Managing Risks of Nitrates to Humans and the Environment

Managing Risks of Nitrates to Humans and the Environment

Edited by

W.S. Wilson, A.S. Ball
Department of Biological Sciences, University of Essex, Colchester, UK

R.H. Hinton
School of Biological Sciences, University of Surrey, Guildford, UK

ROYAL SOCIETY OF CHEMISTRY

The proceedings of a masterclass conference organised by the Agriculture Sector and Toxicology Group of The Royal Society of Chemistry on Managing Risks of Nitrates to Humans and the Environment at the University of Essex on 1–2 September, 1997.

Special Publication No. 237

ISBN 0-85404-768-9

A catalogue record of this book is available from the British Library

Published by The Royal Society of Chemistry,
Thomas Graham House, Science Park, Milton Road, Cambridge CB4 0WF, UK

For further information see our web site at www.rsc.org

Typeset by Computape (Pickering) Ltd, Pickering, North Yorkshire, UK
Printed by MPG Books Ltd, Bodmin, Cornwall, UK

Preface

Numerous scientific papers dealing with the so-called nitrate problem have been published during the past two or three decades. Usually these have taken the form of the Nitrate problem in Agriculture or the Nitrate problem in Water and have been set against the background of establishing limits such as those promulgated by the European Union and the World Health Organization. Frequently criticism is made of the dearth of rigorous scientific evidence or proof to justify these limits such as the European Union's upper limit of 11.3 mg NO_3^--N per litre in drinking water.

In view of the current public demands for thorough-going research in other problem issues such as Bovine Spongiform Encephalopathy (BSE), *E. coli* 0157 and Genetically Modified Organisms (GMOs), it was considered appropriate and opportune to review the wide range of nitrate research being carried out in key spheres such as agriculture, the environment and medicine. Thus, a masterclass conference on Managing the Risks of Nitrates to Humans and the Environment was organized by the Agriculture Sector and Toxicology Group of the Royal Society of Chemistry on 1–2 September 1997 at the University of Essex. Leading scientists in agricultural, environmental and medical aspects of nitrate research along with counterparts of young scientists engaged on research in the chemistry of terminal diseases, especially cancer, were invited to attend. This masterclass arrangement attracted generous financial support from The Angela and Tony Fish Bequest.

The objective of the conference was *inter alia* to investigate the developments whereby nitrate, a simple anion, had become the centrepiece of a furious debate in which farmers and growers were accused of obtaining large profits through the excessive use of nitrogenous fertilizers. The resulting increases in nitrate concentration in natural waters, it was claimed, threatened both the public's health and the environment. The timely and unique conference was asked to assess the foregoing allegations by answering such basic questions as:

- What really happens to nitrogenous fertilizer applied to crops?
- Which environmental problems have nitrate as a primary cause?
- Is nitrate a problem or a solution where our health is concerned?

The answers to these questions proved not only very interesting but rather surprising, particularly where the third question was concerned.

The paper and poster sessions were arranged in three logical sessions and chaired by persons of wide experience. This was augmented by discussion groups and reference to an expert panel and a working party. This procedure was applied to maximize the amount of new knowledge extractable from the work reported.

The book is effectively an account of the proceedings of the conference. It comprises three sections signifying the areas covered. Each section is headed by introductory comments which guide and correlate the constituent chapters. New knowledge in each section proved enlightening to specialists in the other sections. Undoubtedly, the greatest amount of salient knowledge came from the section on medical aspects of nitrates. Most of the total work reported indicated that the time had arrived for a major reassessment of the generally accepted negative attitude to nitrate and, by implication, the need to review the European Union's severe limit for nitrate in drinking water.

The fair balance of new research information contained in this volume will provide scope for persons and organizations involved in policy-making and purveying essential information on nitrates. In this connection active participation of representations from various branches on the Ministry of Agriculture, Fisheries and Food, the Ministry of Health and the Department of the Environment, Transport and the Regions, was significant. Contributions from Agricultural, Horticultural, Environmental, Ecological and Water Research institutes provided full scientific coverage. Representatives from Europe, United States of America and India indicated the universality of the issues discussed. The value and timeliness of the event and the projected publication was acknowledged by Stephen Spivey, Director of Technology Foresight, representing the Office of Science and Technology.

W S Wilson, University of Essex
A S Ball, University of Essex
R H Hinton, University of Surrey

Contents

Acknowledgements

The editors wish to thank the following groups of persons who ensured the production of this book: Conference Organising Committee: Dr A S Ball (University of Essex), Dr P B Barraclough (IACR-Rothamsted Experimental Station), Dr R H Hinton (University of Surrey), Professor D S Powlson (IACR-Rothamsted Experimental Station), Dr R L Richards (John Innes Centre, Norwich) and Dr W S Wilson (University of Essex). Expert Panel: J R Archer (Farming and Rural Conservation Agency), Professor R Kroes (University of Utrecht, Netherlands), Professor P B Tinker (University of Oxford) and Professor R Walker (University of Surrey). Working Party: Professor T M Addiscott (IACR-Rothamsted Experimental Station), Dr T Batey (University of Aberdeen), Professor G Davies (Northeastern University of Boston, USA), Professor M K Garrett (The Queen's University of Belfast), Dr C Leifert (University of Aberdeen), and Professor R Walker (University of Surrey). Thanks are also due to The Angela and Tony Fish Bequest and The Perry Foundation for direct and indirect financial help, to members of the secretarial section of the Department of Biological Sciences, University of Essex and to Janet Freshwater and Kirsteen Ferguson of the Royal Society of Chemistry for their aid in facilitating the publication process.

Section 1

The Nitrate Problem in Agriculture
Introductory Comments

by T. Batey, University of Aberdeen

Although agricultural activity is a major contributor, it is not the only source of nitrates in waters. Rainfall and organic waste of human origin also contribute significant amounts directly and indirectly to waters derived from both anthropic and natural ecosystems. However, the turnover of nitrogen in most agricultural systems is much greater than under natural ecosystems and the potential for loss as nitrate correspondingly higher. As shown by Powlson, the direct contribution of fertilizer N to losses of nitrate is low. During the growing season losses of N are mainly gaseous; it is during autumn and winter that most nitrate is leached from the land. Much is known about the complex dynamics of the N cycle in the soil (Jarvis, Powlson, Stockdale) and on predicting nitrogen losses as nitrate. We know which soils are most vulnerable, the role of rainfall and of crop type and tillage. These factors can be assessed leading to the formulation of a rotation which minimizes the risk of loss (Glendining & Smith). Crops which are likely to contribute high amounts are those which receive large amounts of nitrogen either as fertilizers or as manures and which are relatively shallow rooting, for example potatoes. As discussed by Wilson, there may be scope for reducing the amount of nitrogen applied as fertilizer or altering the timing of applications to crops. Situations at most risk are those where the land is devoid of plant cover or where the growth of crops (and thereby their uptake of nitrogen) has been reduced by drought, pest, disease or the deficiency of another nutrient.

Other questions addressed at this Conference include the role of soluble organic nitrogen. This has been shown in recent work by Murphy et al. to constitute a significant proportion (up to 60%) of total soluble nitrogen in the soil. Some aspects of the effects of increasing levels of atmospheric carbon dioxide are discussed by Ball & Pocock. The possible benefits of nitrate to the health of ruminant livestock are evaluated by Hill.

It is clear from the papers and posters presented that agriculture does not appear to be the bogey which is profligate in its use of fertilizer nitrogen thereby contributing to excessive losses ending up as nitrate in waters. Nevertheless, much still needs to be done to provide a scientific and practical basis to find a system of nitrogen management which continues to provide mankind with crops of satisfactory yield and quality and at the same time minimizes losses as nitrate leaving the land and entering waters.

1

Nitrogen Dynamics in Natural and Agricultural Ecosytems

S. C. Jarvis

INSTITUTE OF GRASSLAND AND ENVIRONMENTAL RESEARCH,
NORTH WYKE RESEARCH STATION, OKEHAMPTON, DEVON,
EX20 2SB, UK

Abstract

Nutrient cycling within ecosystems is influenced by the nature of the element concerned, edaphic and environmental conditions, inputs to the system, and managerial influence. The physico-chemical and biological processes that control cycling are basically the same, whatever the system; some become more influential under different circumstances. For some nutrients, the cycle is relatively 'closed', *i.e.* there are few losses from the system. Most ecosystems, however, whether natural or managed, lose various forms of N. Nitrogen differs from many other nutrients in that it is 'leaky' and can, at various stages within the cycle, change valence and form: this contributes to a large degree of mobility and potential for loss either as gases or into water as NO_3^-, and thus form part of a larger global cycle.

A number of microbiological processes control the flow of N. These contribute to the available pool of 'available' soil mineral N, the form in which it exists and the extent of conversion into gases. In the first instance, mineralization and nitrification are the controlling processes. These are influenced by soil moisture, temperature, pH and other properties, the microbial communities present, and in the case of mineralization, by the 'quality' of the soil organic matter and any returned residues. All of these factors differ between different ecosystems and, depending upon the management, differ between agricultural and natural systems, and contribute to differences in the dynamics of N change. In agricultural systems, there are also large direct inputs into the available pool and large quantities of N recycling within the system. This increases the potential for loss as NO_3^-.

Nitrate leaching varies with soil and rainfall and on the interaction with competitive removal processes, *i.e.* those of uptake/assimilation and denitrification. The extent to which these occur varies with the size of the relevant pools at any one time and their interactions with the soil and weather conditions and plant uptake potential. These obviously differ substantially between different systems (including different natural ecosystems) and will have impact on the amounts of NO_3^- leached from soils into water courses or aquifers.

1 Introduction

Nitrogen (N) cycling processes operate over a wide range of different physical scales from cellular to global and with reaction times from milliseconds to 1000s of years. There can be near instantaneous transfer of N from, for example, inorganic pools in terrestrial or aquatic environments into organic forms or into the atmosphere, and also long-term processes, for example controlling the release of N from stable organic forms in the soil with half-lives of centuries. In a global sense, the important processes which regulate the cycle are those which regulate the fixation or removal of N from the large atmospheric reservoir of dinitrogen gas and those which influence its return, ultimately through denitrification (Fig 1). Within an ecosystem, the processes

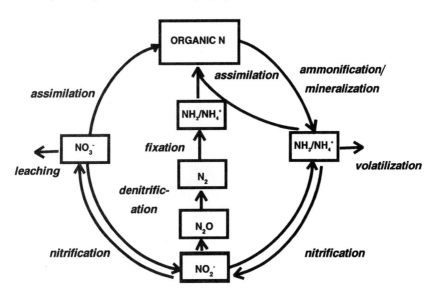

Figure 1   *Global N cycle*

which interact with the global scale are those which introduce N into the system (largely as the result of biological or chemical fixation), assimilation into biomass, ammonification/mineralization (releasing mobile forms) and denitrification (returning stable N_2 back to the atmosphere). These and the other contributory processes which take place, especially in the soil-plant system (Figs 2 and 3), occur in both natural and agricultural environments.

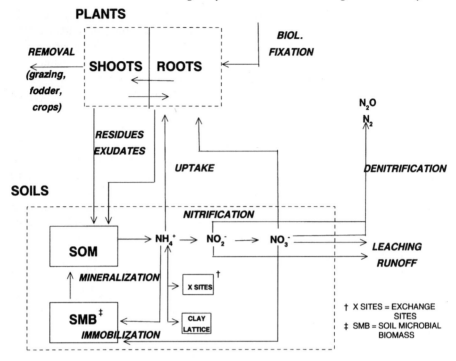

Figure 2   *Nitrogen transformations and transfer in the soil plant system*

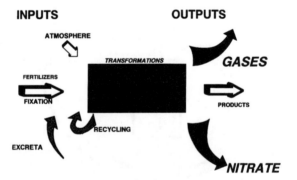

Figure 3   *Inputs and outputs of N for terrestrial ecosystems*

The same basic suite of interactive biological/chemical/physical controls operate in both systems to determine the rate at which N transformations take place. As the result of these transformations, all systems are 'leaky' to some extent or other and N is lost from the system. Nitrogen chemistry is such that a range of different states and chemical forms exist but, because of this, opportunities occur for 'escape' from the system to contribute to the larger cycle (Fig 1). Nitrate (NO_3^-) is a key intermediary in this cycle, in that it is highly mobile and takes part in many of the transfer processes at global (Fig 1) and smaller scales (Fig 2), *i.e.* through assimilation by biomass, denitrification to nitrous oxide (N_2O) and/or N_2, and leaching into aquatic systems (for subsequent participation and contribution to the cycle within another eco-system or removal by denitrification).

In many senses, uptake/assimilation, denitrification and leaching can be seen as competing processes within the soil/plant system. The amount of NO_3^- present in the soil is a reflection of the interaction between removal processes and those which are responsible for its presence, *i.e.* generation through the mineralization/nitrification pathway or by external additions. Again, the same controlling factors (Table 1) for these interactions operate in both natural and agricultural systems. However, by definition, because conditions (*i.e.* extent of input, soil, climate) differ, the equilibria achieved between the interactions differ and as a consequence the rates and dynamics of the transfers and transformations and effects on NO_3^- in soil/plant systems will also differ.

Table 1 *Key processes in controlling N dynamics*

Process	Outcome	Major controlling factors
Uptake into biomass and assimilation	Removal of mobile, mineral N from soil available pools	• environmental (H_2O and temperature) • carbon fixation • soil type and conditions • biomass community structure and populations
Mineralization/ immobilization	Release/removal of mobile mineral N into available pools	• soil type and conditions (temperature and H_2O) • organic matter/residue quality/quantity • system stability and equilibrium
Nitrification	Transfer from relatively immobile (NH_4^+) to highly mobile (NO_3^-) form (some release of N_2O and NO_x)	• substrate (NH_4^+) concentration • soil aerobicity (and other environmental conditions) • nitrifying populations • other soil (*e.g.* pH) conditions
Volatilization	Transfer from terrestrial state to short-lived atmospheric forms (NH_3^o and particulate NH_4^+)	• substrate [NH_4^+/NH_3 (dissolved)] concentration • soil pH • environmental conditions (including windspeed) • enzymic activities
Denitrification	Transfer from terrestrial and aquatic states to atmosphere [N_2O (NO_x) and N_2]	• substrate -(NO_3^-) concentration • anoxia • environmental conditions (temperature) • energy source
'Leaching'	Transfer of mobile NO_3^- (also NO_2^-, dissolved organic forms and some NH_4^+) from terrestrial to aquatic systems	• NO_3^- concentration • hydrological pathways • soil type/conditions • other N cycling processes

The dynamics of N transfer within systems will vary according to many contributory factors (for example, the extent of physical disturbance of the soil system and the degree to which equilibria have been reached). By and large, however, the potential to generate, transform and transfer NO_3^- within and from the system is a reflection of the rate of N input. This increase may take place as the result of transformation processes acting directly on the input itself, leaching or denitrification of added fertilizer N for example. In general, unless management practices are badly timed, *direct* leaching losses from fertilizers are relatively small but those from denitrification can be substantial. The major consequences of N added into any system, whether derived from fertilizer, biological fixation or atmospheric deposition, is to enhance the flows (*via* plant residues, animal excreta for example) through the system. Fertilizer inputs to intensively managed systems can be large (Table 2): it should be emphasized that these are recommended optimum rates of application and are not universally applied. Excesses of N in any system arise when the capacity to be assimilated into biomass is limited: if this happens in combination with the appropriate favourable conditions for particular transformation processes then these will be accelerated and result in increased losses (Fig 3).

Table 2 *Optimum rates of N fertilizer application for intensive systems: values are kg ha^{-1} year^{-1} (from ref. 2)*

Soil fertility	Silage[a]	Grassland for intensive dairying Grazing[b]	
low	420	380	
medium	380	340	
high	340	320	
		Tillage systems	
	Winter wheat	Potatoes	Brussel sprouts
Soil N status (index)[c]			
0	196	230	255
1	136	180	185
2	56	115	115

[a] 4 cuts
[b] rotational grazing. 6+ grazings
[c] on average, for mineral soils

2 Process Influencing NO_3^- Contents in Soils

2.1 Nitrate Generation

2.1.1 Mineralization/immobilization. The microbial activities responsible for the release, in the first instance of NH_4^+ (ammonification/mineralization: Fig 2, Table 1), are the main controlling determinants of the rate of supply of mineral N from internal sources and recycling. In many natural situations where there are no legumes and negligible atmospheric deposition, they will provide the only means for delivering N in a form that is accessible for uptake by plants and so assume considerable importance in ecosystem regulation. The processes involved are complex and interactive (1) and dependent in the first instance on macro-faunal activity to initiate the breakdown process, but then the microbial community determines the dynamics of the release of N into available pools. The net effect in terms of available N – net mineralization – supply is an expression of the balance between gross N mineralization and gross immobilization. Gross mineralization rates depend on the nature of the soil and the history of the background organic matter inputs: Fig. 4 demonstrates an 8–9 fold difference in cumulative gross mineralization in two different soils over a 9 week period. There are also fundamental differences in immobilization rates depending upon the background nature of the system. As illustrated in Table 3, the effects of different grassland managements (from extensive semi-natural to intensive high N input) result in substantial differences in the balance of the patterns in mineralization and immobilization and therefore in the net release of N into the soil available (mineral)-N pool. It is also apparent that even in the semi-natural, low intensity system, microbial activity is releasing substantial amounts of mineral N. This is confirmed by other field based measurements of net mineralization (5) and will be an important feature of other systems in other circumstances, including other semi-natural vegetation types.

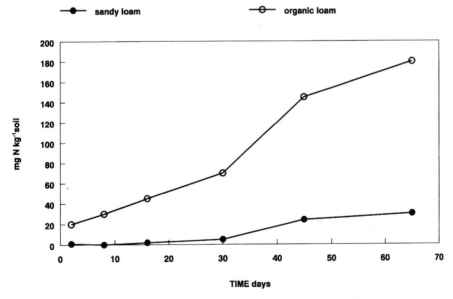

Figure 4   *Cumulative gross mineralization rates from two soil types (3).*

The dynamics of this release depend upon the activities of the soil microbial biomass community structure which will be determined by the nature and extent of the inputs and management (6), the soil type and environmental

Table 3   *Effects of management on gross rates of nitrogen mineralization and immobilization in grassland soils (4). Values are* µg N g^{-1} *dry soil d*$^{-1}$

Treatment	Gross mineralization	Gross immobilization	Balance
1. Semi-natural grassland – no fertilizer or fixed N	9.1	4.8	3.2
2. Mixed grass/clover swards – no fertilizer N	9.7	6.2	4.8
3. Grass sward – 200 kg N ha^{-1} yr^{-1}	11.1	6.5	6.3
SED	1.6	0.8	

conditions. All of these are interrelated. The quality and timing of the residue returns and their incorporation in soil have important effects on mineralization and release of N in mineral pools: in arable soils it is the N derived from these sources during periods when no actively growing crops are present that is likely to be available to be lost by leaching (7). In grassland soils the very large rates of net mineralization (over 300 kg ha^{-1}) (7) is also a major contributory factor because substantial proportions of this occur during periods of maximum rainfall/drainage and at depths which are below the main rooting activity. Both quantities and quality of the residues determine the rate of transfer and although the C:N ratio is considered a major factor on net mineralization influence (Fig. 5), other constituents (*e.g.* phenolics) also play an important role. Organic matter becomes more recalcitrant with respect to N release with time as it becomes more and more degraded. This is of obvious importance in many natural systems, where stable organic materials have been accumulated but do not release N to a high degree. The relationship shown in

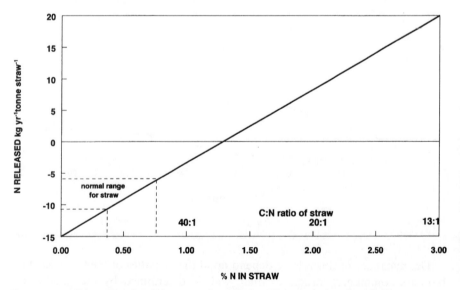

Figure 5   *Effect of residue C:N on N released in arable soils (taken from 8, 7)*

Fig. 5 is for release from cereal straw; although this is too much of a simplification to be applied universally to all systems and vegetation types, the basic principles involved are demonstrated.

Not only does the quality of materials returned to soil differ considerably depending on the particular system but so also do the amounts; Fig. 6 shows that the N returned from the above-ground crop residues at harvest ranges from the equivalent of <20 (field beans) up to >200 (brussel sprouts) kg N ha^{-1}: much of the N in the latter would have a high potential for rapid mineralization under the appropriate conditions. There were also substantial differences in the amounts of mineral N in the soil under the different crops. Nitrogen supplied from both pools could contribute to NO_3^- leached from the soil under these cropping regimes.

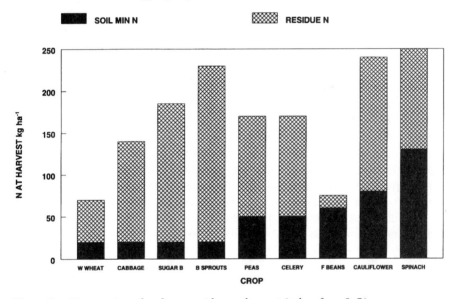

Figure 6   *Nitrogen in soil and crop residues at harvest (taken from 9, 7)*

2.1.2 Nitrification.   The next key phase in the soil N cycle controlling the availability of NO_3^- is regulated by another microbial process, that of nitrification. In the main, this is an autotrophic inorganic oxidative process converting NH_4^+ through a sequence of intermediaries to NO_3^-: nitrite (NO_2^-) is a key, albeit short-lived, product of the process, *i.e.*

$$NH_4^+ \rightarrow NH_2OH \rightarrow [HNO] \rightarrow NO_2^- \rightarrow NO_3^-$$

In some natural environments, organic heterotrophic pathways also exist. Nitrification is an important process in that it provides the mechanism for transformation from a relatively immobile form of N (NH_4^+, which is largely associated with the soil exchange complex – Fig. 2, and not easily removed by leaching) into a highly mobile one. In arable soils, nitrification is usually considered to be non-limiting in the process of generating NO_3^-, but there is

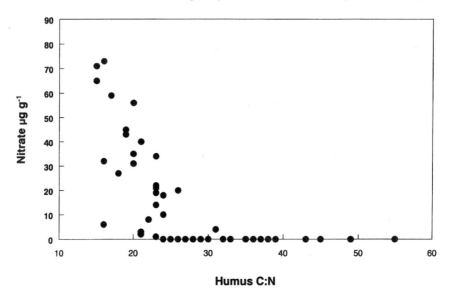

Figure 7   *Effect of humus layer C:N ratio on nitrification in incubated soils (taken from 10, 11)*

evidence of substantial differences in grassland soils which have been influenced by management inputs (12), with enhanced nitrification rates in high N input systems. In natural ecosystems, the quality of the organic residues in terms of C:N contents apparently determines the rate of NO_3^- produced (Fig. 7), although whether this is a direct effect or one which operates through control over mineralization is not clear. Nitrification is an aerobic process, limited by anaerobic conditions and low pH and can be influenced by background history of the site, the effects of which can be expressed over relatively long periods through changes in the microbial community structure which have developed as the result of different rates of substrate (NH_4^+) provision (13).

The processes and microbial activities resulting in the generation of NO_3^- are common to all terrestrial ecosystems, but will be influenced by an interactive and complex set of soil and environmental controls. The rates and dynamics of these processes depend directly, or more likely indirectly, upon N input to the system interacting with external factors. In agricultural systems there are direct inputs to the soil mineral pools through fertilizer additions and, indirectly, enhanced transfer through the organic components which may be derived from either fertilizer or biological fixation.

2.2   Processes Involved in Removal of Mineral N from Soils

2.2.1   Ammonia Volatilization   The next phase in the cycle (Figs 1–3), and which determines the location and fate of the more mobile forms of N, is initially one of competition between the biological processes which operate on or within the soil. Thus ammonia (NH_3) volatilization can remove N from the soil before nitrification, adsorption onto the exchange complex or uptake by biomass can take place. Ammonia volatilization is dependent on NH_3/ NH_4^+ being present in the aqueous phase of the soil matrix, particularly at or near the surface and is enhanced with high pHs, dry soils and high temperatures. Thus, large returns of animal excreta (urine) or addition of urea or NH_3-based fertilizers promote the likelihood of this occurring, some small rates of losses can also occur from leaves of plants if they have high N concentrations in their tissues. Intensive livestock production systems are, because of the large generation/accumulation of excreta, responsible for a rapid rate of loss from agricultural systems and for the increased atmospheric concentrations of NH_3/NH_4^+ species. These have a relatively short half life and are deposited to re-enter the terrestrial N cycle as a mobile species. This transfer can take place over a range of spatial scales and often re-distributes N to other ecosystems which are sensitive to nutrient inputs which may be significantly greater than those under which they established an equilibrium status.

2.2.2 *Competition for the Removal of NO_3^-.* After NO_3^- has been added or generated by nitrification, a number of competing processes, *i.e.* uptake into biomass (above and below ground), denitrification and leaching, operate to eventually determine availability for leaching. Uptake by biomass is a key to this and immobilization into soil microbial biomass determines the net impact of mineralization. Both NO_3^- and NH_4^+ can be utilized by soil microbial biomass, as they can by higher plants. Crop plants have an enormous potential for N uptake – sometimes to luxury levels – and those developed for intensive production systems can operate effectively with very high concentrations of NO_3^- at their root surfaces. Species from natural environments may not have the potential either to utilize large inputs or to be able to survive with them. Part of the skill in agronomic nutrient management is therefore ensuring that the potential for utilization of available N sources, particularly of NO_3^-, meets the likely supply of mineral N. Although NO_3^- may often be the more accessible ion for acquisition by roots (because of its mobility and low competition from other anions), most plants can utilize both ions, under some circumstances NH_4^+ is taken up preferentially.

Under anaerobic conditions, the denitrification process can remove excess NO_3^- from the soil as shown below:

$$NO_3^- \rightarrow NO_2^- \rightarrow [X] \rightarrow N_2O \rightarrow N_2$$

The rate at which it occurs depends on the amounts and location of NO_3^- (and hence inputs to the system), soil moisture conditions and carbon supplies; it is also temperature dependant. Because it is an anaerobic process, denitrification is more likely to occur in poorly drained, fine textured soils. Although denitrification occurs mainly in the top soil, there is some evidence to show that it can take place at depth in the profile and therefore interacts directly with leaching.

The net effect of all these interactive processes is to determine the size of the NO_3^- pool which could be available for transport away from the rooting zone. The balance of what occurs is an expression of the ecosystem, and the particular N cycle determined by the extent of inputs interacting with soil type and climatic conditions. Feedback mechanisms are influenced by the pools of N that arise, then further influence the rates and dynamics of the N transformations and transfers. As a result, there is a continuum of effects within both natural and agricultural systems which influences the availability of NO_3^- for loss.

Table 4 *Summary of recent studies in the fate of ^{15}N applied to grassland and forest systems (from 14)*

Ecosystem	Location	N applied $(g\,m^{-2})$	duration months	Plants	SOM	Extracted	Total soil	Total
					recovered (%)			
Praire grassland	Colorado	2.6	1	52	–	–	28	80
Tussock tundra	Alaska	2.6	1.9	20	84	–	84	100
Mixed grass/ clover	NZ	35.0	0.7	16	6	52	58	74
Grass/legume mixture	Italy	14.0	1.9	24	10	<1	15	39
Mixed conifer	–	0.2	1.3	–	15	17	34	34
Mixed conifer	–	1.0	4.0	–	28	3	52	52

Differential effects of location and ecosystem are illustrated by data in Table 4, although strict comparisons are difficult because the duration of the observations and the inputs were not consistent for the different studies. Nevertheless, if the results for approximately equivalent periods are compared, the effects (as determined by using ^{15}N as a label) of differing process rates have resulted in distinct differences in distribution of the added N in the soil and plant components. One obvious difference is that between the grasslands systems where a much larger percentage of the added labelled N was partitioned into soil organic matter in the tundra system than in other grasslands. The apparent total recovery of the added ^{15}N also differed substantially between comparable managements. Thus, only 39% of the labelled N was recovered from one grass/legume sward but 74% from another, indicating a lower loss of N rate from the latter. There are other similar distinctions which can be drawn between the two woodland types which are shown in Table 4. This again makes the point that, even in apparently relatively similar systems, the dynamics of N transformation are acting and reacting in different ways.

Table 5 *NO₃⁻-N in soil solutions from UK Environmental Change Network Sites: samples were taken 2 x per month and means are shown (values in brackets indicate the number of months of observation) (from 15)*

Site and location		Management	Soil	NO_3^--N $(\mu g\, ml^{-1})$
Drayton	England (Warwickshire)	grazed permanent grass	heavy clay drift	30.72 (12)
Hillsborough	Ireland (County Down)	lowland grassland (grazing/silage)	glacial clay till	1.72 (5)
Wytham Wood	England (Oxfordshire)	ancient woodland	Oxford clay	0.64 (13)
Alice Holt	England (Surrey)	coniferous woodland	–	0.31 (5)
Sourhope	Scotland (Borders)	upland rough grazing	podzol/brown forest soil	0.095 (19)
North Wyke	England (Devon)	extensive grazed grassland	silty clay	0.082 (16)
Rothamsted (Park Grass)	England (Hertfordshire)	unfertilized cut grass	silty loam	0.047 (7)
Glensaugh	Scotland (Grampian)	upland rough grazing	podzol	0.027 (22)
Moorhouse	Pennines (Cumbria)	upland heather moorland	blanket bog	0.003 (27)

The expression of these reactions in terms of NO_3^- present in soil solution is of importance to the potential for leaching. As Table 5 illustrates, the range of concentrations in soil solution is large, *i.e.* from 0.003 μg NO_3^--N l^{-1} in an upland peat to 30 μg NO_3^--N l^{-1} in the intensive dairy pasture system. Intermediate concentrations were present in the other systems, again representing a continuum of effects. One interesting component of Table 5 is that the two woodlands have higher soil NO_3^- concentrations than some of the grasslands which are used for animal production.

2.3   Nitrate Leaching

There are many hydraulic pathways by which NO_3^- in soil solution can be transferred into aquatic systems which depend upon soil type/texture and structure, rainfall volume and pattern. Water movement and transport of nutrients usually either takes place in the vertical plane (*i.e.* piston flow in light textured soil, or by-pass flow in heavy soils), or more or less horizontally (*i.e.* surface, or interflow on sloping land especially with poorly drained soils or when soil surfaces are frozen). For present purposes, any or all of the mechanisms which transport water and its nutrient load from the soil into either surface or subsurface aquifers will be referred to as leaching.

From this it can be inferred that NO_3^- movement into waters is related to concentration in the soil solution (which in turn is dependent upon N input) when drainage occurs: subsequent losses then depend on the further generation or addition of NO_3^- into the soil solution. In many production systems, rates

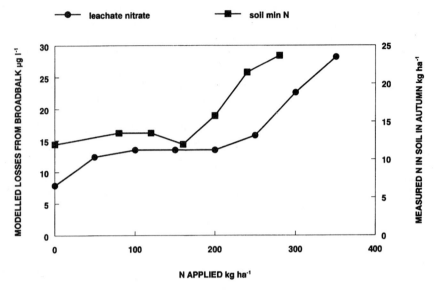

Figure 8   *Nitrate N in soil and drainage waters under cereal production (taken from 7, 16)*

of loss remain relatively constant over a wide range of inputs until there is a marked increase. Fig. 8, which superimposes information from two studies, one in which mineral N in the soil profile was determined in autumn after cropping and the other in which NO_3^- losses were modelled, demonstrates this effect with two very similar patterns. In many such instances, the point of inflection in the curves can be related to the asymptote of the yield response curves to N fertilizer addition. The enhanced soil contents will not necessarily be directly from the fertilizer added, but through increased rates of supply from soil organic matter. Athough in the examples in Fig. 8 the two curves look compatible, it is often difficult to predict accurately the amounts leached from a knowledge of mineral N in the soil profile, because of a continued supply through mineralization.

Mineralization is also a key contributor to the rates of NO_3^- loss from grassland. Results from a long-term study of NO_3^- leaching from grazed grassland on an impermeable silty clay soil in Devon (average rainfall *c.* 1035 mm yr^{-1}) (Table 6) clearly show that losses increase with N input. In this example, increasing fertilizer N input from 0 to 400 kg N ha^{-1} has, on average, increased leaching losses from 5 to >130 kg N ha^{-1} on soils which had artificial drainage. Field drainage had a major influence on leaching and, on average for all fertilized treatments, increased losses by over 3 fold. Whilst some of this effect may be due to a reduction in the removal of NO_3^- by denitrification because of a reduction in anaerobic sites in the drained soils, it is more probable that increased aeration has stimulated mineralization and nitrification. Mineralization/immobilization effects are also demonstrated by the difference in losses between the permanent swards and the reseeds (*i.e.*

Table 6 *Loss of NO_3^--N (kg N ha^{-1} year^{-1}) by leaching from grazed pastures in S.W. England (17)*

Treatment	Range	Mean[a]
200 kg N ha^{-1} long term permanent grass	20.6–54.4	38.4
400 kg N ha^{-1} long term permanent grass	66.8–186.0	133.8
400 kg N ha^{-1} reseeded grass	35.4–68.6	55.7
Undrained soils	19.2–61.9	35.6
Drained soils	79.0–168.8	112.9
No added N long term grass – undrained	1.2–3.4	2.4
No added N long term grass – drained	3.0–8.4	5.1

[a] Values are means of 7 years for fertilized swards, 5 years for the reseed and 3 for unfertilized systems

nearly 2 x more NO_3^--N lost from the old pasture). The effect in this case is probably through changing the equilibrium in the balance of mineralization/immobilization towards more net mineralization in the old sward and a greater release of mineral N. These effects are difficult to predict with confidence and improvement of this predictive capacity is an important requirement for the future development of improved fertilizer recommendation schemes for grassland. Another feature to note is that even those management systems which have no extra N added, have significant, measurable losses.

The determinants of the flows of NO_3^- from land are complex and few, if any, systems are completely leak-free: whilst inputs are one of the major controlling factors, it is the interaction of these with other management, edaphic, ecological and climatic conditions which result in fluxes of mobile N from the soil. Most recent studies have focused on NO_3^- losses from the relatively small scale and there is increasing need for information at the larger scale. As the data in Table 7 indicate, the estimates of rates of loss at the larger scales indicate a wide range of inputs of NO_3^- into surface waters, depending

Table 7 *Survey of rates of NO_3^--N loss from land (from 11)*

Land vegetation type/predominant land use	Loss rate kg (ha^{-1})
Moorland	0.2 – 1.6
Low intensity grassland	0.2 – 2.0
Forest – Sweden	2.0 – 3.8
– USA	4.5
Loch Leven (Scotland) agricultural catchment	7.5
Slapton Ley (S.W. England) agricultural catchment	5.7 – 6.8
River Ouse (England) agricultural catchment	26.4
Swiss Alps agriculture	36.8
Swiss lowland agriculture	84.3
Intensive grassland (England)	56.5
Perthshire – agriculture	58.8
– soft fruit	88.4
– urban	79.5
Intensively managed Danish lake catchment	131.1 – 198.8

upon circumstances. However, one of the current major problems is an inability to extrapolate with confidence from the small experimental to catchment scale because the relationship between land use and soil NO_3^- with soil water quality is not always clear. There are opportunities for many further interactions to occur, as the result of biological, chemical, and hydrological processes, as NO_3^- moves downstream: the net effect of these will determine the final concentration in aquifers. There are further confounding effects of short term temporal variability interacting with immediate antecedent conditions on the land surface, and local hydrological variability. It is clear from Table 7 that NO_3^- was present in all systems and increased as the intensity of land use increased: in some cases this also included significant contributions from non-agricultural sources.

2.4   Transfer of Other Mobile Forms of N

Because of recent environmental pressures and issues, most research attention has focused particularly on NO_3^- moving into waters. Although usually tightly bound to soil particles, NH_4^+-N can enter surface water systems on occasion, usually as the result of a local point source activity (and from both agricultural and non-agricultural sources). Soluble organic forms are known to be present in waters draining from natural ecosystems and recent information suggests that they may make a significant contribution to the N budget of managed land. Preliminary results (Scholefield and Hawkins, pers. comm.) have shown that large proportions of the total N leached from grassland soils were in organic form, thus. *c.* 30% of the total losses of 89–272 kg N ha^{-1} from a high input beef grazing system was as organic N. In grass/clover based grassland, where the total lost in drainage was much lower, > 50% of the loss was in organic N form.

Another inorganic loss route is *via* NO_2^-. This has already been noted as a key intermediary product during nitrification and denitrification processes and although it has a relatively short half life in soil systems, can move into surface water systems *via* drainage (18). The amounts and concentrations involved are small relative to those of NO_3^- but the guide values (19) for the quality of inland waters with respect to fish populations are low (*i.e.* 3 and 9 ppb NO_2^--N for salmonid and coarse fish, respectively): the maximum acceptable concentrations for drinking water is 30 ppb (20). Irish studies (21) have reported NO_2^- concentrations in the rivers within a major lake catchment in the range 100–150 ppb and concluded that land drainage contributed 40% of the river loading and that the remaining 60% arose from transformation of NH_4^+ (which had originated from agricultural sources) at the water/sediment interface of the river system. Other preliminary data from grassland in SW England and the associated river system also indicate that significant (in terms of ecological impact but very low concentrations compared to NO_3^-) amounts of NO_2^- leave agricultural land and that concentrations at certain points in the river system are above the defined maximum (Table 8). Athough there were ranges in the amounts measured in drainage, there was little obvious relationship with N inputs to the grassland in this case.

Table 8 *Nitrite contents of grassland drainage and river waters in S.W. England (22)*

Location and local management	NO_2^--N ($\mu g\, l^{-1}$)	
	Range	Mean
Grassland drainage		
grazed grass – no fertilizer	1–23	7
grazed grass – 280 kg N	1–41	14
grazed grass/clover – no fertilizer	1–18	8
River water (River Taw)		
1. upland extensive grassland	1–3	1
2. lowland intensive grassland	1–25	6
3. lowland intensive grassland	8–28	19
4. lowland intensive grassland	16–50	24

3   General Conclusions

- The dynamics of N cycling, transformation and transfer are complex and highly interactive but with the same basic controls, *i.e.* biological, chemical, and physical (and including hydrological conditions) over the processes which determine whether NO_3^- is present in the soil or is transported or leached into waters from natural and agricultural ecosystems.
- Mineralization/immobilization and nitrification processes are particularly important in determining the extent of leaching. The structure and distribution of soil microbial communities (and the effects of soil mesofauna) are responsible for and responsive to the fluxes of N which influence the availability of NO_3^-. Assimilation by above ground biomass and denitrification are other key determinants. All these interact to different extents in the various ecosystems to produce a continuum of effects which spans natural and agricultural systems.
- Other key controlling factors over dynamics are (i) soils – type (*e.g.* texture), conditions, (*e.g.* structural development), hydraulic status and terrain characteristics (*e.g.* slope) and (ii) location/environment (*e.g.* temperature and moisture). By definition, these will often differ substantially between agricultural and natural soils. As far as agriculture is concerned, other important factors controlling the dynamics and fluxes of N are the off-take from the farm and the efficiency of N capture, the extent of recycling, including mineralization, and also the intensity of animal production and the large amounts of N in excreta and manures. The latter has a particularly large impact on fluxes of N and the dynamics of transfer and the potential for loss (Fig. 9). In natural systems all the above factors operate to a greater or lesser extent but additional factors also come into play and the concept of critical loads for N has been developed (23). This presumes that each ecosystem has a tolerance (which ranges between 5–35 kg N ha^{-1} yr^{-1}) for additional N inputs (*i.e.* largely through atmospheric deposition), which depends on the ecosystem, and above which substantial changes in ecological balance will

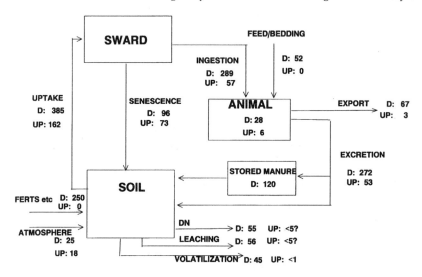

Figure 9   *Annual flows of nitrogen in intensive lowland dairy system (D) (from 24) and an upland semi-natural grazing system (UP) (from 11, 25). Values are kg N ha^{-1} yr^{-1}*

take place. This effect is characterized by changes in soil chemistry, biology, trophic status, acidity and losses and, as a consequence, in ecosystem structure, all of which will have been stimulated by enhanced rates of N dynamics. Additionally, activities undertaken as part of upland 'improvement' also have impact on the natural equilibria for the various components of the N cycle, and those events which influence the stability of systems (*e.g.* afforestation and tree felling) also have impact on N dynamics and transfer. Extreme climatic events, which are more likely to be prevalent in natural ecosystems because of their geographic distribution, such as intensive freeze/thaw changes, snowmelt and burning, will also be of influence. For all situations, any factor which influences soil properties and quality as defined by the requirements of that system will also have importance in the control of N dynamics. Examples are excessive loadings of heavy metals and other pollutants and acid rain.

• The extent of N inputs has a key role in determining N dynamics not only in determining the input/output budgets for the particular system but also the rates at which it is transferred between or accumulated within components of the system (Fig. 9). This will of necessity differ between the extremes of natural and intensive production systems. It is at the point where some of the transformations or transfers occur, either from one chemical state to another or from one component to another, that opportunity for loss is accentuated. Even in the semi-natural grassland system shown in Fig. 8, the amounts of N contained within and transferred between sectors of the system are substantial. It also should be noted that agricultural systems dependent on N from biological fixation operate with the same set of controls over processes as those that are fertilizer based, and where excesses are present, equilibria shift towards loss.

References

1. Jarvis, S. C., Stockdale, E. A., Shepherd, M. A. and Powlson, D. S. (1996) Nitrogen mineralization in temperate agriculture soils: processes and measurement. Adv. Agron., **57**, 188.
2. MAFF. (1995). 'Fertilizer Recommendations for Agricultural and Horticultural Crops (RB209).' HMSO, London, 1995.
3. Monaghan, R. and Barraclough, D. (1996) Contributions to N mineralization from soil macro-organic matter fractions incorporated into 2 field soils. Soil Biol. Biochem., 29, 1215.
4. Ledgard, S. F., Jarvis, S. C. and Hatch, D. J. (1998) Short-term nitrogen fluxes in grassland soils under different long-term nitrogen management regimes. Soil Biol. Biochem., in press.
5. Gill, K., Jarvis, S. C. and Hatch, D. J. (1995) Mineralization of nitrogen in long-term pasture soils: effects of management. Pl. Soil, 1995, **172**, 153.
6. Lovell, R. D. Jarvis, S. C. and Bardgett, R. D. (1995) Soil microbial biomass and activity in long-term grassland: effects of management changes. Soil Biol. Biochem., 1995, **27**, 969.
7. Addiscott, T. M., Whitmore, A. P. and Powlson, D. S. 'Farming, Fertilizers and the Nitrate Problem.' CAB International, Wallingford, 1992.
8. Jenkinson, D. S. In: 'Straw, Soils and Science' (ed. J. Hardcastle) 1985, AFRC, London, p.14.
9. Wehrmann, J. and Scharpt, H. C. In: 'Management Systems to Reduce Impact of Nitrates' (ed. J. C. German) Elsevier, London, p. 147.
10. Kriebitzsch, W. V. (1978) Stickstottnachlieferung in saure Walboden Nordwest-deutschlands. Scr. Geobot., 1978, **14**, 1.
11. INDITE. 'Impacts of Nitrogen Deposition.' 1994, DoE, London.
12. Jarvis, S. C. and Barraclough, D. (1991) Variation in mineral nitrogen content under grazed grassland swards. Pl. Soil, **138**, 177.
13. Willison, T. and Anderson, J. M. (1991) Spatial patterns and controls of denitrification in a Norway spruce plantation. For. Ecol. Manage., **44**, 69.
14. Hart, A.S. C. Firestone, M. K. Paul, E. A. and Smith, J. L. (1993) Flow and fate of soil nitrogen in an annual grassland and a young mixed-conifer forest. Soil Biol. Biochem., **25**, 431.
15. ECN. The UK Environmental Change Network. 1997, World Wide Web. Http:// mwnta.nmw.ac.uk/ecn/index.html.
16. Chaney, K. (1990) Effect of nitrogen fertilizer rate on soil nitrogen content after harvesting winter wheat. J. Agric. Sci., 1990, **114**, 171.
17. Scholefield D., Tyson, K. C., Garwood, E. A., Armstrong, A. C., Hawkins, J. and Stone, A. C. (1993) Nitrate leaching from grazed grassland lysimeters: effects of fertilizer input, field drainage, age of sward and patterns of weather. J. Soil Sci., 1993, **44**, 601.
18. Burns, L. C., Stevens, R. J., Smith, R. V. and Cooper, J. E. (1995) The occurrence and possible sources of nitrite in a grazed, fertilized, grassland soil. Soil Biol. Biochem., 1995, **27**, 47.
19. European Economic Community. EEC 78/695, Brussels, 1978.

20. European Economic Community. EEC 80/77, Brussels, 1980.
21. Smith, R. V., Foy, R. H., Lennox, S. D., Jordon, C., Burns, L. C., Cooper, J. E. and Stevens, R. J. (1995) Occurrence of nitrite in the Lough Neagh River System. J. Environ. Qual., 1995, **24**, 952.
22. Jarvis, S. C. and Dixon, E. (1998), unpublished information.
23. Bull, K.R. (1991) The critical loads/levels approach to gaseous pollutant emission control. Environ. Pollut., 1991, **69**, 105.
24. Jarvis, S. C. (1995) Nitrogen cycling and losses from dairy farms. Soil Use Man., 1995, **9**, 99.
25. Perkins, D. F. In: 'Production Ecology of British Moors and Montane Grasslands.' (eds. O. W. Heal and D. F. Perkins), Springer Verlag, Berlin, 1978, p. 375.

2

Predicting Nitrate Losses from Agricultural Systems: Measurements and Models

E. A. Stockdale

SOIL SCIENCE DEPARTMENT, IACR-ROTHAMSTED, HARPENDEN, HERTS., AL5 2JQ, UK

Abstract

The aim of this article is to set the background for the discussion of the prediction of nitrate losses and the use of such predictions in decision making. The article discusses measurements and models used to predict water flow and nitrate supply, before drawing both together to consider prediction of nitrate loss. The difficulties and advantages of some of the current approaches to predicting nitrate losses from agricultural systems are highlighted.

Models can never be completely independent of measurements. Data are required not only for compilation and evaluation of models but each simulation also requires input data and parameters. The integrated use of both measurements and modelling approaches in the prediction of nitrate losses is therefore advised, with the actual combination of measurements and models controlled by the question to be answered, its scale and the data available. All predictions should be presented in a way that tempers them with an estimate of their reliability, in terms that are readily understood. Where such data are not presented, poor decisions may be made and unacceptable environmental impact or risk may result.

1 Introduction

Nitrate losses from agricultural soils are controlled by interacting biological, chemical and physical factors in the soil and the environment. The main pools, processes and external factors which control nitrate losses from agricultural systems are shown in Figure 1. In this case nitrate losses have been defined as the sum of leaching and runoff losses, while denitrification is included as one of the processes which control nitrate concentration in soil solution. Pools and processes can be separated into those which control water movement in soils

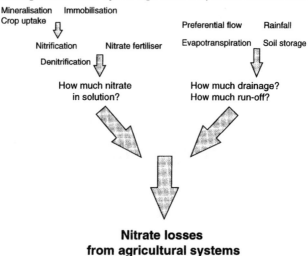

Figure 1   *Processes and environmental factors which interact to control nitrate losses from agricultural systems. These are divided into physical and chemical/ biological processes. This separation has been reflected in approaches to research into nitrate losses*

and those which control the presence of nitrate in the soil solution; this divide is reflected in research. Soil chemists and biologists study nitrate supply in the soil in a range of agro-ecosystems, usually to improve fertiliser recommendations and the efficiency of nitrogen (N) use. Soil physicists study water flow through soils, usually with regard to drainage and irrigation. Nitrate losses can be measured in a variety of ways (Titus and Mahendrappa, 1996). The collection of representative drainage samples and the determination of drainage volumes is difficult and no method is ideal for all soils (Addiscott *et al.*, 1991).

Measured nitrate losses show great variation between agricultural systems (*e.g.* Jarvis, 1998; Powlson, 1998). This is not surprising given the range of climates, crop and livestock enterprises and management practices that they contain. Farmers want to increase profit by increasing N use efficiency. They therefore need information on the magnitude of nitrate losses on a field scale. Water companies have to meet legislative limits for nitrate concentrations in drinking waters. They may therefore need to know the likely nitrate load at an abstraction point, hence requiring predictions of nitrate losses for a ground or surface water catchment. Policy makers may want to know what the impact of changes in farm management practices will be on nitrate losses at a catchment, regional or national scale. There is therefore pressure on scientists to provide answers to various questions related to nitrate losses at a range of scales. Knowledge of the functioning of ecosystems, the effects of management strategies and the remediation of environmental degradation may be pushed to its limits to find answers. Many of the questions asked require scientists to extrapolate or interpolate their knowledge, both in time and space. They may therefore use measurements and models predictively in situations which they have not studied. This may represent prediction *per se*, *i.e.* a forecast or prophecy, but it may also include estimation of processes already past but which were not, or were not able to be, measured.

Scientists use both qualitative assessments and quantitative estimates to predict the timing and amount of nitrate loss. In either case, a robust understanding of the system under study at the appropriate scale is fundamental. Clear definition of the question is also important, as differences in perception can lead to difficulties in analysing and solving problems related to soil behaviour (Bouma, 1993). Measurements, pedo-transfer functions and models may be used either alone or in combination to predict nitrate losses.

Understanding of the system may allow the selection of key pools and/or processes, which can be measured to give an assessment of nitrate losses or the risk that such losses will occur. Measurements may give an indication of the state of a system or its response to external pressure (OECD, 1993). These may be known as indices or indicators. Systems might then be ranked or ordered, and further measurements made in those which are most susceptible to nitrate loss.

Pedo-transfer functions relate basic soil characteristics, such as clay and organic matter content, to data which are difficult or impossible to measure (Bouma and van Lanen, 1987). Simple measurements may therefore be used to calculate more complex properties. Pedo-genetic soil horizons may be used as carriers of information in this way if appropriate measurements were made during soil survey (*e.g.* Batjes, 1997; Bouma, 1989). One or more measured soil properties may therefore be used as inputs to pedo-transfer functions which predict nitrate losses directly or estimate an index.

Models draw together and codify current hypotheses about a system's structure and its controlling factors (Tinker and Addiscott, 1984). They are usually first developed in response to a particular question and at a particular scale. There are many mathematical models of the soil or soil-plant system at a range of scales (*e.g.* Bosatta and Berendse, 1984; de Ruiter *et al.*, 1993; Greenwood and Draycott, 1989; Hansen *et al.*, 1991; Rijtema and Kroes, 1991; Scholefield *et al.*, 1991; Smith *et al.*, 1986). These range in complexity and are usually compiled on computer. Despite this, models represent simplifications of the system; we describe processes in models with various levels of exactitude, all of them imperfect (Addiscott *et al.*, 1991). Models of agricultural systems can be used to predict nitrate losses in response to changes in management or weather patterns or to draw together isolated data to give a prediction for a complete system or region.

The aim of this article is to set the background for the discussion of the prediction of nitrate losses and the use of such predictions in decision-making. This article discusses measurements and models used to predict water flow and nitrate supply, before drawing both together to consider prediction of nitrate loss. The difficulties and advantages of some of the current approaches to predicting nitrate losses from agricultural systems are highlighted. However, a comprehensive review of all possible approaches has not been provided, nor any single approach recommended.

2   Predicting Water Flow

2.1   Measurements

2.1.1 Weather.   Total amounts of rainfall and its intensity and evapotranspiration are the dominant climatic factors controlling the soil water balance and hence the runoff and leaching risk. The presentation of any rainfall data always carries warnings of the extreme variability of the data both temporally and spatially (*e.g.* Collins and Cummins, 1996). Evapotranspiration is much less spatially variable, but data are more difficult to obtain. Estimates of winter drainage for England and Wales show that it is independent of annual precipitation and more variable (Rose, 1991). Differences in soils, slopes and vegetative covers also modify heat and moisture budgets temporally and spatially (Jones, 1976).

Long term weather records have been collected and long term averages can be used to give some indication of the risk of leaching or run-off on a regional basis (Smith and Cassel, 1991). However, the actual variation over any period (*e.g.* day, month) in any year is much greater than mean values indicate. Hence, the use of average weather data to predict leaching risk without a consideration of the likely variability at an appropriate temporal and spatial scale does not give an acceptable degree of sensitivity. The use of recorded weather data over shorter time periods to provide input data for models (*e.g.* using rainfall recorded weekly to generate daily estimates) is also fraught with similar complication (Hutchinson, 1995). The dynamic mathematical modelling of the climate, rather than a statistical modelling approach, is a relatively young field arising from the classic work of Lorenz during the 1960's (Dymnikov and Filatov, 1997) and predictions cannot yet be made reliably over appropriate spatial or temporal scales.

2.1.2 Soil Characteristics.   A range of soil characteristics is known to control water flow: soil permeability, water storage capacity, texture, depth and slope, to name but a few. These characteristics are variable both spatially and temporally within soils. Water storage capacity may vary annually while soil texture, depth and slope can be considered to be stable properties changing only over decades, unless catastrophic erosion events take place (Halvorson *et al.*, 1997). Hydraulic conductivity depends on the size and continuity of the conducting pores and so varies with soil moisture content, reaching a maximum value when the soil is saturated. Measurements under field conditions show that hydraulic conductivity is highly variable both temporally and spatially (Warrick and Nielsen, 1977). The movement of water in soil is reviewed by many textbooks of soil physics (*e.g.* Hanks, 1992; Hillel, 1980 and

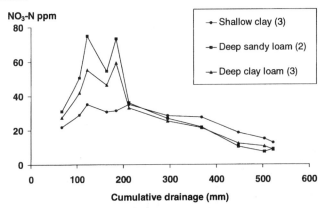

Figure 2   *Nitrate concentration (ppm NO_3^--N) in soil solution, extracted using porous cups installed at 50 cm depth, in the winter following ploughing of a five year grass-clover ley. Drainage was estimated from meteorological data. Total winter drainage was 522 mm. Soils were a shallow clay (leaching risk, 3, moderate), a deep sandy loam (leaching risk, 2, high) and a deep clay loam (leaching risk, 3) occurring within a single field at Tetbury, Gloucs.*

Elrick and Clothier 1990) give an excellent review of the factors controlling solute transport from the microscopic to field scale.

On the basis of site characteristics (*e.g.* slope, vegetation), rainfall (intensity and amount) and the more stable soil properties (depth, texture, stoniness), soil series may be grouped into nitrate leaching risk categories (*e.g.* Smith and Cassel, 1991) or runoff risk categories (*e.g.* Collins and Cummins, 1996). Where nitrate supply is similar then nitrate leaching risk categories can be shown to rank the nitrate losses from soils (Figure 2). However, they give no indication of the timing of nitrate loss. Farmers may be advised to use such classes to assist management decisions such as timing of fertiliser or slurry applications.

Hydraulic conductivity is not widely used as a measurement to predict nitrate losses because of its variability, although attempts have been made to use pedo-transfer functions to calculate unsaturated hydraulic conductivity from measurements of particle size distribution, organic matter content and bulk density (*e.g.* Vereecken *et al.*, 1989). Tietje and Tapkenhinrichs (1993) evaluated a range of pedo-transfer functions, comparing the results to measured data for a broad range of soils, and identified the relationship of Vereecken *et al.* (1989) as the most accurate and widely applicable. However, Vereecken (1995) concluded that without the inclusion of measurements of saturated hydraulic conductivity, which is also very variable under field conditions, pedo-transfer functions did not give accurate estimates of unsaturated hydraulic conductivity.

2.2 Models

Water movement in the unsaturated zone has been a focus for mathematical modelling in soil science for a number of decades. Addiscott and Wagenet

(1985) grouped models of nitrate leaching both according to their purpose and the modelling approaches taken, dividing functional from mechanistic, and separating deterministic from stochastic models. Jury and Flühler (1992) reviewed the modelling of solute and water flow and highlighted the difficulties of applying many models at the field scale.

2.2.1 Mechanistic Models.   Mechanistic models use classical physical theory to describe the flows of water in the soil. In many models the Richards convection-dispersion equation is solved in one-dimension using numerical procedures (Hansen *et al.*, 1991; Johnsson *et al.*, 1987; Wagenet *et al.*, 1989). Mechanistic models of water flow are sometimes known as rate models since they depend mainly on rate parameters (Addiscott *et al.*, 1991). Many of these models have been developed from laboratory studies of soils, often in repacked soil columns (Jury and Flühler, 1992), though undisturbed soil monoliths are also used (Jarvis *et al.*, 1991). Soil water content under field conditions was not accurately simulated by some rate models (de Willigen, 1991), either due to poor parameterisation of soil hydraulic properties from the data available or a failure to account for preferential flow. Ma and Selim (1997) summarised recent progress in mechanistic models of water and solute flow incorporating descriptions of macropore and preferential flow. However, none of the models reviewed could yet be used at the field scale. Jury and Flühler (1992) express strong reservations about the use of the Richards equation to model water flows at the field scale, because many of the process assumptions implicit in the use of the equation are violated.

Jarvis *et al.* (1991) described a mechanistic model to describe water transport in a macroporous soil. The soil porosity was considered to be divided into two interacting flow regions. In macropores, water flows were described by Darcy's equation with unit hydraulic gradient. In micropores, an extended Richards equation simulated convection-dispersion. The boundary condition imposed at the surface controlled the initial entry of water to the two flow regions. Each layer of soil could lose water due to crop uptake, with the surface layer also losing water directly by evaporation. In this case the parameters were fitted for an illitic clay soil (>40% clay) using data collected on water flow from monolith lysimeters. The dynamic pattern of water flow in further lysimeters of the same soil was then well represented by the model with less than 20% of the flow coming from the micropores. This is an example of model development, with this model not yet suitable to answer the questions of farmers or policy-makers.

Transfer function models have been developed for the movement of solutes through soil (Jury, 1982; Jury and Roth, 1990). The arrival of a solute at any depth in soil after the application of an amount of water at the soil surface is characterised by a probability density function (pdf) determined empirically (McCoy *et al.*, 1994). The transfer function framework does not require any process assumptions since the pdf records the characteristics of the medium completely (Jury and Flühler, 1992). Bimodal probability density functions have been used to describe solute flow in an unsaturated structure (Grochulska and Kladivko, 1994; Utermann et al., 1990). However, this approach cannot be used to predict changes in solute concentrations when water flow changes (Jury and Flühler, 1992).

2.2.2 Functional Models. Simpler functional models of water flow consider water to be held in reservoirs of known capacity. The models are known as capacity models (Addiscott et al., 1991) or 'tipping bucket' models. When the capacity of these pools is exceeded then water flows within the soil. Capacity models have been used widely in catchment hydrological modelling (Chiew and McMahon, 1990; Girard, 1975; Vaughan and Corwin, 1994) and to describe water and solute flows within agricultural systems (Groot and de Willigen, 1991; Kersebaum and Richter, 1991). Preferential flow may be included in functional models by including reservoirs of both mobile and immobile water (Whitmore et al., 1991). Addiscott and Whitmore (1991) outlined a functional model which is unusual, since it has both a capacity and rate parameter. The rate parameter is not directly comparable with hydraulic conductivity but is well related to experimental measurements of permeability (Addiscott and Bailey , 1990). The capacity parameter is similar to those used in other capacity models (Addiscott and Whitmore, 1991). Functional models may be more appropriate than mechanistic models for simulation at field and larger scales, since they usually contain fewer parameters. However, their applicability may be limited (de Willigen, 1991).

Chiew and McMahon (1990) used a capacity model to simulate groundwater recharge for a catchment in Southern Australia. Water was held in interception, depression, soil moisture store, groundwater and channel stores within the catchment. Overflow from each store is partitioned to the other stores and leaves the catchment as streamflow and/or deep seepage. The model was parameterised using the streamflow data recorded during 1979–1981. These parameters were used to model streamflow for 1982–1984. The predictions agreed well with measured data. The parameterised model was then used to predict the groundwater recharge from the catchment for the whole period. This is an example of using a model to predict data which could not be measured directly.

3   Predicting Nitrate Supply

3.1   Measurements

Residual mineral nitrogen measured at or soon after harvest may indicate the likely nitrate available for leaching (Chaney, 1990; Sylvester-Bradley and Chambers, 1992). The breakpoint for optimum environmental fertiliser application, *i.e.* residual mineral N similar to unfertilised control plots, is usually close to that of the economic optimum in wheat crops (Glendining et al., 1996). In some vegetable crops the optimum environmental fertiliser application may be significantly lower than the economic optimum and high concentrations of mineral N may therefore be measured in the soil at harvest (Rahn *et al.*, 1992). Residual mineral N levels may also be significant following disastrous yields due to drought, pest or disease (*e.g.* Macdonald *et al.*, 1997). Measurement of mineral N by soil analysis is labour intensive, can be subject to large errors and is difficult to apply on shallow or stony soils. However, the development of rapid field tests for soil mineral nitrogen (Scholefield and Titchen, 1995) may make the use of residual mineral N a feasible index of nitrate supply for leaching.

Crop residues provide a source of mineralisable N. They return between 20 and 145 kg N ha^{-1} in arable crops (Shepherd *et al.*, 1996) and residues can be larger in vegetable crops (Rahn *et al.*, 1992). Where grass swards are cultivated and resown or brought into arable production, large quantities of N are incorporated into the soil (Francis *et al.*, 1992) and large leaching losses commonly result (Whitehead, 1995). Soil organic N may also be mineralised during the autumn and winter period supplying nitrate which may be lost by leaching (Macdonald *et al.*, 1997), even where residual mineral concentrations are low. Indices of soil N supply using biological incubations and simpler chemical extraction techniques are believed to release available N pools preferentially. They might therefore be used to predict the potentially mineralisable N which is at risk of leaching. New methods continue to be developed and these approaches have been extensively reviewed, especially with reference to their potential use in fertiliser recommendation systems (Keeney, 1982; Jarvis *et al.*, 1996). Bundy and Meisinger (1994), however, suggest that such indices can be used only to give a relative indication of the N supply of soils.

^{15}N isotopic dilution methods have allowed the separation of mineralisation and immobilisation and the measurement of gross rates of mineralisation and nitrification (Bjarnason, 1988; Davidson *et al.*, 1991). These methods have greatly extended our understanding of the processes of the soil N cycle and they also give us an opportunity to develop some new process based indices of nitrate leaching risk. Aber (1992) recommended the use of a suite of gross transformation ratios to indicate the degree of N saturation in forest systems. These indices have recently been measured in a range of agricultural systems in south-east England (Willison *et al.*, 1998; Table 1). Ratios of gross nitrification to gross immobilisation greater than a value of 1 indicate that nitrogen mineralised in the system is most likely to be transformed to the more mobile and leachable nitrate ion. Where this ratio is considered alongside the size of the nitrate pool in soil, an index of nitrate supply might be calculated.

3.2 Models

Nitrogen budgets for fields, farms or catchments can give some indication of the likely surplus of nitrogen in the system and therefore that at risk of loss by leaching (Barry *et al.*, 1993; Schröder *et al.*, 1996). N inputs and outputs to the system are estimated by measurement and/or calculation and the N surplus calculated by applying the concept of mass balance (Meisinger and Randall,

Table 1 *Mineral N pool sizes (mg kg^{-1}) and transformation rates (mg kg^{-1} day^{-1}) measured using ^{15}N isotopic dilution techniques in laboratory incubations (Willison et al., 1998). The ratio of gross nitrification and gross immobilisation gives an indication of fate of mineralised N; a ratio greater than 1 indicates that the soil is nitrogen saturated (Aber, 1992)*

	Clay loam Continuous wheat	Clay loam Permanent hay meadow	Peat Drained arable
NH_4 pool size (mg N kg^{-1})	0.2	16.0	2.0
NO_3 pool size (mg N kg^{-1})	6.0	1.5	17.0
Gross mineralisation rate (mg N kg^{-1} day^{-1})	0.6	19.0	3.6
Gross nitrification rate (mg N kg^{-1} day^{-1})	0.1	0.2	3.1
Gross nitrification to NH_4^+ immobilisation ratio	0.1	0.0	2.8

1991). Although the nature and amount of inputs vary among farming systems, regions and even between fields, the mass balance concept provides a framework that can be applied systematically across a range of systems and scales (CLSWC, 1993). The detail with which budgets are compiled varies. Watson and Atkinson (1998) summarise the main three budgeting approaches: economic input-output (EIO) which includes only inputs and outputs with an economic cost; biological input-output (BIO) which includes other inputs such as N fixation and atmospheric deposition; and transfer-recycle input output (TRIO) which includes transfers of N within the system, such as those due to manure management. BIO budgets are the most common (Table 2). However,

Table 2 *Nitrogen surpluses calculated using biological input-output budgets for a conventional and integrated arable farm in the Netherlands (Schröder et al., 1996) and a conventional mixed farm in Canada (Barry et al., 1993). All flows are given in kg N ha^{-1} year^{-1}*

	Netherlands Conventional arable	Netherlands Integrated arable	Canada Conventional mixed
Deposition	39	39	18
Seed	3	3	2
Fixation	18	8	100
Manure	94	95	
Fertiliser	141	94	4
Total inputs	295	240	124
Crop products	135	123	53
Animal products			20
Ammonia volatilisation			17
Total outputs	135	123	90
Surplus	160	117	34

they assume that soil organic matter contents are in equilibrium. A fuller picture of the spatial and temporal flows of N within farm systems is given by TRIO budgets. Such budgets can be used not just to identify systems with a N surplus, but also to highlight components of the system which are particularly leaky.

Process-based models of nitrate supply usually consider soil organic matter as one or more pools of material, where each pool is made up of material which is considered to behave similarly and can be described by a set of unique characteristics (Jarvis *et al.*, 1996). The number and type of pools described varies enormously between models and reflects both the system being modelled and the hypotheses of the modellers. However, many of the complex models appear to be similar. The C and N cycles are closely linked and microbial biomass, readily available and recalcitrant pools of organic matter are often included (Bradbury *et al.*, 1993; Hansen *et al.*, 1991; Parton *et al.*, 1987; van Veen and Frissel, 1981). The rate of release of N from organic matter is affected by external factors such as moisture and temperature and most of the multi-compartmental models use first order kinetics to approximate the decomposition of pools. Monod kinetics have also been introduced to describe microbial growth (Molz *et al.*, 1986). In carbon modelling this pattern has been broken by Bosatta and Agren (1985) who do not break up soil organic matter into discrete pools but describe it as a continuum. However, the mathematics of such an approach is complex.

NCSOIL is an example of a typical process-based nitrate supply model. Molina (1996) gives a full description of the model. Soil organic matter is divided into 3 pools which show increasing resistance to decomposition. Fresh residue and microbial biomass pools are also included. The model timestep is one day or less. Residues and soil organic matter pools are decomposed with first-order rates with respect to their carbon concentrations. The overall shapes of the curves of total and mineral ^{15}N pools simulated by NCSOIL suggest that the model includes the most important processes in the decomposition of ^{15}N labelled wheat straw in laboratory incubations (Hadas *et al.*, 1993). However, quantitative prediction of the pool sizes was not achieved; Hadas *et al.* (1993) suggest that the values used for some of the state variables may be inappropriate for the conditions.

4   Predicting Nitrate Losses

A simple approach to predict nitrate losses could be developed where measured indices of drainage and nitrate supply are coupled. However, this approach is not widely reported in the literature. Pierce *et al.* (1991) use nitrogen budgets to estimate the nitrate available for leaching together with a percolation index. The NCYCLE model developed for UK grassland systems calculates leaching in a similar way (Scholefield *et al.*, 1991). The difference between the inputs to the soil mineral pool and plant uptake is partitioned between denitrification and leaching with reference to soil texture and drainage. Such approaches are limited by the complexity of the processes involved in water movement and N cycling and imprecise data available as inputs (Shaffer et al., 1991). However they provide a rapid and convenient method for screening the N leaching potentials of climate, soil and management combinations.

Simple approaches calculated over a year or cropping cycle give no indication of the likely timing or pattern of the loss which may be as, or more important than, the total amount lost. NLEAP (Shaffer et al., 1991) uses the screening procedure described by Pierce et al. (1991) as a first step. More detailed monthly or event-by-event calculation of N budgets and water flows can then be calculated for sites. NLEAP will then highlight poor management practices associated with N fertiliser, manure, tillage or irrigation. The interactive effects of insect, disease, pest or deficiencies of other nutrients are not taken into account (Shaffer et al., 1991).

More complex models have been developed which use process-based models for both drainage and nitrate supply. Many have been developed not simply to predict nitrate losses but also to simulate the whole soil-plant N cycle and enable fertiliser recommendations to be made. The SUNDIAL model (Bradbury et al., 1993; Smith et al., 1996a) couples an N supply model, based on that described by Jenkinson and Parry (1989), to a functional model of soil water and solute movement (Addiscott and Whitmore, 1991). DAISY (Hansen et al., 1991) uses mechanistic models for both nitrate supply and drainage. NCSOIL forms one module in the larger model NCSWAP which calculates C and N flows in the soil-water-air-plant system (Molina, 1996). ANIMO, a nitrogen turnover model, is coupled to SWATRE (a hydrological model) to simulate nitrogen losses at a field scale (Rijtema and Kroes, 1991). De Willigen (1991) showed that neither the more complex models of soil water nor those of mineralisation simulated measured values were any better than the simpler ones. The data and methods used to derive soil parameters introduce error into predictions as well as poor model structure. It is clear that estimated data in a large model with many parameters will give less reliable results than few measured data in a simple model, so long as both adequately represent the system (Bouma et al., 1996). For use at a catchment scale, complex models may be simplified using the Minimum Information Requirement approach which reduces the uncertainty of data inputs but retains the interacting effects of key envrionmental parameters (Anthony et al., 1996). However, for scientists interested in the processes of nitrate loss, rather than simply the outcome, a model with a physical basis is required (Espeby, 1992).

5   Reliability, Risk and Uncertainty

Where scientists are providing predictions or estimates of nitrate losses in response to the questions asked by farmers, drinking water providers or policy makers, the answers will often be used as a part of a decision-making process. To assess the risk associated with any decision, the possible outcomes and the probability of each outcome should be known. Decisions may also have to be made in uncertain conditions, where the decision maker does not know the probability associated with possible outcomes (Knight, 1921). Predictions of nitrate losses should therefore be accompanied by information about the reliability of estimates, if they are to be used as part of a decision making process (Bouma *et al.*, 1996). Less reliable data have higher variability and so will provide less accurate assessments of nitrate losses. Giving information on the reliability of estimates from either a set of measurements or a model necessitates the consideration of variances, as well as means (Addiscott, 1996). Although the variability of data may be indicated implicitly or even explicitly in scientific reports, if risks are not translated into terms familiar to the end user they might not be recognised or acknowledged (Linthurst *et al.*, 1995). Where easily comprehended information about the reliability of predictions is not provided, poor decisions may be made and unacceptable risks might result.

The reliability of prediction is, in large part, dependent on the choice made in selecting a method to predict nitrate losses. A bad choice of method will use an approach which requires substantial extrapolation. There is no scale independent uniformitarianism of patterns and processes in the natural world (Wiens, 1989). Using models, pedo-transfer functions or measurements at scales, other than those at which they were designed, is therefore difficult. Scaling measurements up is relatively easy for normally distributed data (Stockdale *et al.*, 1997). However, where distributions are skewed, errors in both the mean and the variance of the distribution may be introduced (Halvorson *et al.*, 1997). Changing the scale at which a model is applied is not easy (Lefelaar, 1990). Smith (1995) showed that changing the scale of a model may result in changes in its scope, the heterogeneity of input values and the data requirements. Further development of robust guidelines for scaling needs close collaboration with disciplines such as statistics.

Any model, even the most detailed, represents a grossly generalised picture of reality. Therefore when exclusive emphasis is placed on modelling to characterise soil behaviour, 'pseudo-realities' can result (Bouma, 1993). It is therefore important that all methods to predict nitrate losses, whether measurements or models, are rigorously evaluated against measured data, especially where the predictions are to be used in decision making (Addiscott and Powlson, 1989). The evaluation of models has recently been extensively reviewed (Whitmore, 1991; Smith *et al.*, 1996b). Simple statistical tests can be used to compare the measured sampling distribution for nitrate losses to any indicator sampling distribution.

5.1   Reliability of Measurements

The sampling variance of a measurement has been used as a criterion of its reliability since the beginnings of statistics (Kendall and Stuart, 1973). The determination of the variance and the form of the sampling distribution for any predictive measurement is therefore as important as the determination of its mean (Arrouays *et al.*, 1993). Soil variables may appear to vary spatially in well defined trends, vary erratically or appear constant depending on the scale of measurement (Parkin, 1993). Temporal patterns may relate to daily or seasonal environmental fluctuation in temperature and moisture and may also be affected by sample depth or soil type. In general, soil variables which vary most in time, such as moisture content or soil microbial activity, also have relatively high spatial variability. Spatial and temporal patterns may also interact (Halvorson *et al.*, 1997). Where the likely variability is known then standard formulae can be used to determine the number of samples needed reliably to estimate the mean of any soil property and its sampling distribution (Cochran and Cox, 1977). However, the number of samples required may sometimes be very large. Soil variability even within mapped soil units meant that 100 samples may have been insufficient to define the risk of pesticide leaching narrowly (Nofziger *et al.*, 1996).

Measurements are not only required as indices of nitrate leaching but also to provide input to pedo-transfer functions and models. Error associated with either measuring or estimating basic soil characteristics will express itself in the pedo-transfer function (Vereecken *et al.*, 1992). Pedo-transfer functions should reflect the measurement error of the parameters being used and the effects of spatial variability when simulations are made for areas of land (*e.g.* Petach *et al.*, 1991; Bouma and Hack-ten Broeke, 1993). Some data used as inputs for agricultural models, *e.g.* crop areas and fertiliser use, are collected through by census (MAFF *et al.*, 1996). These data are generalised and often not geo-referenced. The accuracy of the data used as inputs is of key importance in determining the error associated with modelling (Heuvelink *et al.*, 1989).

5.2   Reliability of Models

Deterministic models presume that a certain set of events lead to a uniquely definable outcome, while stochastic models presuppose the outcome to be uncertain and are structured to accommodate this uncertainty. Most of the models used in predicting nitrate losses are deterministic. However, it is possible to assess the reliability of the predictions of a deterministic model if we know the likely variability in model parameters or model input data, *e.g.* weather. A series of independent and equally likely weather sequences might be used as data input for a model (*e.g.* Nofziger *et al.*, 1996) and the prediction expressed as the probability of exceeding a given threshold or the fraction of years in which the drainage water concentration exceeds that limit. Knowledge of soil variability might be used to vary the model parameters in a series of simulations (Finke, 1993) and estimate the risk of exceeding drainage water nitrate concentrations of 50 mg nitrate l^{-1}. This is often called sensitivity analysis.

Parameters used in models may be spatially correlated and hence their statistical properties need to be defined in terms of the Theory of Regionalised Variables (Matheron, 1965). Using geostatistical algorithms, such as kriging, to estimate normally-distributed capacity parameters is not difficult. However, kriging rate parameters is more difficult because of the skewed nature of their distribution (Webster and Addiscott, 1991). If we wish to use models at a large scale on a grid basis with validly averaged parameters, we need to look for models that are linear with respect to their parameters and whose parameters are not too variable in space (Addiscott, 1993). There are some simple models (*e.g.* capacity models for water movement) that probably fulfil both these conditions. However, the application of geostatistics in this way assumes implicitly that the spatial structure of the parameters is stable in time. The water status of a Vesuvian soil showed that water tension and volumetric water content are essentially multi-dimensional when observed in space and time (Comegna and Vitale, 1993).

Failure to account for the variation in model parameters, as well as their means, can lead to false results (Addiscott and Wagenet, 1985). Parameters obtained by optimisation within models may indirectly compensate for deficiencies in the model, including spatial, temporal and management effects and their intrinsic variability (van Es, 1993). If we carry out sensitivity analyses we must therefore ask, not only how sensitive the model is to changes in the value of its parameters, but also how sensitive it is to changes in the variability of its parameters (Addiscott, 1993).

6   Conclusions

Prediction of nitrate losses is possible using either measurements or models. However, the timing of nitrate losses is more difficult to predict than the total loss over a year or season and such predictions almost certainly require the use of modelling approaches. The use of information already collected (*e.g.* soil maps) or easily compiled (*e.g.* nutrient budgets) may allow a rapid identification of areas with high leaching risk at a catchment or regional level. Focused measurements of residual mineral or mineralisable N, or soil physical properties, may allow fields, or parts of fields, to be identified according to leaching risk and managed accordingly. The selection of an appropriate model depends on the available data as well as the scale of interest. At catchment and regional scales (and perhaps even at field scales) simple functional models may be most appropriate, due to their lower data requirements and the lower spatial variability of their parameters. However, functional models cannot be extrapolated reliably. Mechanistic models are important as research tools and may be used during a diagnosis phase to try and understand the mechanisms of nitrate losses and their main controlling factors. The guidelines given above suggest the integrated use of both measurement and modelling approaches in the prediction of nitrate losses, with the combination of measurements and models controlled by the question, scale and data available.

Models can never be completely independent of measurements. Data are required not only for compilation and evaluation of models but each simulation also requires input data and parameters. This provides an opportunity to develop simple methods that can be used by farmers to measure key pool sizes or parameters. Simple measurements may only provide an index for the parameter of interest or integrate across several pools or parameters. However, where models are developed to accommodate such measurements then the site-specificity and accuracy of simulations may be improved. Diagnostic measurements made during the simulation (*e.g.* field measurements of mineral N) might also be used to allow internal adjustment of models. Measurements and models should be linked in this way as soon as possible.

Although scientists can make predictions, they are often reluctant to do so. They are aware of the vulnerability of a single estimate to misinterpretation. Neither measurements nor models are currently used widely as part of fertiliser recommendation systems or environmental impact assessment. Predictions should be presented in a way that tempers them with an estimate of their reliability, in terms that are readily understood. Such presentational techniques should be brought quickly into use.

Scientists, farmers, environmentalists, policy makers and all those who seek to understand and predict nitrate losses need to be aware that models are not the whole answer. Measurements and models must go hand in hand.

'Ask not the computer simulation alone, for it may portray a make-believe world. Ask nature itself too . . .' (Hillel, 1993).

Acknowledgements

IACR-Rothamsted acknowledges the grant-aided support of the Biotechnology and Biological Sciences Research Council. The author acknowledges financial support from the Ministry of Agriculture Fisheries and Food to carry out part of this work and the help of many colleagues in the Soil Science department, particularly Dan Murphy and Toby Willison.

References

Aber J. D., 1992, Nitrogen cycling and nitrogen saturation in temperate forest ecosystems., *Tree*, **7**, 20.

Addiscott T. M., 1993, Simulation modelling and soil behaviour., *Geoderma*, **60**, 15.

Addiscott T.M., 1996, Measuring and modelling nitrogen leaching: parallel problems., *Plant Soil*, **181**, 1.

Addiscott T. M. and N. J. Bailey, 1990, Relating the parameters of a leaching model to the percentage of clay and other soil components., In: 'Field scale solute and water transport through soil' Roth, K., Flühler, H., Jury, W. A. and Parker, J. C. (eds.) BirkenhäuserVerlag, Basel, p. 209.

Addiscott T. M. and D. S. Powlson, 1989, Nitrogen., *New Scientist*, **29 April**, 28.

Addiscott T. M. and R .J. Wagenet, 1985, Concepts of solute leaching in soils: a review of modelling approaches., *J. Soil Sci.*, **36**, 411.

Addiscott T. M. and A. P. Whitmore, 1991, Simulation of solute leaching in soils of differing permeabilities., *Soil Use Manag.*, **7**, 94.

Addiscott T. M., A. P. Whitmore and D. S. Powlson, 1991, 'Farming, fertilizers and the nitrate problem', CAB International, Wallingford, UK. p. 55

Anthony S., P. Quinn and E. Lord, 1996, Catchment scale modelling of nitrate leaching, *Aspects Appl. Biol.*, **46**, Modelling in applied biology: spatial aspects. p23.

Arrouays D., I. Vion and J. L. Kicin, 1993, Spatial analysis and modelling of topsoil carbon storage in temperate forest humic loamy soil of France., *Soil Sci.*, **159**, 191.

Barry D. A. J., D. Goorahoo and M. J. Goss, 1993, Estimation of nitrate concentrations in groundwater using a whole farm nitrogen budget., *J. Environ. Qual.*, **22**, 767.

Batjes N. H., 1997, A world dataset of derived soil properties by FAO-UNESCO soil unit for global modelling., *Soil Use Manag.*, **13**,. 9.

Bjarnason S., 1988, Calculation of gross nitrogen immobilization and mineralization in soil., *J. Soil Sci.*, **39**, 393.

Bosatta E. and G. I. Agren., 1985, Theoretical analysis of decomposition of hetrogeneous substrates., *Soil Biol. Biochem.*, **17**, 601–610.

Bosatta E. and F. Berendse, 1984, Energy or nutrient regulation of decomposition: implications for the mineralization-immobilization response to perturbations., *Soil Biol.Biochem.*, **16**, 63.

Bouma J, 1989, Using soil survey data for quantitative land evaluation. In: B. A. Stewart (ed.), Advances in Soil Science, 9, Springer, New York., p. 225.

Bouma J., 1993, Soil behaviour under field conditions: differences in perception and their effects on research., *Geoderma*, **60**, 1.

Bouma J. and M. J. D. Hack-ten Broeke, 1993, Simulation modelling as a method to study land qualities and crop productivity related to soil structure differences., *Geoderma*, **57**, 51.

Bouma J. and J. A. J. van Lanen, 1987, Transfer functions and threshold values: from soil characterisitics to land qualities. In: 'Quantified land evaluation' K. J. Beck (ed.), Proceedings of a ISSS and SSSA Workshop, Washington D.C. International Institute of Aerospace Survey and Earth Sciences Publication no. 6. Entschede, The Netherlands., p. 106.

Bouma J., H. W. G. Boooltink, A. Stein and P. A. Finke, 1996, Reliability of soil data and risk assessment of data applications., In: 'Data reliability and risk assessment in soil interpretations'. W.D. Nettleton *et al.* (eds.) SSSA Special Publication no. 47., p. 63.

Bradbury N. J., A. P. Whitmore, P. B. S. Hart and D. S. Jenkinson, 1993, Modelling the fate of nitrogen in crop and soil in the years following application of ^{15}N-labelled fertilizer to winter wheat.,.*J. Agric. Sci. (Camb.)*, **121**, 363.

Bundy L. G. and J. J. Meisinger, 1994, Nitrogen availability indices., In: 'Methods of soil analysis. Part 2.' SSSA Book Series 5. SSSA, Madison., p. 951.

Chaney K., 1990, Effect of nitrogen fertilizer rate on soil nitrate nitrogen content after harvesting winter wheat., *J. Agric. Sci. (Camb.)*, **114**, 171.

Chiew F. H. S and T. A. McMahon, 1990, Estimating groundwater recharge using a surface watershed modelling approach., *J. Hydrol.*, **114**, 285.

Cochran W. G. and G. M. Cox, 1977, 'Experimental Designs'. Wiley, New York., p.17.

Collins J. F. and T. Cummins, 1996, 'Agroclimatic atlas of Ireland', AGMET (Joint Working Group on Applied Agricultural Meteorology)., p. 96.

Comegna V. and C. Vitale, 1993, Space-time analysis of water status in a volcanic Vesuvian soil., *Geoderma*, **60**, 135.

Committee on Long Range Soil and Water Conservation (CLSWC), 1993, 'Soil and Water Quality: an agenda for agriculture.' National Research Council, National Academy Press. Washington., p. 431.

Davidson E. A., S. C. Hart, C. A. Shanks and M. K. Firestone, 1991, Measuring gross nitrogen mineralisation, immobilization, and nitrification by ^{15}N isotopic pool dilution in intact soil cores., *J. Soil Sci.*, **42**, 335.

de Ruiter P. C., J. C. Moore, K. B. Zwart, L. A. Bouwan, J. Hassink, J. Bloemde, J. A. Vos, J. C. Y. Marinissen, W. A. M. Didden, G. Lebbink and L. Brussaard, 1993,

Simulation of nitrogen mineralization in the below-ground food webs of two winter wheat fields., *J. Appl. Ecol.*, **30**, 95.

de Willigen P, 1991, Nitrogen turnover in the soil-crop system: comparison of fourteen simulation models., *Fert. Res.*, **27**, 141.

Dymnikov V. P. and A. N. Filatov, 1997, 'Mathematics of climate modelling'. Birkhäuser, Boston, p. 1.

Elrick D. E. and B. E. Clothier, 1990, Solute transport and leaching., In: 'Irrigation of Agricultural crops' B. A. Stewart and D. R. Nielsen (eds.) Agronomy Series volume 30. ASA, CSSA, SSSA. Madison, Wisconsin., p. 93.

Espeby B., 1992, Coupled simulations of water flow from a field investigating glacial till slope using a quasi-two dimensional water and heat model with bypass flow., *J. Hydrol.*, **131**, 105.

Finke P. A., 1993, Field scale variability of soil structure and its impact on crop growth and nitrate leaching in the analysis of fertilizing scenarios., *Geoderma*, **60**, 89.

Francis G. S., R. J. Haynes, G. P. Sparling, D. J. Ross and P. H. Williams, 1992, Nitrogen mineralization, nitrate leaching and crop growth following cultivation of a temporary leguminous pasture in autumn and winter., *Fert. Res.*, **33**, 59.

Girard G., 1975, 'Modèle global ORSTOM 1974. Première application du modèle à discrétisation spatiale sur le bassin versant de la crique Grégoire en Guyanne. Atelier hydrologique sur les modèles mathématiques', ORSTOM., p. 1.

Glendining M. J., D. S. Powlson, P. R. Poulton, N. J. Bradbury, D. Palazzo and X. Li, 1996, The effects of long term applications of inorganic nitrogen fertilizer on soil nitrogen in the Broadbalk Wheat Experiment. *J. Agric. Sci. (Camb.)*, **127**, 347.

Greenwood D. J and A. Draycott, 1989, Experimental validation of an N-response model for widely different crops, *Fert. Res.*, **18**, 153.

Grochulska J. and E. J. Kladivko, 1994, A two region model of preferential flow of chemicals using a transfer function approach., *J. Environ. Qual.*, **23**, 498–507.

Groot J. J. R. and P. de Willigen, 1991, Simulation of nitrogen balance in the soil and winter wheat crops., *Fert. Res.*, **27**, 261.

Hadas A., S. Feigenbaum, M. Sofer, J. A. E. Molina and C. E. Clapp, 1993, Decomposition of nitrogen-15–labeled wheat and cellulose in soil: modeling tracer dynamics., *Soil Sci. Soc. Am. J.*, **57**, 996.

Halvorson J. J., J. L. Smith and R. I. Papendick, 1997, Issues of scale for evaluating soil quality., *J. Soil Water Conserv.*, **52**, 26–30.

Hanks R. J., 1992, 'Applied soil physics. Soil water and temperature applications. 2nd Edition'. Springer-Verlag. New York., p. 63.

Hansen S., H. E. Jensen, N. E. Nielsen and H. Svendsen, 1991, Simulation of nitrogen dynamics and biomass production in winter wheat using the Danish simulation model DAISY., *Fert. Res.*, **27**, 245.

Heuvelink G. B. M., P. A. Burrough, and A. Stein, 1989, Propagation of errors in spatial modelling with GIS, *Int. J. Geog. Inform. Syst.*, **3**, 303.

Hillel D., 1980, 'Fundamentals of soil physics'. Academic Press. London, p. 166.

Hillel D., 1993, Science and the crisis of the environment., *Geoderma*, **60**, 377.

Hutchinson M. F., 1995, Stochastic space-time weather models form ground based data. *Agric. For. Meteorol.*, **73**, 237.

Jarvis N. J., P.-E. Jansson, J. P. Dik and I. Messing, 1991, Modelling water and solute transport in macroporous soil. I. Model description and sensitivity analysis., *J. Soil Sci.*, **42**, 59.

Jarvis S. C., 1998 (this volume) pp. 2–20.

Jarvis S. C., E. A. Stockdale, M. A. Shepherd and D. S. Powlson, 1996, Nitrogen mineralization in temperate agricultural soils: processes and measurement., *Adv. Agron.*, **57**, 188–235.

Jenkinson D. S. and L. C. Parry, 1989, The nitrogen cycle in the Broadbalk wheat experiment- a model for the turnover of nitrogen through the soil microbial biomass., *Soil Biol. Biochem.*, **21**, 535.

Johnsson H., L. Bergström, P. E. Jansson and K. Paustian, 1987, Simulated nitrogen dynamics and losses in a layered agricultural soil., *Agric. Ecosyst. Environ.*, **18**, 333.

Jones M. E., 1976, In: Topographic climates: soils, slopes and vegetation. In: 'The Climate of the British Isles.' T. J. Chandler and S. Gregory (eds.) Longman. London., p. 288.

Jury W. A., 1982, Simulation of solute transport using a transfer function model., *Water Resourc. Res.*, **18**, 363.

Jury W. A. and H. Flühler, 1992, Transport of chemicals through soil: mechanisms, models and field applications, *Adv. Agron.*, **47**, 141.

Jury W. A. and K. Roth, 1990, 'Transfer functions and solute movement through soil. Theory and applications'. Birkhäuser Verlag, Berlin, p. 1.

Keeney D. R., 1982, Nitrogen availability indices., In:'Methods of Soil Analysis Part 2' A. L. Page, R. H. Miller and D. R. Keeney (eds.) ASA, Madison., p. 711.

Kendall M. G. and A. Stuart, 1973, 'The advanced theory of statistics. Volume 2. Inference and relationship. Third edition.' Griffin, London., p. 1.

Kersebaum K. C. and J. Richter, 1991, Modelling nitrogen dynamics in a plant-soil system with a simple model for advisory purposes., *Fert. Res.*, **27**, 273.

Knight F. H., 1921, 'Risk, uncertainty and profit'. Houghton Mifflin, Boston., p.1.

Lefelaar P. A., 1990, On scale problems in modelling: an example from soil ecology., 'Theoretical production ecology: reflections and prospects', R. Rabbinge, J. Goudriaan, H. van Keulen, F. W. T Penning de Vries, and H. H. van Laar, (eds.) Simulation Monographs No. 34, PUDOC., Wageningen, , p.301.

Linthurst R. A., P. Bourdeau and R. G. Tardiff, (eds.), 1995, 'Methods to assess the effects of chemicals on ecosystems.' SCOPE 53. Wiley, Chichester. p. 13.

Ma L. and H. M. Selim, 1997, Transport of chemicals through soil- mechanisms, models and field applications., *Adv. Agron.*, **58**, 95.

Macdonald A. J., P. R. Poulton, D. S. Powlson and D. S. Jenkinson. 1997, Effects of season, soil type and cropping on recoveries, residues and losses of ^{15}N-labelled fertilizer applied to arable crops in spring. *J. Agric. Sci. (Camb.)*, **129**, 125.

Matheron G., 1965, 'Les Variables Régionalisées et Leur Estimation' Masson, Paris. p.1.

McCoy E. L., C. W. Boast, R. C. Stehouwer and E. J. Kladivko, 1994, In: 'Soil processes and water quality' R. Lal and B. A. Stewart, (eds.) CRC. Boca Raton. USA, p. 303.

Meisinger J. J. and G. W. Randall, 1991, Estimating Nitrogen budgets for soil-crop systems., In: 'Managing nitrogen for groundwater quality and farm profitability', R. F. Follett, D. R. Keeney and R. M. Cruse (eds.). SSSA, Madison, Wisconsin., p. 85.

Ministry of Agriculture Fisheries and Food, The Scottish Office Agriculture. Environment and Fisheries Department, Department of Agriculture for Northern Ireland and the Welsh Office. 1996, 'The Digest of Agricultural Census Statistics 1995.' The Stationery Office, London. p. 3–2.

Molina J. A. E., 1996, Description of the model NCSOIL., In: 'Evaluation of soil organic matter models. Using existing long-term datasets' D.S. Powlson, P. Smith and J. U. Smith (eds.) NATO ASI series Vol. 9., Springer, Berlin, p. 269.

Molz F. J., M. A. Widdowson and L. D. Benefield, 1986, Simulation of microbial growth dynamics coupled to nutrient and oxygen transport in porous media., *Water Resourc. Res.*, **22**, 1207.

Nofziger D. L., J.-S. Chen and A. G. Hornsby, 1996, Uncertainty in pesticide leaching risk due to soil variability., In: 'Data reliability and risk assessment in soil interpretations', W.D. Nettleton *et al.* (eds), SSSA Special Pub. 47. SSSA, Madison. p. 99.

OECD, 1993, 'OECD Core set of indicators for environmental performance reviews. A synthesis report by the Group on the State of the Environment.' OECD, Paris, p.5.

Parkin T. B., 1993, Spatial variability of microbial processes in soil – a review., *J. Environ. Qual.*, **22**, 409.

Parton W. J., D. S. Schimel, C. V. Cole and D. S. Ojima, 1987, Analysis of factors controlling soil organic matter levels in Great Plains grassland., *Soil Sci. Soc. Am. J.*, **51**, 1173.

Petach M. C., R. J. Wagenet and S. D. De Gloria, 1991, Regional water flow and pesticide leaching using simulations with spatially distributed data., *Geoderma*, **48**, 245.

Pierce F. J., M. J. Shaffer and A. D. Halvorson, 1991, Screening procedure for estimating potentially leachable nitrate-nitrogen below the root zone., In: 'Managing nitrogen for groundwater quality and farm profitability'. R. F. Follett, D. R. Keeney and R. M. Cruse (eds.). SSSA, Madison, Wisconsin., p. 259.

Powlson D. S., 1998 (this volume, pp. 42–57).

Rahn C. R., L. V. Vaidyanathan and C. D. Paterson, 1992, Nitrogen residues from brassica crops., *Aspects Appl. Biol.*, **30**, Nitrate in Farming Systems, 263.

Rijtema P. E. and J. G. Kroes, 1991, Some results of nitrogen simulations with the model ANIMO., *Fert. Res.*, **27**, 189–198.

Rose D. A., 1991, The variability of winter drainage in England and Wales., *Soil Use Manag.*, **7**, 115–121.

Scholefield D. and N. M.. Titchen, 1995, Development of rapid field test for soil mineral nitrogen and its application to grazed grassland., *Soil Use Manag.*, **11**, 33.

Scholefield D., D. R. Lockyer, D. C. Whitehead and K. C. Tyson, 1991, A model to predict transformations and losses of nitrogen in UK pastures grazed by beef cattle., *Plant Soil*, **132**, 165.

Schröder J. J., P. van Asperen, G. J. M. van Dongen and F. G. Wijnands, 1996, Nutrient surpluses on intrgrated arable farms., *Eur. J. Agron.*, **5**, 181.

Shaffer M. J., A. D. Halvorson and F. J. Pierce, 1991, Nitrate leaching and economic analysis package (NLEAP): Model description and application. In: 'Managing nitrogen for groundwater quality and farm profitability'. R. F. Follett, D. R. Keeney and R. M. Cruse (eds.). SSSA, Madison, Wisconsin., p. 285.

Shepherd M. A., E. A. Stockdale, D. S. Powlson and S. C. Jarvis, 1996, The influence of organic nitrogen mineralization on the management of agricultural systems in the UK., *Soil Use Manag.*, **12**, 76.

Smith J. L., B. L. McNeal, H. H. Cheng and G. S Campbell, 1986, Calculation of microbial maintenance rates and net nitrogen mineralization in soil at steady-state., *Soil Sci. Soc. Am. J.*, **50**, 332.

Smith J. U., 1995, Models and scale: up- and down- scaling., In: 'Models in action.' P.A. Stein, F. W. T. Penning de Vries and P. J. Schotman (eds.) AB-DLO. Haren., p. 25.

Smith J. U., N. J. Bradbury and T. M. Addiscott, 1996a, SUNDIAL: a PC-based system for simulating nitrogen dynamics in arable land., *Agron. J.*, **88**, 38.

Smith J. U., P. Smith and T. M. Addiscott, 1996b, Quantitative methods to evaluate and compare soil organic matter (SOM) models., In: 'Evaluation of soil organic matter models. Using existing long-term datasets' D.S. Powlson, P. Smith and J. U. Smith (eds.) NATO ASI series Vol. 9., Springer, Berlin, p. 181.

Smith S. J. and D. K. Cassel, 1991, Establishing nitrate leaching in soil., In: 'Managing nitrogen for groundwater quality and farm profitability', R. F. Follett, D. R. Keeney and R. M. Cruse (eds.). SSSA, Madison, Wisconsin, p. 165.

Stockdale E. A., J. G. Gaunt and J. Vos, 1997, Soil-plant nitrogen dynamics: what concepts are required?, *Eur. J. Agron.*, **7**, 145.

Sylvester-Bradley R. and B. J. Chambers, 1992, The implications of restricting use of fertilizer nitrogen for the productivity of arable crops, their productivity and potential pollution by nitrate., *Aspects Applied Biol.* **30**, Nitrate in Farming Systems, 85.

Tietje O. and H. Tapkenhinrichs, 1993, Evaluation of pedo-transfer-functions., *Soil Sci. Soc. Am. J.*, **57**, 1088.

Tinker P. B. and T. M. Addiscott, 1984, 'The Nitrogen Requirements of Cereals. Reference Book 385'. ADAS, Ministry of Agriculture, Fisheries and Food., p.265.

Titus B. D. and M. K. Mahendrappa, 1996, 'Lysimeter systems designs used in soil research: a review', Canadian Forest Service, Canada, p. 14.

Utermann J., E. J. Kladivko and W. A. Jury, 1990, Evaluating pesticide migration in tile-drained soils with a transfer function model., *J. Environ. Qual.*, **19**, 707.

van Es H. M.,1993, Evaluation of temporal, spatial, and tillage-induced variability for parameterization of soil infiltration., *Geoderma*, **60**, 187

van Veen J. A. and M. J. Frissel, 1981, 'Simulation of nitrogen behaviour of soil-plant systems.' (M.J. Frissel and J. A. Van Veen eds.) PUDOC, Wageningen., p. 126.

Vaughan P. J. and D. L. Corwin, 1994, A method of modelling vertical fluid flow and solute transport in a GIS context., *Geoderma*, **64**, 139.

Vereecken H., 1995, Estimating the unsaturated hydraulic conductivity from theoretical models using simple soil properties., *Geoderma*, **65**, 81.

Vereecken H., J. Diels, J. van Orshoven, J. Feyen and J. Bouma, 1992, Functional evaluation of pedotransfer functions for the estimation of soil hydraulic properties., *Soil Sci. Soc. Am. J.*, **56**, 1371.

Vereecken H., J. Maes and J. Feyen,. 1989, Estimation of the soil moisture retention characteristic from texture, bulk density, and carbon content., *Soil Sci.*, **148**, 1.

Wagenet R. J., J. L. Hutson and J. W. Biggar, 1989, Simulating the fate of a volatile pesticide in unsaturated soil: a case study with DBCP., *J. Environ. Qual.*, **18**, 78.

Warrick A. W. and D. R. Nielsen, 1977, In: 'Applications of soil physics.' Hillel, D. (ed.) Academic Press, New York., p. 319.

Watson C. A. and D. Atkinson, 1998, Using nitrogen budgets to indicate nitrogen use efficiency and losses from whole farm systems: a comparison of three methodological approaches , *Nut. Cycl. Agro-Ecosyst.*, in press.

Webster R. and T. M. Addiscott, 1991, Spatial averaging of solute and water flows in soil., In: 'Field-scale Water and Solute Flux in Soils' K. Roth, H. Flühler, W. A. Jury & J. C. Parker, (eds.), Birkhauser, Basel. p. 163.

Whitehead D. C., 1995, 'Grassland Nitrogen' CAB International. Wallingford.., p. 147.

Whitmore A. P., 1991, A method for assessing the goodness of computer simulation of soil processes., *J. Soil Sci.*, **42**, 289.

Whitmore A. P., K. W. Coleman, N. J. Bradbury and T. M. Addiscott, 1991, Simulation of nitrogen in soil and winter wheat through organic matter., *Fert. Res.*, **27**, 283.

Wiens J. A., 1989, Spatial scaling in ecology., *Funct. Ecol.*, **3**, 385.
Willison T. W., D. V. Murphy, J. C. Bakerand K. W. T. Goulding, 1998, Gross nitrogen transformations in soils under different land uses, *J. Environ. Qual.*, in press

3

Fate of Nitrogen from Manufactured Fertilizers in Agriculture

D. S. Powlson

SOIL SCIENCE DEPARTMENT, IACR-ROTHAMSTED, HARPENDEN, HERTS AL5 2JQ, UK

1 Introduction

In virtually all situations the use of land for agricultural production leads to greater loss of nitrogen (N) by nitrate leaching and other mechanisms than if the same land were kept under natural or semi-natural vegetation. In part this is because N inputs to an agricultural system as inorganic fertilizers, manures, plant residues or biological N fixation are almost always greater than to natural ecosystems. As a broad generalization it is also true that any effort to increase agricultural production by increasing N inputs, from any source, will increase the risk of larger N losses occurring. The actual losses that occur are dependent on soil type, weather and a range of site-specific management factors.

It is often assumed that inorganic fertilizer is the main source of nitrate moving from agriculture to natural waters but a large body of research conducted over recent decades demonstrates that this is a great over-simplification (Addiscott *et al.*, 1991). Nitrate derived from the mineralization of organic N contained in soil organic matter, plant residues or organic manure is often of greater significance. Fertilizers do have important impacts on N loss though the effects are often indirect. Some decrease in the input of inorganic fertilizer, or its partial replacement with organic or biologically fixed N, is possible but in the foreseeable future it is difficult to envisage large-scale agriculture in western Europe that is not heavily dependent on inorganic fertilizer. It is therefore entirely appropriate that its fate is scrutinized closely but, equally, all sources of nitrate loss must be identified and the controlling factors understood if mitigation measures are to be effective.

Losses of N other than nitrate leaching also have major environmental impacts. Nitrous oxide (N_2O) evolved from soil through the processes of denitrification (or, under some circumstances, from nitrification) is both a powerful greenhouse gas and is involved in stratospheric ozone depletion

(Bouwman, 1990). Ammonia (NH_3) which can be evolved in large quantities from animal manure can cause both nutrient enrichment and soil acidification when redeposited, especially in sensitive semi-natural ecosystems. It is vital that agricultural management strategies designed to decrease nitrate leaching do not cause an increase in other loss pathways.

2 Losses of Fertilizer N During the Growing Season

Experiments with ^{15}N-labelled fertilizers are valuable for quantifying losses of fertilizer N during the growing season, though not necessarily for identifying the processes causing loss. Fertilizer labelled with ^{15}N is applied to small plots under field conditions, the crop and soil are sampled at final harvest or at intermediate stages and the total quantity of ^{15}N that can be accounted for in both crop and soil is measured. In experiments in southeast England in which ^{15}N-labelled fertilizer was applied in spring to a range of arable crops (winter wheat, winter oilseed rape, potatoes, sugar beet) at various rates up to 240 kg N ha^{-1}, losses varied from 5 to 35% of the N applied (Macdonald *et al.*, 1997; Powlson *et al.*, 1992). Losses were influenced by the state of the crop (sowing date and whether or not diseases were prevalent), soil type and weather. A major factor was rainfall during the first few weeks following fertilizer application. In most cases this could be represented by the linear regression:

$$L_{70} = 11.9 + 0.32R_3, = 0.62$$

where L_{70} = percentage of labelled fertilizer N *not* recovered at harvest in crop or soil to a depth of 70 cm and R_3 = cumulative rainfall (in mm) in the 3 weeks following ^{15}N-labelled fertilizer application in spring. There was a similar but closer relationship between N loss and rainfall if only the experiments with winter wheat were considered.

The two major processes causing N loss, nitrate leaching and denitrification, could both be increased by rainfall soon after fertilizer application when much of the applied N is still present in soil as nitrate. Addiscott and Powlson (1992) used the SLIM leaching model (Addiscott and Whitmore, 1987) to calculate the maximum likely contribution of leaching to the total loss of fertilizer-derived N applied to winter wheat. On average, leaching represented about one-third of the loss but often was much less. In two cases out of the 13 experiments examined, when there was unusually high rainfall, it was a greater proportion. The overall conclusion was that conditions favouring a major loss of N through nitrate leaching in spring, following fertilizer application, were uncommon under the climatic conditions of eastern England and probably for much of Europe. Addiscott and Powlson (1992) presumed that denitrification was the major process causing loss but other gaseous pathways may also contribute, including emission of N from crop foliage as ammonia (Schorring *et al.*, 1989) or possibly oxidized forms and N_2O or NO from nitrification (Hutchinson and Davidson, 1993; Yamulki *et al.*, 1997).

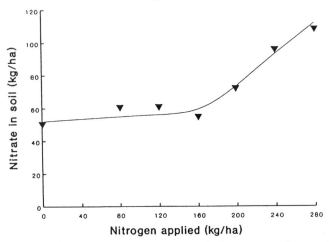

Figure 1   *Relationship between the nitrogen applied to winter wheat and the amount remaining unused in soil. The 'surplus nitrate' curve (Chaney, 1990)*

3   Losses of Nitrate During Winter

In a maritime climate, any nitrate remaining in soil after harvest, plus that mineralized in autumn, is at considerable risk of being lost by leaching during the subsequent winter. From the standpoint of water quality it is therefore essential to aim at management practices that decrease this quantity. In continental climates, where soil usually freezes for a considerable period in winter, nitrate accumulated in the soil during autumn will remain in the frozen soil and become subject to leaching or runoff when the soil thaws in spring. It is necessary to be aware of this difference in timing when transferring results between maritime and continental areas.

Figure 1, from Chaney (1990), shows the amounts of nitrate remaining in soil after harvest where a wide range of fertilizer rates were applied to winter wheat. For rates up to 160 kg N ha^{-1}, residual nitrate did not differ markedly from that in the unfertilized control treatment; above 160 kg N ha^{-1} it increased sharply. Figure 2 from Glendining *et al.* (1996) shows a similar situation with winter wheat on the long-term Broadbalk Experiment at Rothamsted. Residual nitrate in soil after harvest was about 50 kg N ha^{-1} in the unfertilized control and in all N rates up to 144 kg N ha^{-1}. At higher N rates it rapidly increased to about twice this value. Maximum yield was essentially attained with 144 kg N ha^{-1}; thus, in this case, production goals could be achieved with an input of N fertilizer that left no more residual nitrate than wheat given no fertilizer N at all. These two examples show that there is not necessarily a direct relationship between the amount of N fertilizer applied to winter wheat and the quantity of residual nitrate in soil. However it would be incorrect to conclude that there is never a conflict between attaining a profitable yield and leaving undesirably large nitrate residues. This was

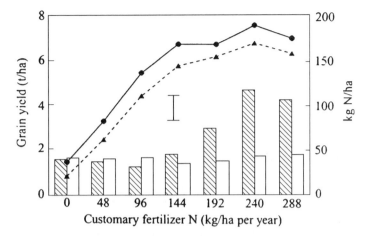

Figure 2 *Inorganic N (kg ha^{-1}) in soil (0–100 cm) at harvest in 1990, customary fertilizer N applied (▨) or withheld (□). Grain yield (-•-, 85% DM t ha^{-1}) and N in crop (-▲-, grain plus straw, kg N ha^{-1}) for areas where fertilizer N was applied at the customary rate. Error bar indicates ± S.E., for soil inorganic N content, 14 D.F.*

demonstrated in other years on the Broadbalk Experiment (Glendining *et al.*, 1996). In a year when weather conditions constrained maximum grain yield to less than 6 t ha^{-1}, nitrate residues of 100 kg N ha^{-1} remained even where only 144 kg N ha^{-1} was applied; in another year when yields followed a similar pattern to those in Figure 2 residual nitrate did not increase sharply even at the highest N rate of 288 kg N ha^{-1}.

Measuring how much of the residual nitrate in soil after harvesting crops is labelled with ^{15}N gives an indication of the direct contribution of N from fertilizer to nitrate exposed to leaching during winter: with cereals it is usually small. For example, Figure 3 (from Macdonald *et al.*, 1989) shows that less than 10% of the nitrate in soil after winter wheat was derived from fertilizer. Typically the fertilizer-derived quantity is less than 3 kg N ha^{-1} out of the 50 kg N ha^{-1} or more of total nitrate-N commonly found in soils to a depth of 1 m in autumn. Data from ^{15}N experiments has to be interpreted with some caution as there are reasons why the contribution of fertilizer-derived N may be underestimated; see Jenkinson *et al.* (1985) for a discussion of this. But even if the contribution were twice that commonly measured, a vast overestimate of the possible error, the direct contribution of fertilizer N to residual nitrate would still be modest. Such results emphasize the importance of mineralization of soil organic matter as a source of the nitrate leached during winter from agricultural soils.

In the Brimstone Experiment, situated on a cracking clay soil of the Denchworth Series in Oxfordshire, UK, nitrate movement to field drains has been continuously monitored from 0.24 ha hydrologically separated plots for almost 20 years (Cannell *et al.*, 1984; Catt *et al.*, 1992). In some years there was

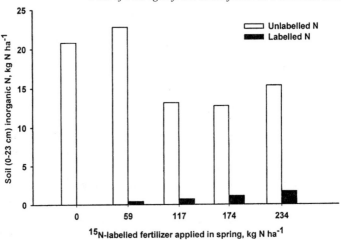

Figure 3   *Unlabelled and labelled inorganic N remaining in soil (0–23 cm) at harvest of winter wheat*

a short-lived peak of high nitrate concentration if drainage occurred after N fertilizer application in spring but the major losses of nitrate were consistently during winter. Losses during winter were not related to N fertilizer applied in the previous spring but reflected the production of nitrate from mineralization of soil organic matter or crop residues.

Nitrate residues following some crops are greater than after cereals. For example, larger residues were found after potatoes and a greater proportion was derived from the fertilizer applied in spring (Macdonald *et al.*, 1997). Many horticultural crops, such as leafy brassicas, leave much larger nitrate residues than cereals (Rahn *et al.*, 1992) for a combination of reasons. They generally have a shallow root system that is inefficient at capturing nitrate and are often given very high rates of N fertilizer to compensate for this inefficiency because of the high value of the product. In addition, the crop residues have a high N concentration and their decomposition in soil after harvest leads to a large production of nitrate. Legumes such as peas, beans or lucerne (alfalfa) generally leave large nitrate residues even though they are given little or no N fertilizer (*e.g.* Macdonald *et al.*, 1997) because they preferentially absorb biologically fixed N, leaving in the soil much of the nitrate derived from mineralization.

Any agronomic practice that increases soil organic matter content or involves a greater return of crop residues inevitably increases the quantity of N that will eventually be mineralized and at risk of being leached. This causes a dilemma as increasing soil organic matter content is a major way of enhancing soil quality in terms of fertility, physical properties, biodiversity and long-term sustainability. Where arable crops are grown in rotation with pasture, mineralization of the accumulated roots and organic matter can lead to much greater residues of nitrate than in a continuous arable situation for at least two years after ploughing up the ley (Macdonald *et al.*, 1989). Applications of animal manure, important both as a means of recycling nutrients and of improving soil quality, can also lead to major increases in N mineralization and nitrate loss from both grassland and arable crops (Jarvis *et al.*, 1995).

4   Indirect Effects of N from Fertilizers

Although the *direct* contribution of inorganic fertilizers on winter nitrate leaching is usually small, indirect effects are significant. Excessively large applications lead to high nitrate residues: the most common reason for this is an underestimation of nitrate production from mineralization, emphasizing the importance of more precise prediction of this quantity. Larger inputs of N from inorganic fertilizer lead to larger quantities being retained in organic forms in soil or returned as crop residues. Eventually this extra N is mineralized to nitrate, some of which is absorbed by crops but some is lost. For example, in the Broadbalk Experiment at Rothamsted, annual nitrate production has increased by up to two-fold in the plot receiving 144 kg N ha^{-1} in fertilizer annually for 150 years compared to the unfertilized plot (Glendining *et al.*, 1996). This is reflected by an increase in the mean concentration of nitrate-N in drainage water from 2 mg l^{-1} from the unfertilized plot to 7 mg l^{-1} from the plot receiving 144 kg N ha^{-1} (Powlson and Goulding, 1995). This trend is seen in many long-term experiments worldwide with increases in mineralization rate of 20–50% of the value in unfertilized soil often detected within 20–30 years (Glendining and Powlson, 1995).

5   Agricultural Management Strategies to Decrease Nitrate Loss

5.1   Crop Cover During Winter

The presence of plants during winter is beneficial as any nitrate absorbed will no longer be at risk of leaching. Many experiments show that early sowing of a crop in autumn leads to greater N uptake and decreased leaching (Widdowson *et al.*, 1987) but the actual decrease is often less as commercial cereal crops may only absorb about 20 kg N ha^{-1} before winter. As the cultivations required for autumn sowing stimulate mineralization to some extent, the N absorbed by the crop may barely exceed the extra nitrate produced. However, less leaching is generally found in the presence of an autumn-sown crop than where the soil is bare for many months before sowing a spring crop (*e.g.* Webster *et al.*, 1992). In the Brimstone Experiment, nitrate leaching has been monitored in various rotations: apart from grass that was unfertilized and ungrazed, the least total leaching over 5 years was from continuous winter cereals. However, the 50 mg l^{-1} nitrate EU limit was exceeded on many occasions, illustrating the difficulty of meeting this limit even where management strategies regarded as 'good practice' were followed (Catt *et al.*, 1992).

Growing a cover crop or 'catch crop' during winter decreases leaching in the period when soil would otherwise be bare before sowing a spring crop. If conditions for growth are favourable, cover crops can absorb over 50 kg N ha^{-1} and greatly decrease leaching (Christian *et al.*, 1992) but effectiveness is highly dependent on sowing date and weather conditions in autumn (Davies *et al.*, 1996). Another possibility for achieving the necessary rapid establishment is to undersow a cover crop before the previous crop is harvested. A more radical approach is to regard crop cover as an integral part of arable management and use an undersown cover crop even when the following crop is to be sown in autumn.

Cover crops are either incorporated into soil in spring or grazed by animals before establishing the next crop. In both cases, N derived from the cover crop will later be mineralized to nitrate but as yet there has been rather little research on the rate and timing of N release (Macdonald *et al.*, 1996). There is preliminary evidence from the Brimstone Experiment that N mineralized from an incorporated cover crop caused increased nitrate leaching in the subsequent winter.

5.2 More Accurate Matching of Fertilizer N Applications to Crop Requirements

For many years, advice to farmers on N fertilizer applications has been based on an ability to predict the following quantities:

(1) N becoming available to the crop from soil – essentially the differences between N mineralized (plus nitrate residues from the previous crop) and N lost during winter plus N from atmospheric deposition.
(2) N requirement of the crop.
(3) Average efficiency of recovery of fertilizer N.

This approach has been gradually refined and is the basis of the recommendations used in the UK contained in MAFF Reference Book RB209 (Ministry of Agriculture, Fisheries and Food, 1997) and systems related to it. In situations where large nitrate residues are expected the system is supplemented by measurements of mineral N in soil (ammonium plus nitrate, termed N_{min}) to a depth of 90 cm or more. In several other European countries, N_{min} measurements have long formed a central part of the N fertilizer recommendation system (*e.g.* Dilz *et al.*, 1982). In recent years there has been a strong move to provide advice on a field-specific basis using dynamic models of N cycle processes that use input information that is readily available to farmers or their advisers. For example, in the UK the SUNDIAL system (*Simulation of Nitrogen Dynamics in Arable Land*; Bradbury *et al.*, 1993; Smith *et al.*, 1996) was recently launched for arable crops and WELL-N (Rahn *et al.*, 1996) has been in use for some years for horticultural crops. NCYCLE (Scholefield *et al.*, 1991) is based on an annual N budget and is used for giving fertilizer advice for grassland.

The increased use of model-based fertilizer recommendation systems should decrease the risk of crops being either under-fertilized, causing an economic loss to the farmer, or over-fertilized and leaving an excessive quantity of nitrate in soil. It is correct to use the word 'risk' as no system can guarantee to calculate the optimum fertilizer application (whether a biological or economic definition of optimum is used) for at least two reasons. First, weather and other factors can influence crop growth well after the time that the latest application of N must be given. Second, the interactions between the various N cycle processes, and the ways they are influenced by weather and management factors, are complex and there is always scope for a model to fail to capture some aspect in a given situation.

A critical aspect of improved fertilizer recommendation, whether based on computer models or any other approach, is the more precise prediction of N mineralization from soil organic matter, crop residues and animal manure. Sylvester-Bradley *et al.* (1997) found that differences in the supply of N from soil was the largest source of variation in the optimum requirement for fertilizer N in 48 field trials in the UK. Consequently identifying more closely the complex interactions between factors controlling the quantity and timing of mineralization is a major research objective; see Jarvis *et al.* (1996) for a review. Initial fertilizer recommendations made in early spring and based on a dynamic N cycle model may be refined by diagnostic measurements on crop or soil as the season progresses (Stockdale *et al.*, 1997). Another possible approach is to make measurements of N_{min}, or other soil or crop N pools, at 'benchmark' sites to check model simulations in the light of actual weather conditions each year.

5.3 Forms and Timing of Fertilizer Applications

Numerous experiments have been conducted to compare the agronomic effectiveness of different chemical forms of N fertilizer (*e.g.* ammonium, nitrate, urea) and their timing. Using urea rather than ammonium nitrate decreases the peak concentration of nitrate in soil and would be expected to decrease the risk of direct N loss by leaching or denitrification, but results are inconsistent. Applying fertilizer in several small doses, rather than a single large dose, decreases the risk of a large proportion of the total N application coinciding with weather conditions likely to cause loss. This is already widely practised but there is scope to manage the timing of applications more closely, either by using the simple relationship between rainfall and loss to adjust the size of later applications (Powlson *et al.*, 1992) or by using a dynamic N cycle model or diagnostic measurement to track the progress of crop N uptake and the supply of N from soil.

Slow release fertilizers or the use of chemicals to inhibit nitrification or urease activity are often claimed to be of major potential value in decreasing nitrate leaching. This is unlikely because, as discussed above, most leaching occurs during winter and affects nitrate derived from mineralization. Whilst these approaches may well have agricultural value, they will not solve the nitrate problem. By contrast, adding nitrification inhibitors to animal slurry applied in autumn may decrease nitrate leaching.

It has been clearly demonstrated in many field experiments that N fertilizer applied to autumn-sown cereals at around the time of sowing is used very inefficiently; for example, Powlson *et al.* (1986) found losses of 39–78% using ^{15}N. The nitrate is in soil during winter when crop uptake of N is low and the risk of leaching is high. Similarly large leaching losses were found in the Brimstone Experiment when autumn N was sometimes applied in the early years (Cannell *et al.*, 1984). There is no agronomic justification for applying fertilizer N to cereals in autumn so it is surprising that 10% of cereals in England and Wales still receive an average of 30 kg N ha^{-1} at this time (Burnhill *et al.*, 1996).

Substituting a part of the total N fertilizer dose given to crops with foliar applications is a possible means of decreasing leaching or denitrification by avoiding contact of the fertilizer with soil. There are some benefits from foliar applications but experiments with ^{15}N-labelled fertilizers indicate substantial losses through other mechanisms, probably ammonia volatilization (Poulton *et al.*, 1990). Thus a major shift from soil to foliar applications may substitute one environmentally sensitive loss with another.

5.4   Spatially Variable Fertilizer Applications Within Fields

There can be significant variations in N cycle dynamics or crop growth within a field, often because of variations in soil type, water availability, or other factors. The technology to vary fertilizer applications within a field in accordance with a predetermined map now exists, as does the facility for constructing yield maps from the output of a combine harvester. There are considerable difficulties in correctly interpreting spatial data within fields and formulating variable fertilizer recommendations but it is worthwhile to attempt it. The aim is to concentrate higher N fertilizer rates in areas where yield potential is greatest and to decrease applications in areas that consistently give lower yields, thus decreasing nitrate residues.

5.5   Expanded Field Margins

It is often observed that the edges of fields consistently give lower yields than the field as a whole. Whilst there will be multiple reasons for this, it is thought that soil compaction caused by machinery turning is often a factor. Establishing expanded field margins, perhaps growing grass, and not given fertilizer or pesticides, may be a practical means of decreasing total leaching losses to water with a relatively small impact on profitability. Field margins are also claimed to offer other environmental benefits such as refuge for beneficial insects that can play a part in integrated pest management strategies.

5.6 Animal Manures

The conversion of N into animal products is inefficient so much N is returned to the soil, either directly in the excreta of grazing animals or indirectly as slurry or manure (Jarvis, 1993). In part this N can be recycled and used to substitute for inorganic fertilizer but animal-based systems are inherently leaky for several reasons. Manures are very variable in composition so it is difficult to predict the quantity or timing of N release. If organic manures are applied to arable crops this is often done in autumn so much nitrate can be produced before winter and then leached. In grazed pastures N in urine and faeces is mineralized during summer and autumn leading to a large accumulation of nitrate before winter.

Several approaches are being developed that can decrease nitrate loss but the impacts on ammonia and nitrous oxide emissions must also be assessed. Restrictions on times of application within Nitrate Vulnerable Zones are intended to minimize winter leaching losses and there are now serious attempts to apply slurry in spring, even where an autumn-sown crop is already established (Smith and Chambers, 1993). Attempts are being made to develop practical on-farm tests to assess the composition of individual batches of manure or slurry and some progress has been made in modelling N release from manures (Chambers *et al.*, 1998). This is of great importance as surveys show that farmers regularly undervalue the nutrient contribution from manures, so inorganic fertilizer applications are only decreased slightly (Chalmers *et al.*, 1992). A scheme for using the NCYCLE model together with a rapid field test for nitrate in soil has been developed (Titchen and Scholefield, 1992); in one case using this 'tactical N' approach decreased residual nitrate in soil by over 30%.

5.7 Organic Farming

It is often assumed that nitrate leaching from organic farms will be less than from conventional farms because of the absence of inputs from N fertilizer. This is not necessarily the case but there is a paucity of information on which to make an assessment. Organic farms are almost always based on a rotation comprising a grazed grass/clover pasture, managed to maximize biological N fixation, followed by arable crops. Leaching is usually considerable immediately after ploughing pasture (Johnston *et al.*, 1994) but this can be offset by low leaching during the period under pasture, especially on organic farms where stocking density is generally lower. It is difficult to make direct comparisons between organic and conventional farms as confounding factors always influence the results. One comparative study (E.I. Lord, C.E. Stopes, L. Philipps, unpublished) showed slightly smaller overall losses from organic farms than from conventional mixed farms selected to be as similar as possible in other respects. Measurements of leaching on an organic farm (Philipps *et al.*, 1998) were in the range measured on other mixed farms but a modelling study based on the data suggested that leaching would have been greater if the farm had been managed conventionally. There is significant scope to decrease leaching on organic and other mixed farms by adjusting the timing of ploughing the pasture phase, limiting grazing shortly before ploughing and appropriate selection of the first arable crop. Thus organic farming appears to offer some potential for decreased nitrate leaching.

5.8   Planning Crop Rotations

Crops differ greatly in the quantity of nitrate and readily mineralizable organic N left in soil after harvest and also in their ability to capture inorganic N left from the previous crop. Consequently a change in the sequence of crops grown in a rotation can have a significant effect on the quantity of nitrate leached during winter. A new decision support system has been developed, based on the SUNDIAL model, to simulate N transformations in a large number of crop rotations in order to assist in the design of rotations that optimize N use and minimize losses (Smith and Glendining, 1996). In one example (Smith *et al.*, 1998) 26 possible variations of a 6–course rotation were compared for a number of typical arable farms in different regions of England having a defined set of soil types, weather conditions and management constraints. The decrease in calculated N loss between the best and worst rotations averaged 22 kg N ha^{-1} yr^{-1}. This approach offers a powerful means of exploring ways of minimizing N loss within farming systems. Another useful approach is to construct nutrient budgets for whole farms to identify systems or practices that lead to large imbalances (Watson and Atkinson, 1998; Watson and Stockdale, 1998).

5.9   Buffer strips

The low-lying areas of land that often lie between farmland and water courses can act as buffers by removing nutrients from the waters moving through them (Burt and Haycock, 1993; Correll, 1997). Usually called Buffer Strips or Buffer Zones, their use has been established for some time in Europe and the USA, especially to trap eroded sediments. Buffer strips may remove nutrients by many mechanisms, but denitrification and nutrient uptake into plants growing in the strip are dominant for nitrate removal (Haycock and Pinay, 1993).

Much work on buffer zones in the USA refers to very large areas adjoining rivers, sometimes the entire floodplain. Winter flooding and the generally higher water table of a floodplain are ideal conditions for denitrification. A loss of nitrate from water of up to 90% can sometimes occur as water moves through a large buffer zone. In Europe there is a tendency to regard buffer zones as narrow strips, perhaps a few metres wide, alongside a watercourse. It does not follow that these will be as effective but there is sufficient encouraging data for the idea to be pursued (Haycock *et al.*, 1997).

If denitrification is a major pathway of nitrate removal it will be necessary to establish the quantity of N_2O produced and calculate the impact on national scale emissions if buffer strips were to be used extensively.

5.10 Drainage Manipulation

Temporarily restricting water flow to field drains in clay soils causes the cracks that develop in such soils during summer to close more quickly than normal, thus decreasing the extent to which a pulse of nitrate moves through drains to surface waters by by-pass flow. It also creates sub-soil conditions conducive to denitrification, thus decreasing the amount of nitrate at risk to leaching. The idea is currently being tested at the Brimstone Experiment (Catt, 1996) where flow restriction is achieved by installing a large rotatable U-bend in the collector pipe carrying water from the mole drains. This is raised at the start of the drainage period and lowered later in the winter to permit the degree of drainage required for crop growth. If the idea is to be widely used in practice it may be realised by allowing present field drains to decay somewhat and designing new drains to a lower specification for flow rates. As with buffer zones, it will be necessary to check the impact on N_2O production.

5.11 Catchment Planning

Some crops and rotations leave greater nitrate residues than others. It is virtually impossible to prevent all drainage or runoff water from every field of productive farmland from exceeding the EU 50 mg l^{-1} nitrate limit at all times. It is therefore necessary to consider water resources at the catchment and regional scale, recognizing that high nitrate water from some areas will be diluted by lower nitrate water from others. To do this, models are being developed that can simulate N dynamics over large areas, taking account of nitrate movement to both groundwaters and surface waters. It is necessary to simulate the impacts of different mixtures of managements on both total nitrate (and phosphate) losses and peak concentrations. It is also necessary to ensure that models working at the catchment or regional scale are capable of providing reasonable simulations of reality despite using coarse scale and imprecise input data. There is promising progress in the development of minimum information requirement modelling systems (*e.g.* Anthony *et al.*, 1996) but the problem of errors being propagated and magnified as models are scaled up (Gaunt *et al.*, 1997) has to be borne in mind so that erroneous conclusions are not drawn.

6 Conclusions

When Addiscott *et al.* (1991) reviewed some of the medical literature seven years ago it appeared that health reasons for lowering the limit for nitrate in public drinking water supplies in western Europe were largely unfounded. There are, however, genuine environmental reasons for wishing to decrease leakage of combined nitrogen from agriculture to water and the atmosphere. Also, as price support for agricultural production declines, it will be of increasing economic importance to minimize the wastage of nutrients. Using current knowledge of the nitrogen cycle it is possible to achieve significant decreases in nitrate loss to water with minimal effects on the productivity or profitability of agriculture. However, to meet the 50 mg l^{-1} limit for drinking water, and many surface waters, consistently it is likely that more drastic changes will be necessary. These will inevitably have significant impacts on productivity and profitability. Developing the scientific understanding for management practices in which organic materials such as animal manures and crop residues are used more rationally represents a major challenge. Another challenge is to decrease the loss of nitrate to water without increasing the evolution of environmentally undesirable gases, especially N_2O. The continued development and up-dating of models and diagnostics that form the basis for decision aids in an increasingly knowledge-based agricultural industry seems the only logical way forward. Scientific solutions will decrease the frequency and probability of adverse environmental impacts arising from agriculture but cannot eliminate them completely as many management decisions have to be made on the basis of incomplete knowledge.

Acknowledgements

The author thanks his many colleagues at IACR-Rothamsted for providing information, especially Margaret Glendining, Andy Macdonald, Jo Smith, Paul Poulton, John Catt, Nabeel Mirza and Liz Stockdale. Much of the work referred to was funded by the UK Ministry of Agriculture, Fisheries and Food. IACR-Rothamsted receives grant-aided support from the Biotechnology and Biological Sciences Research Council of the United Kingdom.

References

Addiscott, T.M. & Powlson, D.S. 1992. Partitioning losses of nitrogen fertilizer between leaching and denitrification. *Journal of Agricultural Science, Cambridge* **118**, 101–107.

Addiscott, T.M. & Whitmore, A.P. 1987. Computer simulation of changes in soil mineral nitrogen and crop nitrogen during autumn, winter and spring. *Journal of Agricultural Science, Cambridge* **109**, 141–157.

Addiscott, T.M., Powlson, D.S. & Whitmore, A.P. 1991. *Farming, Fertilizers and the Nitrate Problem*. Wallingford: CAB International.

Anthony, S., Quinn, P. & Lord, E. 1996. Catchment scale modelling of nitrate leaching. *Aspects of Applied Biology* **46**, 22–32.

Bouwman, A.F. (1990) Introduction. In: *Soils and the Greenhouse Effect* (Ed. A.F. Bouwman), Wiley, Chichester, pp. 25–32.

Bradbury, N.J., Whitmore, A.P., Hart, P.B.S. & Jenkinson, D.S. 1993. Modelling the fate of nitrogen in crop and soil in the years following application of ^{15}N-labelled fertilizer to winter wheat. *Journal of Agricultural Science, Cambridge* **121**, 363–379.

Burnhill, P., Chalmers, A. & Fairgrieve, J. 1996. *British Survey of Fertiliser Practice 1995*, HMSO, London, 76 pp.

Burt, T.P. & Haycock, N.E. 1993. Controlling losses of nitrate by changing land use. In *Nitrate. Processes, Patterns and Management* (Eds. T.P. Burt, A.L. Heathwaite & S.T. Trudgill), Wiley, Chichester, pp. 341–367.

Cannell, R.Q., Goss, M.J., Harris, G.L., Jarvis, M.G., Douglas, J.T., Howse, K.R. & Le Grice, S. (1984) A study of mole drainage with simplified cultivation for autumn-sown cereals on a clay soil. I. Background, experiment and site details, drainage systems, measurement of drainflow and summary of results 1978–80. *Journal of Agricultural Science, Cambridge* **102**, 583–594.

Catt, J.A. 1996. The Brimstone Experiment. *IACR Report for 1995*, pp. 41–42.

Catt, J.A., Christian, D.G., Goss, M.J., Harris, G.L. & Howse, K.R. 1992. Strategies to reduce nitrate leaching by crop rotation, minimal cultivation and straw incorporation in the Brimstone Farm Experiment, Oxfordshire. *Aspects of Applied Biology* **30**, *Nitrate and Farming Systems*, 255–262.

Chalmers, A.G., Dyer, C.J., Leech, P.K. & Elsmere, J.I. 1992. Fertilizer use on farm crops, England and Wales 1991. *Survey of Fertiliser Practice*, MAFF Publications, London.

Chambers, B., Lord, E., Nicholson, F. & Smith, K. 1998. Predicting nitrogen availability and losses following land application of manures. In: *International Workshop on Environmentally Friendly Management of Animal Farm Wastes*, Sapporo, Japan, November 1997 (in press).

Chaney, K. 1990. Effect of nitrogen fertilizer rate on soil nitrogen content after harvesting winter wheat. *Journal of Agricultural Science* **114**, 171–176.

Christian, D.G., Goodlass, G. & Powlson, D.S. 1992. Nitrogen uptake by cover crops. *Aspects of Applied Biology* **30**, *Nitrate and farming systems*, 291–300.

Correll, D.L. 1997. Buffer zones and water quality protection: general principles. In *Buffer Zones: Their Processes and Potential in Water Protection* (Eds N.E. Haycock, T.P. Burt, K.W.T. Goulding & G. Pinay), Quest Environmental, Harpenden, pp. 7–20..

Davies, D.B., Garwood, T.W.D. & Rochford, A.D.H. 1996. Factors affecting nitrate leaching from a calcareous loam in East Anglia. *Journal of Agricultural Science, Cambridge* **126**, 75–86.

Dilz, K., Darwinkel, A., Boon, R. & Verstraeten, L.M.J. (1982) Intensive wheat production as related to nitrogen fertilization, crop protection and soil nitrogen: experience in the Benelux. *Proceedings of the Fertilizer Society* **211**, 93–124.

Gaunt, J.L., Riley, J., Stein, A. & Penning de Vries, F.W.T. 1997. Requirements for effective modelling strategies. *Agricultural Systems* **54**, 153–168.

Glendining, M.J. & Powlson, D.S. 1995. The effects of long continued applications of inorganic nitrogen fertilizer on soil organic nitrogen – a review. In *Soil management, experimental basis for sustainability and environmental quality* (Eds. R. Lal and B.A. Stewart), CRC Lewis Publishers, London, pp. 385–446.

Glendining, M.J., Powlson, D.S., Poulton, P.R., Bradbury, N.J., Palazzo, D. & Li, X. 1996. The effects of long-term applications of inorganic nitrogen fertilizer on soil nitrogen in the Broadbalk Wheat Experiment. *Journal of Agricultural Science, Cambridge* **127**, 347–363.

Haycock, N.E. & Pinay, G. 1993. Groundwater nitrate dynamics in grass and poplar vegetated riparian buffer strips during the winter. *Journal of Environmental Quality* **22**, 273–278.

Haycock, N.E., Burt, T.P., Goulding, K.W.T. & Pinay, G. 1997. *Buffer Zones: Their Processes and Potential in Water Protection*, Quest Environmental, Harpenden.

Hutchinson, G.L. & Davidson, E.A. 1993. Processes for production and consumption of gaseous nitrogen oxides in soil. In *Agricultural Ecosystem Effects on Trace Gases and Global Climate Change*. ASA Special Publication Number 55, Madison, USA, pp. 79–93.

Jarvis, S.C. 1993. Nitrogen cycling and losses from dairy farms. *Soil Use and Management* **9**, 99–105.

Jarvis, S.C., Scholefield, D. & Pain, B.F. 1995. Nitrogen cycling in grazing systems. In *Nitrogen Fertilization in the Environment* (Ed. P. Bacon), Marcel Dekker, New York, pp. 381–419.

Jarvis, S.C., Stockdale, E.A., Shepherd, M.A., Powlson, D.S. 1996. Nitrogen mineralisation in temperate agricultural soils: processes and measurement. *Advances in Agronomy* **57**, 187–235.

Jenkinson, D.S., Fox, R.H. & Rayner, J.H. 1985. Interactions between fertilizer nitrogen and soil nitrogen – the so-called 'priming' effect. *Journal of Soil Science* **36**, 425–444.

Johnston, A.E., McEwen, J., Lane, P.W., Hewitt, M.V., Poulton, P.R. & Yeoman, D.P. 1994. Effects of one to six year old ryegrass-clover leys on soil nitrogen and on the subsequent yields and fertiliser requirements of the arable sequence winter wheat, potatoes, winter wheat, winter beans (*Vicia faba*) grown on a sandy loam soil. *Journal of Agricultural Science, Cambridge* **122**, 73–89.

Macdonald, A.J., Powlson, D.S., Poulton, P.R. & Jenkinson, D.S. 1989. Unused fertiliser nitrogen in arable soils – its contribution to nitrate leaching. *Journal of the Science of Food and Agriculture* **46**, 407–419.

Macdonald, A.J., Barraclough, D., Gibbs, P. & Ayaga, G.O. 1996. Effects of crop residue incorporation on gross nitrogen mineralization rates in an arable soil. In *Transactions of the 9th Nitrogen Workshop, Braunschweig, September 1996*, pp. 265–266.

Macdonald, A.J., Poulton, P.R., Powlson, D.S. & Jenkinson, D.S. 1997. Effects of season, soil type and cropping on recoveries, residues and losses of ^{15}N-labelled fertilizer applied to arable crops in spring. *Journal of Agricultural Science, Cambridge* **129**, 125–154.

Ministry of Agriculture, Fisheries and Food. 1997. *Fertiliser Recommendations for Agricultural and Horticultural Crops (RB209)*, The Stationery Office, London.

Philipps, L., Stockdale, E. & Watson, C. 1998. Nitrogen leaching losses from mixed organic farming systems in the UK. Workshop on 'Mixed Farming Systems in Europe', Wageningen, 25–28 May 1998 (in press).

Poulton, P.R., Vaidyanathan, L.V., Powlson, D.S. & Jenkinson, D.S. 1990. Evaluation of the benefit of substituting foliar urea for soil-applied nitrogen for winter wheat. *Aspects of Applied Biology* **25**, *Cereal Quality II*, 301–307.

Powlson, D.S. & Goulding, K.W.T. 1995. Agriculture, the nitrogen cycle and nitrate. In *Nitrate control policy, agriculture and land use, Proceedings of the fourth professional environmental seminar, 11 October 1994, Cambridge* (Ed. J. North). Cambridge Environmental Initiative, pp. 5–25.

Powlson, D.S., Hart, P.B.S., Pruden, G. & Jenkinson, D.S. 1986. Recovery of ^{15}N-labelled fertilizer applied in autumn to winter wheat at four sites in eastern England. *Journal of Agricultural Science, Cambridge* **107**, 611–620.

Powlson, D.S., Hart, P.B.S., Poulton, P.R., Johnston, A.E. & Jenkinson, D.S. 1992. Influence of soil type, crop management and weather on the recovery of ^{15}N-labelled fertilizer applied to winter wheat in spring. *Journal of Agricultural Science, Cambridge* **118**, 83–100.

Rahn, C.R., Vaidyanathan, L.V. & Paterson, C.D. 1992. Nitrogen residues from brassica crops. *Aspects of Applied Biology* **30**, *Nitrate and farming systems*, 263–270.

Rahn, C.R., Greenwood, D.J. & Draycott, A. 1996. Prediction of nitrogen fertiliser requirement with the HRI WELL-N computer model (Eds. O. Van Cleemput *et al.*) *Progress in Nitrogen Cycling Studies*, Kluwer, Dordrecht, pp. 255–258.

Scholefield, D,., Lockyer, D.R., Whitehead, D.C. & Tyson, K.R. 1991. A model to predict transformations and losses of nitrogen in UK pastures grazed by beef cattle. *Plant and Soil* **132**, 165–177.

Schorring, J.K., Nielsen, N.E., Jensen, H.E. & Gottschau, A. 1989. Nitrogen losses from field-grown spring barley plants as affected by rate of nitrogen application. *Plant and Soil* **116**, 167–175.

Smith, J., Glendining, M. & Smith, P. 1998. Optimisation of crop rotations with respect to nitrogen use efficiency. In: *World Congress of Soil Science*, Montpellier, France, 1998 (in press).

Smith, J.U. & Glendining, M.J. 1996. A decision support system for optimising the use of nitrogen in crop rotations. *Aspects of Applied Biology* **47**, *Rotations and cropping systems*, 103–110.

Smith, J.U., Bradbury, N.J. & Addiscott, T.M. 1996. SUNDIAL: A PC-based version of the Rothamsted nitrogen turnover model. *Agronomy Journal* **88**, 38–42.

Smith, K.A. & Chambers, B.J. 1993. Utilizing the nitrogen content of organic manures on farms – problems and practical solutions. *Soil Use and Management* **9**, 105–112.

Stockdale, E.A., Gaunt, J.L. & Vos, J. 1997. Soil-plant nitrogen dynamics: what concepts are required? In: *Perspectives for Agronomy – Adopting Ecological Principles and Managing Resource Use* (Eds. M.K. van Ittersum & S.C. van de Geijn), Elsevier, Amsterdam, pp. 201–215.

Sylvester-Bradley, R., Davies, D.B., Dyer, C., Rahn, C. & Johnson, P.A. 1997. The value of nitrogen applied to wheat during early development. *Nutrient Cycling in Agroecosystems* **47**, 173–180.

Titchen, N.M. & Scholefield, D. 1992. The potential for a rapid soil mineral N test for tactical applications of N fertilizer to grassland. *Aspects of Applied Biology* **30**, *Nitrate and farming systems*, 223–229.

Watson, C.A. & Atkinson, D. 1998. Using nitrogen budgets to indicate nitrogen use efficiency and losses from whole farm systems: A comparison of three methodological approaches. *Nitrogen Cycling in Agroecosystems* (in press).

Watson, C.A. & Stockdale, E.A. 1998. Using nutrient budgets as management tools in mixed farming systems. Workshop on 'Mixed Farming Systems in Europe', Wageningen, 25–28 May 1998 (in press).

Webster, C.P., Macdonald, A.J., Poulton, P.R. & Christian, D.G. 1992. The effectiveness of winter cover crops in minimising nitrogen leaching loss. In *Proceedings of the Second Congress of the European Society of Agronomy*, Warwick University, August 1992, pp. 380–381.

Widdowson, F.V., Penny, A.P., Darby, R.J., Bird, E. & Hewitt, M.V. 1987. Amounts of NO_3^--N and NH_4^+-N in soil, from autumn to spring, under winter wheat and their relationships to soil type, sowing date, previous crop and N uptake at Rothamsted, Woburn and Saxmundham, 1979–85. *Journal of Agricultural Science, Cambridge* **108**, 73–95.

Yamulki, S., Harrison, R.M., Goulding, K.W.T. & Webster, C.P. 1997. N_2O, NO and NO_2 fluxes from a grassland: effect of soil pH. *Soil Biology and Biochemistry* **29**, 1199–1208.

4

Using a Rotational Modelling System to Explore the Effect of Straw Incorporation on the Efficiency of Nitrogen Use

Margaret J. Glendining and Jo U. Smith

SOIL SCIENCE DEPARTMENT, IACR-ROTHAMSTED, HARPENDEN, HERTS, AL5 2JQ, UK

Abstract

A decision support system has been constructed around the nitrogen (N) turnover model SUNDIAL which allows farmers and policy makers to explore how arable rotations respond to practical strategies for reducing N losses. It automatically derives all crop rotations allowed within an imposed set of farming constraints and presents the N dynamics for each rotation. Total N losses (by leaching and gaseous losses) and crop N offtake were simulated for a six year arable rotation based on two winter wheat crops, spring barley, winter oilseed rape, winter beans and set-aside (cropped with industrial oilseed rape). The simulation was run using three basic soil types (sand, loam and clay), for three different productivity levels (low, medium and high), and using typical weather data from three different geographical regions in England (Southwest, Central and East Anglia), to look at the effects of changing the order of the crops in the rotation and incorporating straw on N losses and crop N offtake. The simulations suggest that straw incorporation decreases potential N losses to the environment, but that crops are generally unable to make use of the saved N. The average N offtake following straw incorporation is lower than when straw is not incorporated. Changing the sequence of crops in the rotation has a greater effect, significantly reducing N losses and increasing mean crop N offtake. On average, the best rotations lost 319 kg N ha^{-1} throughout the six year rotation, compared to the worst rotations, which lost 464 kg N ha^{-1}, a saving of 145 kg N ha^{-1} achieved merely by changing the order of the crops.

1 Introduction

The maximum concentration of nitrate-N permitted in drinking water by the 1980 European Community Directive on the Quality of Water Intended for Human Consumption is set at 50 mg dm^{-3}. This regulation constrains farming practices likely to result in excess nitrate leaching. Since the UK ban on straw burning in 1993 it has been estimated that an extra 6 x 10^6 tonnes of cereal straw has to be disposed of each year. A large proportion of this straw will be incorporated into the soil at harvest. In the short-term, the decomposition of such crop residues with a high C:N ratio will lead to net immobilization of inorganic N, which may decrease nitrate leaching in autumn and winter. In the longer term, the continued incorporation of straw may increase soil organic matter content, with a potential increase in the amount of N available for mineralization (Powlson, 1993). Depending on the timing of this mineralization, the additional N may be available to subsequent crops, so reducing their fertilizer N requirement, or at risk to loss by leaching or denitrification. The objective of this study is to assess the likely influence of incorporating straw on N losses and crop N offtake over whole rotations.

2 Methods

2.1 General Methodology

Much of the research to improve knowledge about factors controlling nitrate leaching has concentrated on developing principles on a single crop or in a single year. In practice, farmers operate within more complex systems involving rotations of crops and planning timescales of several years. A decision support system (DSS) has been constructed around the dynamic N turnover model SUNDIAL (Bradbury, *et al.*, 1993; Smith, *et al.*, 1996) that will allow farmers or policy makers to explore how arable rotations respond to practical strategies for decreasing N losses.

SUNDIAL (Figure 1) is a dynamic model of nitrogen turnover in the soil/crop system. It incorporates descriptions of all the major processes of N turnover on a weekly basis. Unlike many other N models, N dynamics in SUNDIAL are driven by the carbon (C) cycle. Nitrogen may be added to the soil/crop system as inorganic fertilizer, organic manure or by atmospheric deposition. Nitrate and ammonium are taken up by the crop in proportion to the expected yield of the crop, and the cumulative temperature since sowing. Nitrogen and C are then returned to the soil, not only at harvest as stubble, straw and other crop residues, but also throughout the growing season as root exudates, dead leaves and fragments of roots. The amount of C and N returned to the soil from straw is calculated from the grain yield using the harvest index, and the proportion of dry matter, C and N in the straw. Decomposition of the crop residues is represented by partitioning the C and N into biomass and humus according to the soil type. The C:N ratios of these organic matter pools are assumed to remain constant, and are set at 8.5. If the C:N ratio rises, due

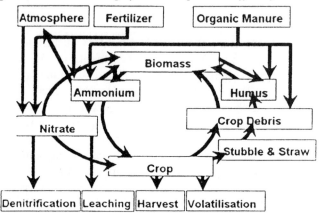

Figure 1   *Structure of the SUNDIAL model showing the nitrogen pools and their interactions.*

to a higher C:N compostion of the crop residue, N is immobilized first from ammonium and then from nitrate in the soil. If the C:N ratio falls, N is mineralized to ammonium. Ammonium may then be nitrified to nitrate, and nitrate may be lost by denitrification or leaching. A more detailed description of the model structure is given by Bradbury *et al.* (1993).

SUNDIAL is designed to be used in a 'carry-forward' mode, with one year's run providing the inputs for the next. This allows it to be used to investigate more complex systems involving rotations of crops and planning timescales of several years. By attaching SUNDIAL to a systematic tool for deriving crop rotations, or *scenario generator*, all crop rotations allowed within an imposed set of farming constraints may be obtained and the N dynamics of each rotation investigated. Constraints are initially defined by restrictions on cropping and management practices, for example, due to pest and disease considerations, or EC regulations. The N dynamics of all the resulting permutations of crops are automatically simulated and the optimum management strategies selected. Smith and Glendining (1996) and Smith *et al.* (1997) give further details of the development of the DSS. In this paper we present the results of simulations used to explore the influence of straw incorporation on N losses.

2.2   The Farming Constraints Imposed on the Simulations

The simulations are for a six-course arable rotation based on two winter wheats, spring barley, winter oilseed rape, set-aside (cropped with industrial oilseed rape) and winter beans. The following constraints are imposed in this example: no more than two cereals may be grown in succession, in order to reduce the effects of take-all (*Ophiobolus graminae*); the first cereal must always be winter wheat, to maximize profit; oilseed rape cannot be grown in succession, to reduce the risk of soil borne pests and diseases, such as *Sclerotinia sclerotiorum* and club root (*Plasmodiophora brassicae*); there should be at least a four-year break between crops of beans, to reduce the risk of problems from soil-borne fungi and nematodes. There are restrictions as to which crops may follow set-aside (MAFF, 1997); all the crops selected for this simulation are allowed to follow set-aside; if they were not allowed, the system would prevent the user from selecting them.

The crops are grown according to standard management practices, taken from Nix (1995) and MAFF (1994). These cover sowing and harvest dates, rates of fertilizer N applied and dates of the fertilizer applications.

The area of the farm under each crop is assumed to remain approximately constant throughout the rotation, so as to maintain an approximately constant marketable yield of each crop. This is achieved by applying the rotation over subunits of equal area on the farm. The user defines subunits by grouping the fields according to convenience for crop management.

2.3 Criteria for Optimum Management

Criteria for optimum management may be entered in the DSS by the user. The system then determines whether the results of each SUNDIAL simulation meet the entered criteria, and rejects all rotations that do not. In this example, the *best-case* rotations are defined as the rotations in which crop productivity is maintained, while the total losses of N to the environment (both leaching and gaseous losses) are minimized. Conversely, the *worst-case* rotations are defined as the rotations resulting in the highest total losses of N to the environment.

2.4 Running the Simulations

The DSS generates a complete list of possible rotations from the given starting ratios, and excludes any rotations that do not obey the above rules. The remaining permutations provide the inputs for SUNDIAL. The simulations were run for the following conditions:

2.4.1 Soil. The simulations were run using three basic soil types: sand, loam and clay. It was assumed that the crop roots were not limited by any impenetrable layer, and so the depth of the soil was given as 150cm. The soil is assumed to be in long-term arable cultivation, with no period under grass in the last ten years.

2.4.2 Weather. SUNDIAL requires weekly cumulative rainfall, evapotranspiration over grass and average air temperature as inputs. The simulations were run using weather data from three different geographical regions in England (Table 1): South-west Region (meteorological station Plymouth); Central Region (meteorological station Rothamsted); and East Anglian Region (meteorological station Wattisham).

Table 1   *Average annual weather data used in the simulations*

	South-west region	Central region	East-Anglian region
Rainfall (mm yr^{-1})	960	695	610
Air temperature (°C)	13.6	9.4	12.6
Evapotranspiration over grass (mm yr^{-1})	560	485	507

2.4.3 Cropping.   The simulations were run assuming low, medium and high productivity, and the associated crop yields were obtained from crop statistics for England and Wales (Nix, 1995; Soffe, 1995). Straw was incorporated after all the crops grown. For comparison, the simulations were also run with no straw incorporation. No organic manure was applied to any of the crops.

2.5   Statistical Analysis of Results

The output values of total N losses and crop N offtake were analysed using the Analysis of Variance procedure of Genstat (Genstat 5 Committee, 1993), using either N losses or crop N offtake as the dependent variables, with combinations of region, soil types, productivity, with or without straw and best or worst-case rotation as explanatory variables. This allows the potential improvement of N use attributable to the ordering of the crops to be examined and the effect of straw incorporation to be assessed.

3   Results and Discussion

The results of all the simulations are summarised in Tables 2–4. Averaged over all three soil types, productivity levels and regions, changing the sequence of crops in the rotation significantly decreases N losses (Table 2) and increases mean crop N offtake (Table 3). If straw is not incorporated during the

Table 2   *Mean total N losses* [kg N ha^{-1} (6 yr)$^{-1}$]

Rotation	With Straw	No Straw	Difference	F prob	S.E.D
Best-Case	319	361	−42	0.001	2.53
Worst-Case	464	475	−11	0.001	2.53
Difference	−145	−114			

Table 3   *Mean crop N offtake* [kg N ha^{-1} (6 yr)$^{-1}$]

Rotation	With Straw	No Straw	Difference	F prob	S.E.D
Best-Case	741	833	−92	0.001	3.05
Worst-Case	589	713	−124	0.001	3.05
Difference	152	121			

Table 4 *Value of N saved by optimizing the crop sequence and incorporating straw*

	Value of saved N[a] ($£ ha^{-1} yr^{-1}$)	Percentage of Saved N taken up by crops
Optimized Sequence (Straw)	7.70	105
Optimized Sequence (No Straw)	6.06	105
Straw Incorporation (Best-Case)	2.23	
Straw Incorporation (Worst-Case)	0.58	

[a] Cost of N = $£0.3188 kg^{-1}$

rotation, the worst-case rotations lose 114 kg N ha^{-1} more than the best-case rotations over the six years. This represents a decrease in the environmental impact of arable agriculture and a potential saving in N, merely by changing the order of the crops. Given a price of N of £0.3188 kg^{-1}, if all the saved N could replace fertilizer N, the total value of N saved over the six year rotation would be equivalent to £6.06 ha^{-1} yr^{-1} (Table 4). If straw is incorporated throughout the rotation, changing the sequence of the crops has an even greater effect. The potential savings increase to £7.70 ha^{-1} yr^{-1}. In both cases, all of the saved N is exploited by the crops in the rotation (Table 3), and so the N saving translates into a real increase in gross margins for the farmer.

Analysis of the N losses associated with straw incorporation indicates that this change in management has a smaller potential for reducing N losses than changing the order of the crops (at least with this combination of crops). In this simulation, straw incorporation decreases N losses by an average of 42 kg N ha^{-1} over the six year rotation in the best-case rotations, and by only 11 kg N ha^{-1} in the worst-case rotations. As the straw is incorporated, some N is immobilized into the soil microbial biomass, leaving a little less N at risk to loss. This N is generally not available to the subsequent crops in the rotation. The average N offtake following straw incorporation is generally lower than when straw is not incorporated (Table 3). In the most N efficient rotations, the saved N remains in the soil throughout the six years of the rotation, resulting in smaller N losses in all regions, soil types and productivity levels. However, over the six years of the least N efficient rotations, straw incorporation results in greater total N losses on sandy soils and when using weather data from the East Anglian region, than when straw is not incorporated, and is of no benefit under conditions of low productivity and with weather data from the Central region.

The results of these simulations suggest that straw incorporation has some potential to decrease N losses from arable rotations, especially on clay and loam soils and under conditions of moderate to high productivity. This saved N is generally retained in the soil organic matter, rather than being available to subsequent crops. Straw contains less N than is required for its microbial decomposition, and more N is immobilized, so that the average N offtake following straw incorporation is less than when straw is not incorporated. In the longer term, straw incorporation may increase losses of N, as the additional soil organic matter is mineralized. In this simulation this appears to have already occurred during the six year simulation in the worst case rotations on sandy soils and using weather data from the East Anglian region.

Acknowledgments

We are grateful to Nicky Bradbury, David Jenkinson, Gordon Dailey, Penny Leech, Tom Addiscott, Gill Tuck and Pete Smith for their contributions to the development of SUNDIAL. Funding for the development of the decision support system is provided by the Ministry of Agriculture, Fisheries and Food, United Kingdom. IACR-Rothamsted receives grant-aided support from the Biotechnology and Biological Sciences Research Council of the United Kingdom.

References

Bradbury N. J., A.P. Whitmore, P.B.S. Hart and D.S. Jenkinson 1993. Modelling the fate of nitrogen in crop and soil in the years following application of ^{15}N-labelled fertilizer to winter wheat. *J. Agric. Sci.,Cambs* **121**:363–379.

Genstat 5 Committee 1993. *Genstat 5 Release 3 Reference Manual.* Oxford: Clarendon Press.

MAFF 1994. *Fertilizer recommendations for agricultural and horticultural crops (RB209), 6th Edition.* London: HMSO.

MAFF 1997. Arable Area Payments Scheme Explanatory Guide: 1988 Update. Ministry of Agriculture, Fisheries and Food, PB 3196. **AR 30.**

Nix J. 1995. *Farm management pocket book, 26th Edition.* Wye College, University of London: Wye College Press, 216 pp.

Powlson D.S. 1993 Understanding the soil nitrogen cycle. *Soil Use and Management* **9**: 86–94.

Smith J.U. and M.J. Glendining 1996. A decision support system for optimising the use of nitrogen in crop rotations. *Aspects of Applied Biology* **47**, *Rotations and cropping systems:*103–110.

Smith J.U., N.J. Bradbury and T.M. Addiscott 1996. SUNDIAL: A PC-based version of the Rothamsted nitrogen turnover model. *Agron. J.* **88**:38–42.

Smith J.U., M.J. Glendining and P. Smith 1997. The use of computer simulation models to optimise the use of nitrogen in whole farm systems. *Aspects of Applied Biology* **50**, *Optimising cereal inputs: Its scientific basis*: 147–154.

Soffe R. J. (Ed.) 1995. *Primrose McConnell's, The agricultural notebook, 19th Edition.* Oxford: Blackwell Science Limited.

5

Assessing the Importance of Soluble Organic Nitrogen in Agricultural Soils

D. V. Murphy, S. Fortune, J. A. Wakefield, E. A. Stockdale,
P. R. Poulton, C. P. Webster, W. S. Wilmer, K. W. T. Goulding
and J. L. Gaunt

SOIL SCIENCE DEPARTMENT, IACR-ROTHAMSTED, HARPENDEN,
HERTS., AL5 2JQ, UK

Abstract

Soluble organic nitrogen (SON) is not generally considered to be a significant
pool of nitrogen (N) in agricultural soils. Given the importance of SON in
forest and natural systems, this view is challenged with examples from current
research. Methods to collect and measure SON are described and possible
roles for SON in N transformations are discussed. Gaps in the understanding
of N availability and losses in agricultural systems are highlighted.

1 Introduction

The plough layer of arable soils often contains more than 3000 kg of N per
hectare (Stevenson, 1982; Streeter and Barta, 1988). However, most of this N
is not available to plants and is composed of a continuum of complex organic
forms. Organic N in soil is commonly divided, conceptually, into a number of
pools (Paul and Juma, 1981) which may include organic N which is virtually
inert to further decomposition (Hsieh, 1992) as well as N present in the living
bodies of the soil microbial biomass (Jenkinson and Powlson, 1976; Figure
1). The mineral N pool makes up only a small proportion of the total N in
the soil (Harmsen and Kolenbrander, 1965; Bremner, 1965), usually about
1% in arable soils (Jarvis *et al.*, 1996; Figure 1), except after recent fertiliser
additions. However, mineral N cycles rapidly. It is supplied by mineralisation
of soil organic matter, as well as additions of fertilisers, manure and atmospheric deposition, and depleted through uptake by plants and micro-organisms (immobilisation), and through denitrification and leaching. Although the
pool of microbial N in soil is only 3 to 5% of total N (Figure 1),

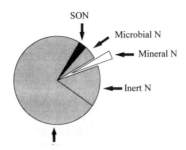

SON

Microbial N

Mineral N

Inert N

Slowly decomposable organic N

Figure 1   *Pie diagram showing the relative size of soil nitrogen pools within agricultural soils*

mineralisation, immobilisation and denitrification are all microbially mediated processes and the flux of N through the microbial biomass is large compared to its size at a given time. Microbial N is therefore often determined alongside mineral N in studies of N dynamics in agricultural systems (Murphy *et al.*, 1998c). Concerns over the environmental and health impacts of nitrate (NO_3^--N) leached from soils, as well as the importance of ammonium (NH_4^+-N) and NO_3^--N in crop nutrition, has focused attention on the study of mineral N in agricultural soils.

In contrast, few measurements of SON have been made in agricultural soils. The size of the SON pool (determined by EUF; see section 2.3) in arable soils is only about 0.3 to 1% of total organic N (Mengel, 1985) and it has generally been considered to be an insignificant source of N for plant uptake (Harper, 1984). However, rapidly cycling pools (*e.g.* NH_4^+-N) may be very small (Davidson *et al.*, 1990) and since this fraction of N is likely to reflect easily mineralisable soil N (Németh *et al.*, 1988) it is likely to be an important pool in N transformation pathways and plant uptake.

SON has been identified as a key pool in soil-plant N cycling in forest systems (Qualls and Haines, 1991), arctic tundra (Atkin, 1996) and in subtropical wet heathland (Schmidt and Stewart, 1997). Indeed SON represents a major input of N to lakes in forested watersheds (Wissmar, 1991) suggesting that leached SON could be a major N loss route from at least some soils. However, little is known about the form and function of SON and the role that it plays in soil N cycling, particularly within mineralisation, immobilisation and N loss pathways.

2   Measurement of SON

2.1   Sampling of SON in the Field

Spatially separate pools of mineral N and/or SON may occur in soil (*e.g.* within mobile or immobile water). Hence, depending on our question the sampling method should remove N selectively from the pool of N which supplies the process of interest. For example, with leaching studies it is necessary to preferentially sample mobile water to determine the NO_3^--N concentration of soil solution which is subject to leaching. Samples of soil solution can be collected in a number of ways (Addiscott *et al.*, 1991; Goulding and Webster, 1992; Titus and Mahendrappa, 1996). All of the methods have limitations especially where soils show heterogeneous flow patterns due to preferential flow through soil macropores.

Suction cups composed of hydrophilic material containing small pores can be used to collect soil solution directly when a negative pressure is created within the cups (Grossmann and Udluft, 1991). Such samplers work best in sandy soils (Webster *et al.*, 1993) which have a relatively small range of pore sizes and are not subject to cracking. On these soils the solution composition measured by suction cups and lysimeter drainage is in close agreement (Webster *et al.*, 1993). However, use of suction cups on poorly drained, heavy clay soils (Hatch *et al.*, 1997) is inappropriate for determining leaching. On these soils immobile water is preferentially sampled (Hatch *et al.*, 1997) and macropores can be the dominant drainage route, especially after heavy rains (Grossmann and Udluft, 1991) or irrigation. However, suction cups may still be suitable to determine the pool of water that the root utilises in plant uptake studies.

When using suction cups it is difficult to determine the soil volume from which the solution was removed and the reproducibility of both volume of water removed and the concentration of NO_3^--N is often unsatisfactory (Addiscott *et al.*, 1991). Suction cups with ceramic heads are frequently used to obtain soil solution for analysis of NO_3^--N (Webster *et al.*, 1993; Poss *et al.*, 1995; Hatch *et al.*, 1997) but their use is questionable for measuring phosphate (Hansen and Harris, 1975) and also we expect for SON due to sorption on surfaces. Teflon cup samplers (Zimmermann *et al.*, 1978) are becoming popular especially in phosphorus leaching studies (Bottcher *et al.*, 1984) and are not believed to interfere with the sampling of SON. However, they are more expensive than ceramic samplers decreasing their widespread application.

Soil solution samples can also be obtained as drainage water from field drains (Lawes *et al.*, 1882; Tyson *et al.*, 1997), lysimeters (Smolander *et al.*, 1995; Titus and Mahendrappa, 1996), or from hydrologically isolated plots (Cannell *et al.*, 1984; Vinten and Redman, 1990). In this case, mobile water is sampled preferentially and the methods are widely used to study leaching. However, field drains may only partially intercept drainage through soil and also require constant (*i.e.* automatic) sampling. Goulding and Webster (1992) concluded that lysimeters are the only reliable method to measure total water and NO_3^- loss, although results from suction cups were acceptable in freely draining soils. However, lysimeters can give problems through shrinkage from the walls and large scale lysimeters or hydrologically isolated plots, which allow agricultural operations to be carried out normally, are very expensive to set up (Cannell *et al.*, 1984). No comparison of methods has yet been conducted for SON but the process of assessing SON concentration in drainage flow against suction cups and extraction from the soil profile is being undertaken.

2.2   Extraction of SON from Soil

Soluble forms of N can be extracted from soils after shaking with water. However, such extractions cause the dispersion of clays and it may be difficult to obtain clean solutions for analysis (Young and Aldag, 1982). A range of salt solutions have been used for soil extraction, most commonly KCl, $CaCl_2$ and K_2SO_4. Salt extracts may disturb adsorption equilibria on soil surfaces and release organic N which was not dissolved in soil solution.

The SON pool in soils cannot be measured directly but instead is determined by subtracting the mineral N concentration from the total soluble N (TSN) concentration. Kjeldahl digestion was first used to determine TSN in sea water and has also been used for soil solutions (Beauchamp *et al.*, 1986). This method is based on the reduction of N to NH_4^+-N in an acid solution, and has been described in detail (Bremner and Mulvaney, 1982). However, the method is slow and cumbersome and high N contents within blank samples decreases the accuracy and sensitivity of the procedure (Smart *et al.*, 1981).

The development of simple, rapid and automated methods by which TSN can be routinely analysed has encouraged more measurements of SON to be made in recent years. Persulfate ($K_2S_2O_8$) oxidation was originally used for sea water analysis (D'Elia *et al.*, 1977; Koroleff, 1983) but has been modified to determine TSN in fresh water (Solórzano and Sharp, 1980) and soil extracts (Ross, 1992; Cabrera and Beare, 1993; Sparling *et al.*, 1996). $K_2S_2O_8$ oxidation is based on the principle that in the presence of a strong oxidising agent both NH_4^+-N and SON are converted to NO_3^--N. Complete oxidation is achieved by autoclaving the soil extract (Cabrera and Beare, 1993; Williams *et al.*, 1995; Sparling *et al.*, 1996) or by ultraviolet digestion. Both approaches are suitable for the processing of large batches of samples (100 per day) and only require common laboratory equipment. This technique has become popular in recent years for determining microbial biomass-N from the difference in TSN between chloroform fumigated and non-fumigated soils (Ross, 1992; Sparling and Zhu, 1993; Murphy *et al.*, 1998a). By also measuring the mineral N content in the non-fumigated soil SON can be estimated by difference (Jensen *et al.*, 1997; McNeill *et al.*, 1998). Alternatively, KCl salt extracts are routinely used to measure mineral N in soil to assist with fertiliser recommendations (Shepherd *et al.*, 1996; Wilson *et al.*, 1996) and it is possible to use samples of the same soil extract for determination of both TSN and mineral N.

Smart *et al.* (1981) determined that $K_2S_2O_8$ oxidation was more precise than Kjeldahl digestion in samples collected from a range of aquatic habitats. Studies comparing Kjeldahl digestion and $K_2S_2O_8$ oxidation have found no significant difference in the amount of TSN determined in water extracts of forest litter (Yu *et al.*, 1994) or 0.5 M K_2SO_4 soil extracts (Cabrera and Beare, 1993; Sparling *et al.*, 1996). However, in 1 M KCl soil extracts, TSN was overestimated at low concentrations by $K_2S_2O_8$ oxidation, and underestimated at higher concentrations when compared to Kjeldahl digestion (Cabrera and Beare, 1993). More recently Merriam *et al.* (1996) have proposed a high-temperature catalytic oxidation technique to measure TSN and found that results compare well to $K_2S_2O_8$ oxidation.

Experiments have shown the extractant to have only a slight influence on the size of both the mineral N and SON pools in a sandy loam soil (Figure 2). Results suggest, that on this soil type, extractions collected using routine procedures for mineral N (2 *M* KCl) or microbial biomass N (0.5 *M* K_2SO_4) are also suitable for the determination of SON pool size. However, this needs to be examined across a range of soil types with differing texture and cation exchange capacity.

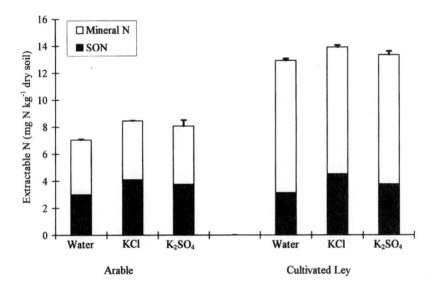

Figure 2 *Effect of extractant on the size of the mineral N and SON pools in a sandy loam soil under long term arable cropping or cultivated grass ley which was under 1st year of wheat*

2.3 Separation of TSN into Mineral and Organic Fractions

Electro-ultrafiltration (EUF) is based on applying an electric field to soil suspensions to separate fractions of soluble N by forced diffusion through membrane filters (Németh, 1979; Németh, 1985). The method removes both a mineral ($EUF-NO_3$) and organic ($EUF-N_{org}$) fraction of N from the soil solution and is considered to remove all forms of N that are available for either plant uptake or microbial transformations over the short term. The advantage of this technique is that the rate of nutrient release can be determined whereas soil extraction techniques only determine pool sizes

(Németh, 1985). However, EUF is labour intensive, costly, and results in greater variability compared to soil extraction (Houba *et al.*, 1986). Houba *et al.* (1986) found EUF and soil extraction by 0.01 *M* $CaCl_2$ to be highly correlated and concluded that the two techniques were interchangeable. However, Feng *et al.* (1990) found that the EUF-N_{org} fraction was larger than the pool of N extracted with $CaCl_2$ one week after the incorporation of ground and dried rape tops to soil.

3   Size of the SON Pool in Soil

In forest systems increased levels of atmospheric deposition (Goulding *et al.*, 1998) have elevated the N content of soil and brought many forest soils close to N saturation (Aber *et al.*, 1989; Aber, 1992; Currie *et al.*, 1996; Koopmans *et al.*, 1997). Sizeable pools of SON have been measured in forest floor leachates (Yavitt and Fahey, 1986; Stevens and Wannop, 1987; Qualls *et al.*, 1991; Currie *et al.*, 1996). Qualls *et al.* (1991) found that 94% of the dissolved N leaching through a deciduous forest soil was present in organic form. Yu *et al.* (1994) also found that SON was the dominant form of N in a coniferous forest soil after extraction with water.

Smith (1987) measured SON pools as large or larger than mineral N pools in air dried agricultural soils and suggested that this reflected the presence of freshly decomposed plant material and/or organic matter components disrupted during sample preparation. Recent studies with fresh soil have confirmed that the size of the SON pool can be as large as that of mineral N under agricultural cropping systems (Jensen *et al.*, 1997; McNeill *et al.*, 1998). Jensen *et al.* (1997) showed that 0.5 *M* K_2SO_4 extractable SON varied seasonally between 8 to 20 kg SON-N ha^{-1} in a coarse sand and 15 to 30 kg SON-N ha^{-1} in a sandy loam (0–15 cm); with the minimum during winter and the maximum in late summer. McNeill *et al.* (1998) showed that SON comprised 55–66% of the TSN under wheat (20 mg TSN-N kg^{-1}) and pasture (26 mg TSN-N kg^{-1}) on a loamy sand (0–10 cm) which was dry during summer months. Within the long-term Woburn Ley-Arable experiment (Johnston, 1973) SON in the 0–25 cm layer ranged from 2 mg SON-N kg^{-1} under continuous arable cropping to 5 mg SON-N kg^{-1} after eight years of grass ley and accounted for between 33 and 60% of the TSN (Figure 3). Within fields of an organic farm SON in the 0–25 cm layer accounted for 80% of TSN (Figure 4). At this site SON ranged from 7 to 13 mg SON-N kg^{-1} and increased with increasing number of previous years under grass/clover ley. Clearly SON is a significant pool within agricultural soils.

4   Role of SON in N Transformations

Understanding and predicting N release from soil organic matter is essential for improving fertiliser recommendations (Stockdale *et al.*, 1997). To predict the supply of N from soil it is necessary to determine the total pool of N which is easily mineralisable and likely to be released during crop growth. Smith *et al.*

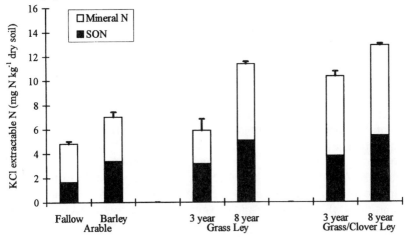

Figure 3   *Size of the 2* M *KCl extractable mineral N and SON pools in treatments from the Ley-Arable Experiment at Woburn, Bedfordshire. Some plots have been in continuous arable cropping since 1938 (Johnston, 1973), others alternate between 3 or 8 year grass or grass/clover leys and 2 years arable cropping*

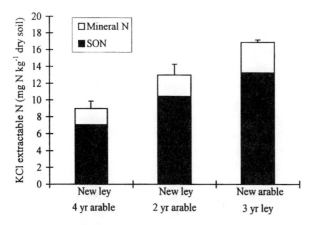

Figure 4   *Size of the 2* M *KCl extractable mineral N and SON pools in treatments from Duchy Home Farm, Tetbury, Gloucestershire which is a working organic farm. Arable plots (previously 2 or 4 yr arable) had recently been converted to new leys. The ley plot (previously 3 yr ley) had been converted to a new arable treatment*

(1980) had recognised the importance of SON in fertiliser predictions when they concluded that 'the use of salt solutions for periodic mineral-N extractions during mineralisation incubations should include estimates of the total N leached or some justification for not considering the amounts of organic N leached'. Németh (1985) and Recke and Németh (1985) also concluded that it is necessary to measure both the EUF-NO_3 and EUF-N_{org} fractions of soil N when predicting N fertiliser requirements. The EUF method has been applied to sugar beet (Wiklicky, 1982; Recke and Németh, 1985; Sheehan, 1985), grape production (Eifert *et al.*, 1982) and agricultural crops (Appel and Mengel, 1990; Saint Fort *et al.*, 1990; Linden *et al.*, 1993) to aid in the prediction of fertiliser requirements. For example, Rex *et al.* (1985) used EUF-NO_3 plus EUF-N_{org} to indicate the yield potential of a soil and found a highly significant correlation between grain yield of winter wheat and EUF-N content in the soil. However, Smith (1987) found that mineral N production alone can provide a representative picture of the potential soil N supply. Appel and Xu (1995) have subsequently showed that the decline in the size of the ^{15}N labelled EUF-N_{org} fraction does not necessarily correspond to the production of ^{15}N-labelled mineral N. Determination of the SON pool size alongside that of mineral N may improve subsequent fertiliser recommendation. However, the processes and conditions which promote and consume SON are not, as yet, well defined.

4.1   Mineralisation

Mineralisation studies have usually concentrated on changes in the size of the soil mineral N pool while largely overlooking SON. However, the distinction between organic matter mineralised through to NH_4^+-N and decomposition, which results in the production of SON, is important since the form of N released determines which microorganisms and plants can utilise the N. Recently Appel and Mengel (1993) have suggested that $CaCl_2$ extractable SON is a reliable indicator of the pool of organic N which is available for mineralisation in sandy soils. DeLuca and Keeney (1994) measured a decline in the size of the soluble amino N pool in soil under tallgrass prairie which coincided with an increase in microbial biomass and N mineralisation. Kielland (1995) has suggested that rapid turnover of amino acids in arctic tundra soils result in high rates of gross N mineralisation. Similarly, Murphy *et al.* (1998b) have suggested that changes in the size of the SON pool in a loamy sand under pasture and wheat may result in changes in gross N mineralisation rates. However, no direct link between SON turnover and gross N mineralisation rates has been established. Similarly, there is no direct evidence as to the extent to which the SON pool, recovered by EUF or $CaCl_2$ extracts, is mineralised during crop growth (Appel and Mengel, 1993). Appel and Xu (1995) attempted to relate the mineralisation of EUF-N_{org} to the appearance of mineral N by using ^{15}N-labelled rape residue as a tracer. They found that while EUF was able to selectively extract organic N derived from the rape, the decline in the size of the EUF-N_{org} fraction was not sufficient to account for all of the production of ^{15}N-labelled mineral N. Appel and Xu (1995) also found that a large (but unknown) fraction of the EUF-N_{org} derived from the rape was not easily mineralisable.

Smith (1987) showed that during long term aerobic incubation of soil with periodic leaching (Stanford and Smith, 1972) SON was produced between leaching intervals; although the majority of leached N was removed as NO_3^--N. Smith (1987) determined the mineralisation potential of the leachates in the absence of soil and concluded that the leached SON was not 'exceptionally susceptible' to mineralisation. This may imply that the SON would be stable in water courses. However, since the mineralisation potential was determined outside of the soil the findings of Smith (1987) do not indicate a rate of turnover for this pool *in situ* within soil. Also the size of the pool of SON does not necessarily indicate its importance in N turnover.

No strong relationship was found between the size of the 2 *M* KCl extractable SON pool and the amount of potentially available N (PAN; determined by anaerobic incubation) in soil collected from a range of soil types and landuses (Figure 5). In fact both the mineral N and SON pools increased after anaerobic incubation, with the ratio between mineral N and SON being

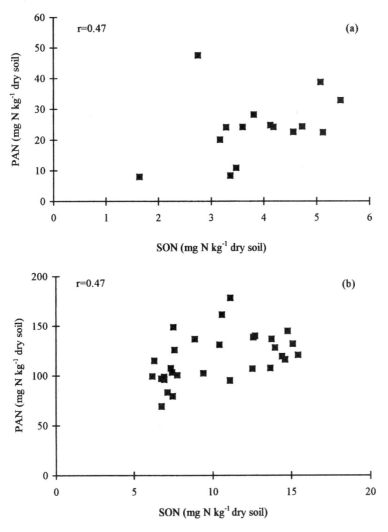

Figure 5   *Correlation between the amount of 2* M *KCl extractable SON and potentially available N (PAN) for soil collected from (a) the Ley-Arable experiment at Woburn and (b) Duchy Home Farm*

the same at the beginning and at the end of the incubation 7 days later (Figure 6). The absence of any change in this ratio does not indicate that SON is not the source of mineral N production, but results suggest that as an index PAN would rank soils in the same order whether SON was included or not. However, if the magnitude of PAN is thought to reflect the amount of the N that will become available to plant uptake, then inclusion of the SON pool may be warranted.

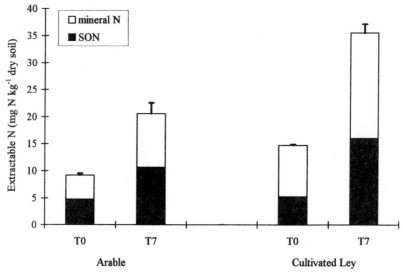

Figure 6   *Amount of 2 M KCl extractable mineral N and SON before (T0) and after (T7) anaerobic incubation of soil from arable and cultivated ley plots*

To date there is still no direct evidence as to what extent the organic molecules within the SON pool are actually mineralised. Breakdown of organic matter can result in the production of organic compounds which are soluble but recalcitrant to further microbial decomposition (Smolander *et al.*, 1995). This may mean that only a fraction of the SON pool is likely to be mineralised. Determination of the composition of the SON pool from a forest floor revealed that humic substances and hydrophilic acids were the main fractions involved with transportation of N into the mineral soil (Qualls *et al.*, 1991). However, a large component of the EUF-N_{org} fraction is composed of amino acids, proteins and other hydrolysable N compounds which are assumed to be easily mineralised (Németh *et al.*, 1988; Warman and Isnor, 1989; Appel and Mengel, 1993). Németh *et al.* (1988) determined the chemical composition of the EUF-N_{org} fraction from three arable and two forest soils. They found that in the arable soils about 3% of the EUF-N_{org} fraction was free amino acids, 23–55% was amino N (hydrolysable N) and the remainder was non-hydrolysable N. In the forest soils free amino acids comprised a greater percentage of the total fraction (Németh *et al.*, 1988). Until recently the characterisation of dissolved organic matter (DOM) into functional or structural groups has been restricted by suitable analytical techniques. Modern techniques including fluorescence spectroscopy (Tam and Sposito, 1993; Erich and Trusty, 1997), fourier-transform infrared (Candler *et al.*, 1988; Gressel *et al.*, 1995a), ^{13}C-nuclear magnetic resonance spectra (Candler *et al.*, 1988; Novak *et al.*, 1992) and ultraviolet-visible spectra (Candler and Van Cleve, 1982; Cronan *et al.*, 1992) are now being employed in the investigation of DOM composition. Methods are still being evaluated and modified from the analysis of humic substances (Gressel *et al.*, 1995a) but recent results (Gressel *et al.*, 1995b) suggest that these methods are likely to increase our understanding of the availability and transformation of DOM in the future.

4.2 Immobilisation

Microbial immobilisation of N has been shown to be predominantly of NH_4^+-N (Jansson and Person, 1982; Recous *et al.*, 1988; Shen *et al.*, 1989) although NO_3^--N can also be assimilated when carbon is available (Azam *et al.*, 1986; Recous *et al.*, 1988). The classical Mineralisation-Immobilisation Theory (MIT) of microbial assimilation assumes that all N uptake is from the mineral N pool. However, micro-organisms can also utilise low molecular weight, SON compounds (Molina *et al.*, 1983; Barak *et al.*, 1990; Barraclough, 1997) suggesting that MIT may not be an absolute. Hart *et al.* (1994) calculated the substrate C:N ratio that was being used by heterotrophic micro-organisms in a forest soil and showed that it was similar to the C:N ratio of the K_2SO_4 extractable organic pool. They have also shown that the K_2SO_4 extractable SON pool declines when the $CHCl_3$-labile N pool (*i.e.* microbial-N) increases and suggest that extractable SON is a major source of N for micro-organisms in forest soil (Hart and Firestone, 1991; Hart *et al.*, 1994). Recently Barraclough (1997) has shown that, in a soil under winter wheat, only 44% of added leucine-N and 82% of glycine-N was mineralised through to NH_4^+-N while all of the amino-N had disappeared from the soil, presumably due to direct assimilation by the microbial biomass. These findings suggest that conventional views of N transformation within soil may be an over-simplification and provides evidence that direct assimilation of SON may be occurring within agricultural soils.

Plant residues contain significant water soluble fractions of C and N (Mengel and Kirkby, 1978) and water soluble fractions of straw have been shown to be relatively rich in N (Jensen *et al.*, 1997). N from plant residues are rapidly assimilated by the microbial biomass, and Jensen *et al.* (1997) found no significant increase in the amount of SON in the soil after 4 and 8 t ha^{-1} inputs of oilseed rape straw compared to unamended soil. In contrast, studies of nutrient dynamics on tropical soil amended with green manures have shown that concentrations of SON increase substantially at the same time as there is immobilisation of mineral N (Mulongoy and Gasser, 1993). Hence, where only mineral N pools are measured false conclusions about the magnitude or direction of nutrient fluxes might be drawn. The role of soluble organic compounds within plant residues in the immobilisation process is unclear and warrants further investigation.

4.3   Leaching

Ecosystem studies have shown that concentrations of SON in forest floor leachates exceed concentrations of mineral N (Yu *et al.*, 1994) and SON has been found to be a dominant source of N to lakes in forested watersheds (Wissmar, 1991). Stevens and Wannop (1987) studied the composition of TSN from lysimeter leachate under the organic horizon and from ceramic cup solution in the soil mineral layers after clearfelling of Sitka spruce in North Wales. They found that within the organic horizons SON was more than 90% of TSN but that at depth in the soil profile $NO_3^- $-N predominated, suggesting transformation of SON to NO_3^--N at depth. However, results may also reflect low recoveries of SON at depth from ceramic cups (see section 2.1).

The current perception is that in temperate climates, NO_3^--N produced by mineralisation is the main source of N in drainage water from agricultural soils (Jarvis *et al.*, 1996; Macdonald *et al.*, 1989). In the light of the data reviewed above we challenge this perception and suggest that SON may also be a vehicle for significant loss of N from the soil. Significant concentrations of SON have been found in drainage waters leaving grassland lysimeters in Devon, U.K. (J. M. B. Hawkins, IGER, North Wyke, pers. comm.). Substantial amounts of SON have also been found at depth (0–90 cm) when extracted from soil under wheat and permanent pasture (Figure 7). Given the concentrations of SON extracted from soils it seems likely that there may be as much, or more, SON leached from the soil compared to NO_3^--N. To date it is unclear whether SON leaving soils can be transformed to NO_3^--N in surface or groundwaters.

Losses of N from soil may therefore be greater than previously considered, and leached SON may partially explain the imbalance which is sometimes seen in studies looking at the recovery of ^{15}N-labelled fertiliser (Glendining *et al.*, 1997). This is supported by Németh (1985) who found that fertiliser N was recovered in both EUF-NO_3 and EUF-N_{org} fractions and concluded that it is necessary to measure the easily mineralisable EUF-N_{org} fraction in fertiliser recovery studies. However, Appel and Mengel (1992) did not measure an increase in the size of the EUF-N_{org} fraction following fertiliser addition.

4.4   Plant Uptake

The release of N through the mineralisation of SOM and plant residues has long been identified as an important source of N for plant uptake (Scarsbrook, 1965; Paul and Voroney, 1980). Plant uptake is usually considered to be dominantly as NH_4^+-N and NO_3^--N, with NO_3^--N being favoured when it is available in abundance (Hageman, 1984; Streeter and Barta, 1988). Plants are able to take up urea directly in the absence of hydrolysis (Harper, 1984), but more slowly than mineral N (Kirkby and Mengel, 1970). Bollard

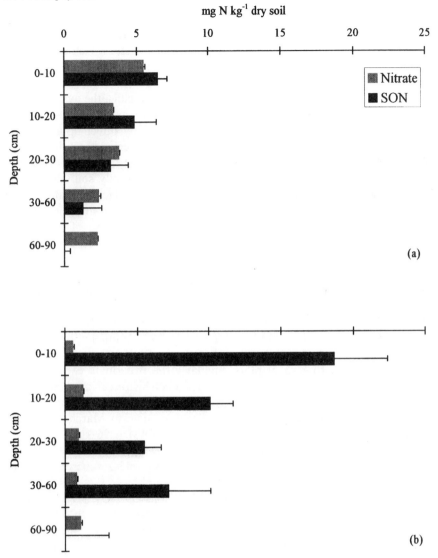

Figure 7   *Distribution of 2 M KCl extractable NO₃⁻-N and SON within the soil profile (a) under continuous arable and (b) permanent pasture plots on a sandy loam soil located at Woburn Farm, Bedfordshire*

(1959) demonstrated that many plants are able to use SON compounds as their sole N source. In some natural environments (*e.g.* temperate and arctic heathlands) mineralisation rates are too low to supply all of the N required by plants and uptake of SON is a major source of plant available N (Chapin *et al.*, 1993; Kielland, 1994; Mercik and Németh, 1985). Schmidt and Stewart (1997) found that over one year of measurements NH_4^+-N and amino acids were the dominant forms of low molecular weight N under subtropical heathland. It is well established that arctic plants are capable of utilising SON either directly (Chapin *et al.*, 1993) or in association with ectomycorrhizae (Atkin, 1996; Kielland, 1994), although there is still a lack of information on the contribution of SON to the total intake of N by mycorrhizal plants.

It has usually been thought that, within agricultural soils, the availability of organic N compounds to plant roots is minimal (Harper, 1984). It is not clear whether agricultural crops benefit from SON either by direct uptake or *via* microbial immobilisation of organic molecules and then remineralisation of NH_4^+-N. Appel and Mengel (1990; 1992) found that the size of the organic N pool extracted by EUF or $CaCl_2$ did not decline in sandy soils during the growth of rape plants, although they pointed out that the lack of change in the size of the SON pool does not indicate that the fraction is not turning over and supplying N to plants. Further research employing the use of ^{15}N tracer studies is required to determine what proportion of extractable SON fractions are utilised by the plant.

Although it is well known that amounts of NO_3^--N in the soil are influenced by numerous factors including plant uptake, residue returns and previous crop history (Addiscott *et al.*, 1991) the major factors influencing SON pool sizes and plant uptake have not been determined. Mercik and Németh (1985) have shown that long term (since 1923) annual applications of inorganic N (90 kg N ha^{-1}) increased both the EUF-NO_3 and EUF-N_{org} fractions compared to zero N plots. Similarly the Broadbalk experiment at Rothamsted has treatments which have received historical applications of inorganic N and/or farmyard manure (FYM) since 1844 and is being used to investigate N losses from arable land. Soils were taken in March 1997 from plots receiving either 0 or 144 kg N ha^{-1} y^{-1} or FYM + 96 kg N ha^{-1} y^{-1} and from wheat grown either continuously or following potatoes. As might be expected 2 *M* KCl extractable mineral N and SON pool sizes increased under treatments in the order 0 kg N ha^{-1} y^{-1} < 144 kg N ha^{-1} y^{-1} < FYM + 96 kg N ha^{-1} y^{-1} (Figure 8). Greater concentrations of NO_3^--N was found where wheat followed potatoes (Figure 8). Merick and Németh (1985) also found that leaching of EUF-NO_3 was greater under potato monoculture compared to rye monoculture. On Broadbalk less SON was present where wheat followed potatoes (8–20% of TSN) compared to continuous wheat (26–29% TSN), especially at depth (Figure 8). At present we are unsure whether this was related to: (i) a differences in the quality and quantity of residue inputs to soil under potatoes compared to wheat, or (ii) a difference in the utilisation of SON between plant species.

5   Conclusion

Concerns over the environmental and health impacts of NO_3^--N leached from soils, as well as the importance of NH_4^+-N and NO_3^--N in crop nutrition, have focused attention on the study of mineral N in agricultural soils. In contrast the SON pool has received little attention despite its being as large as

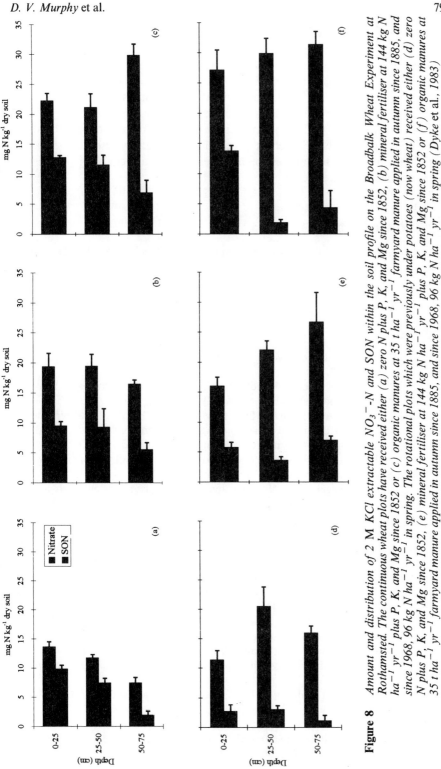

Figure 8 *Amount and distribution of 2 M KCl extractable NO_3^--N and SON within the soil profile on the Broadbalk Wheat Experiment at Rothamsted. The continuous wheat plots have received either (a) zero N plus P, K, and Mg since 1852, (b) mineral fertiliser at 144 kg N ha^{-1} yr^{-1} plus P, K, and Mg since 1852 or (c) organic manures at 35 t ha^{-1} yr^{-1} farmyard manure applied in autumn since 1885, and since 1968, 96 kg N ha^{-1} yr^{-1} in spring. The rotational plots which were previously under potatoes (now wheat) received either (d) zero N plus P, K, and Mg since 1852, (e) mineral fertiliser at 144 kg N ha^{-1} yr^{-1} plus P, K, and Mg since 1852 or (f) organic manures at 35 t ha^{-1} yr^{-1} farmyard manure applied in autumn since 1885, and since 1968, 96 kg N ha^{-1} yr^{-1} in spring (Dyke et al., 1983)*

the mineral N pool. The role of SON in N transformations is unclear. The absence of a relationship between SON and other indices of N availability suggest that this measurement does not reflect the size of the mineralisable N pool. However, information on its role in N transformations is sparse and often contradictory. The presence of significant amounts of SON within the soil profile suggests that it is subject to leaching losses and may be of importance in water pollution. At present there are many unanswered questions concerning SON. The recent development of simple methods to measure SON and improved techniques to characterise SON are enabling these issues to be studied in greater detail.

Acknowledgements

IACR receives grant-aided support from the UK Biotechnology and Biological Sciences Research Council and the authors also acknowledge funding from the UK Ministry of Agriculture, Fisheries and Food.

References

Aber J. D., 1992, Nitrogen cycling and nitrogen saturation in temperate forest ecosystems, *Tree*, **7**, 220–224.

Aber J. D., K. J. Nadelhoffer, P. Steudler and J. M. Melillo, 1989, Nitrogen saturation in northern forest ecosystems, *Bio Science*, **39**, 378–386.

Addiscott T. M., A. P. Whitmore and D. S. Powlson, 1991, 'Farming, fertilizers and the nitrate problem', Rothamsted Experimental Station, Hertfordshire, p. 1.

Appel T. and K. Mengel, 1990, Importance of organic nitrogen fractions in sandy soils, obtained by electro-ultrafiltration or $CaCl_2$ extraction, for nitrogen mineralization and nitrogen uptake of rape, *Biol. Fertil. Soil.*, **10**, 97–101.

Appel T. and K. Mengel, 1992, Nitrogen uptake of cereals grown on sandy soils as related to nitrogen fertilizer application and soil nitrogen fractions obtained by electro-ultrafiltration (EUF) and $CaCl_2$ extraction, *Eur. J. Agron.*, **1**, 1–9.

Appel T. and K. Mengel, 1993, Nitrogen fractions in sandy soils in relation to plant nitrogen uptake and organic matter incorporation, *Soil Biol. Biochem.*, **2**, 685–691.

Appel T. and F. Xu, 1995, Extractability of ^{15}N-labelled plant residues in soil by electro-ultrafiltration, *Soil Biol. Biochem.*, **27**, 1393–1399.

Atkin O. K., 1996, Reassessing the nitrogen relations of Arctic plants: a mini-review, *Plant Cell Environ.*, **19**, 695–704.

Azam F., K. A. Malik and F. Hussain, 1986, Microbial biomass and mineralization-immobilization of nitrogen in some agricultural soils, *Biol. Fertil. Soil.*, **2**, 157–163.

Barak P., J. A. E. Molina, A. Hadas and C. E. Clapp, 1990, Mineralization of amino acids and evidence of direct assimilation of organic nitrogen, *Soil Sci. Soc. Am. J.*, **54**, 769–774.

Barraclough D., 1997, The direct or MIT route for nitrogen immobilization: A ^{15}N mirror image study with leucine and glycine, *Soil Biol. Biochem.*, **29**, 101–108.

Beauchamp E. G., W. D. Reynolds, D. Brasche-Villeneuve and K. Kirby, 1986, Nitrogen mineralization kinetics with different soil pretreatments and cropping histories, *Soil Sci. Soc. Am. J.*, **50**, 1478–1483.

Bollard E. E., 1959, 'Utilisation of nitrogen and its compounds by plants', Symposia of the Society of Experimental Biology XIII, Academic Press, New York, p. 304.

Bottcher A. B., L. W. Miller and K. L. Campbell, 1984, Phosphorus adsorption in various soil-water extraction cup materials: Effect of acid wash, *Soil Sci.*, **137**, 239–244.

Bremner J. M., 1965, 'Methods of soil analysis, part 2', C. A. Black (ed.), ASA, Madison, Wisconsin, Agronomy 9, p. 1179.

Bremner, J. M. and C. S. Mulvaney, 1982, 'Methods of Soil Analysis', A. L. Page (ed.), p. 599.

Cabrera M. L. and M. H. Beare, 1993, Alkaline persulfate oxidation for determining total nitrogen in microbial biomass extracts, *Soil Sci. Soc. Am. J.*, **57**, 1007–1012.

Candler R. and K. Van Cleve, 1982, A comparison of aqueous extracts from the B horizon of a birch and aspen forest in interior Alaska, *Soil Sci.*, **134**, 176–180.

Candler R., W. Zech and H. G. Alt, 1988, Characterization of water-soluble organic substances from a Typic Dystrochrept under spruce using GPC, IR, ^{1}H NMR, and ^{13}C NMR spectroscopy, *Soil Sci.*, **146**, 445–452.

Cannell R. Q., M. J. Goss, G. L. Harris, M. G. Jarvis, J. T. Douglas, K. R. Howse, and S. Le Grice, 1984, A study of mole drainage with simplified cultivation for autumn-sown crops on a clay soil. 1. Background, experiment and site details, drainage systems, measurement of drainflow and summary of results, 1978–80, *J. Agric. Sci.*, **102**, 539–559.

Chapin III F. S., L. Moilanen and K. Kielland, 1993, Preferential use of organic nitrogen for growth by a non-mycorrhizal arctic sedge, *Nature*, **316**, 150–153.

Cronan C. S., S. Lakshman and H. H. Patterson, 1992, Effects of disturbance and soil amendments on dissolved organic carbon and organic acidity in red pine forest floors, *J. Environ. Qual.*, **21**, 457–463.

Currie W. S., J. D. Aber, W. H. McDowell, R. D. Boone and A. H. Magill, 1996, Vertical transport of dissolved organic C and N under long-term N amendments in pine and hardwood forests, *Biogeochem.*, **35**, 471–505.

Davidson E. A., M. J. Stark and M. K. Firestone, 1990, Microbial production and consumption of nitrate in an annual grassland, *Ecology*, **71**, 1969–1975.

D'Elia C. F., P. A. Steudler and N. Corwin, 1977, Determination of total nitrogen in aqueous samples using persulfate digestion, *Oceanogr.*, **22**, 760–764.

DeLuca T. H. and D. R. Keeney, 1994, Soluble carbon and nitrogen pools of prairie and cultivated soils: Seasonal variation, *Soil Sci. Soc. Am. J.*, **58**, 835–840.

Dyke G. V., B. J. George, A. E. Johnston, P. R. Poulton and A. D. Todd, 1983, 'The Broadbalk Wheat Experiment 1968–1978: yields and plant nutrients in crops grown continuously and in rotation', Rothamsted Experimental Station Report for 1982, Part 2, p. 5.

Eifert J., M. Várnai and L. Szöke, 1982, Application of the EUF procedure in grape production, *Plant Soil*, **64**, 105–113.

Erich M. S. and G. M. Trusty, 1997, Chemical characterization of dissolved organic matter released by limed and unlimed forest soil horizons, *Can. J. Soil Sci.*, **77**, 405–413.

Feng K., D. Hua-Ting and K. Mengel, 1990, Turnover of plant matter in soils as assessed by electro-ultrafiltration (EUF) and $CaCl_2$ extracts, *Agribiol. Res.*, **43**, 337–347.

Glendining M. J., P. R. Poulton, D. S. Powlson and D. S. Jenkinson, 1997, Fate of ^{15}N-labelled fertilizer applied to spring barley grown on soils of contrasting nutrient status, *Plant Soil*, **195**, 83–98.

Goulding K. W. T. and C. P. Webster, 1992, Methods for measuring nitrate leaching, *Aspects Applied Biol.*, **30**, 63–69.

Goulding K. W. T., N. J. Bailey, N. J. Bradbury, P. Hargreaves, M. T. Howe, D. V. Murphy, P. R. Poulton, and T. W. Willison, 1998, Nitrogen deposition and its contribution to nitrogen cycling and associated soil processes, *New Phytol.*, **139**, 49–58.

Gressel N., A. E. McGrath, J. G. McColl and R. F. Powers, 1995a, Spectroscopy of aqueous extracts of forest litter. I: Suitability of methods, *Soil Sci. Soc. Am. J.*, **59**, 1715–1723.

Gressel N., J. G. McColl, R. F. Powers and A. E. McGrath, 1995b, Spectroscopy of aqueous extracts of forest litter. II: Effects of management practices, *Soil Sci. Soc. Am. J.*, **59**, 1723–1731.

Grossmann J. and P. Udluft, 1991, The extraction of soil water by the suction-cup method: A review, *J. Soil Sci.*, **42**, 83–93.

Hageman R. H., 1984, 'Nitrogen in crop production', R. D. Hauck (ed.), ASA, CSSA, SSSA, Madison, p. 67.

Hansen E. A. and A. R. Harris, 1975, Validity of soil-water samples collected with porous ceramic cups, *Soil Sci. Soc. Am. Proc.*, **39**, 528–536.

Harmsen G. W. and G. J. Kolenbrander, 1965, 'Soil nitrogen', W. V. Bartholomew and F. E. Clark (eds.), ASA, Madison, Agronomy 10, p. 43.

Harper J. E., 1984, 'Nitrogen in crop production', R. D. Hauck (ed.), ASA, CSSA, SSSA, Madison, p. 165.

Hart S. C. and M. K. Firestone, 1991, Forest floor-mineral soil interactions in the internal nitrogen cycle of an old-growth forest, *Biogeochem.*, **12**, 103–127.

Hart S. C., G. E. Nason, D. D. Myrold and D. A. Perry, 1994, Dynamics of gross nitrogen transformations in an old-growth forest: The carbon connection, *Ecology*, **75**, 880–891.

Hatch D. J., S. C. Jarvis, A. J. Rook, and A. W. Bristow, 1997, Ionic contents of leachate from grassland soils: A comparison between ceramic suction cup samplers and drainage, *Soil Use Manag.*, **13**, 68–74.

Houba V. J. G., I. Novozamsky, A. W. M. Huybregts and J. J. Van der lee, 1986, Comparison of soil extractions by 0.01 M $CaCl_2$, by EUF and by some conventional extraction procedures, *Plant Soil*, **96**, 433–437.

Hsieh Y. P., 1992, Pool size and mean age of stable soil organic carbon in cropland, *Soil Sci. Soc. Am. J.*, **56**, 460–464.

Jansson S. L. and J. Person, 1982, 'Nitrogen in Agricultural soils', F. J. Stevenson (ed.), ASA, CSSA, SSSA, Madison, pp. 229–252.

Jarvis S. C., E. A. Stockdale, M. A. Shepherd and D. S. Powlson, 1996, Nitrogen mineralization in temperate agricultural soils: Processes and measurement, *Adv. Agron.*, **57**, 187–235.

Jenkinson D. S. and D. S. Powlson, 1976, The effects of biocidal treatments on metabolism in soil – 1. Fumigation with chloroform, *Soil Biol. Biochem.*, **8**, 167–177.

Jensen L. S., T. Mueller, J. Magid and N. E. Nielsen, 1997, Temporal variation of C and N mineralization, microbial biomass and extractable organic pools in soil after oilseed rape straw incorporation in the field, *Soil Biol. Biochem.*, **29**, 1043–1055.

Johnston A. E., 1973, 'The effects of ley and arable cropping systems on the amount of soil organic matter in the Rothamsted and Woburn Ley-Arable experiments', Rothamsted Experimental Station Report for 1972, Part 2, pp. 131–159.

Kielland K., 1994, Amino acid absorption by arctic plants: Implications for plant nutrition and nitrogen cycling, *Ecology*, **75**, 2373–2383.

Kielland K., 1995, Landscape patterns of free amino acids in arctic tundra soils, *Biogeochem.*, **31**, 85–98.

Kirkby E. A. and K. Mengel, 1970, 'Nitrogen nutrition of the plant', E. A. Kirkby (ed.), University of Leeds, Leeds, p. 35.

Koopmans C. J., D. van Dam, A. Tietema, and J. M. Verstraten, 1997, Natural ^{15}N abundance in two nitrogen saturated forest ecosystems, *Oecologia*, **111**, 470–480.

Koroleff F., 1983, 'Methods of seawater analysis, 2nd edition', K. Grasshoff (ed.), Verlag Chemie, Weinheim, pp. 151–157.

Lawes J. B., J. H. Gilbert and R. Warington, 1882, On the amount and composition of the rain and drainage-waters collected at Rothamsted. Part III. The drainage waters from land cropped and manured, *J. Royal Agric. Soc. England*, **18**, 1–71.

Linden B., I. Lyngstrad, J. Sippola, J. D. Nielsen, K. Soegaard and V. Kjellerup, 1993, Evaluation of the ability of three laboratory methods to estimate net nitrogen mineralization during the growing season, *Swedish J. Agric. Res.*, **23**, 161–170.

Macdonald A. J., D. S. Powlson, P. R. Poulton and D. S. Jenkinson, 1989, Unused fertiliser nitrogen in arable soils – its contribution to nitrate leaching, *J. Sci. Food Agric.*, **46**, 407–419.

McNeill A. M., G. P. Sparling, D. V. Murphy, P. Braunberger and I. R. P. Fillery, 1998, Changes in extractable and microbial C, N, and P in a Western Australian wheatbelt soil following simulated summer rainfall, *Aust. J. Soil Res.*, in press.

Mengel K., 1985, Dynamics and availability of major nutrients in soils, *Adv. Soil Sci.*, **2**, 65–131.

Mengel K. and E. A. Kirkby, 1978, 'Principles of Plant Nutrition', International Potash Institute, Switzerland, p. 314.

Mercik S. and K. Németh, 1985, Effects of 60-year N, P, K and Ca fertilization on EUF-nutrient fractions in the soil and on yields of rye and potato crops, *Plant Soil*, **83**, 151–159.

Merriam J., W. H. McDowell and W. S. Currie, 1996, A high-temperature catalytic oxidation technique for determining total dissolved nitrogen, *Soil Sci. Soc. Am. J.*, **60**, 1050–1055.

Molina J. A. E., C. E. Clapp, M. J. Shaffer, F. W. Chichester and W. E. Larson, 1983, NCSOIL, a model of nitrogen and carbon transformations in the soil: Description, calibration, and behaviour, *Soil Sci. Soc. Am. J.*, **47**, 85–91.

Mulongoy K. and M. O. Gasser, 1993, Nitrogen-supplying capacity of leaves of *Dactyladenia barteri* (Hook ex olw) and *Leucaena leucocephala* (Lam.) de Wit in two soils of different acidity from southern Nigeria, *Biol. Fertil. Soils*, **16**, 57–62.

Murphy D. V., G. P. Sparling and I. R. P. Fillery, 1998a, Stratification of microbial biomass C and N and gross N mineralisation with soil depth in two contrasting Western Australian agricultural soils, *Aust. J. Soil Res.*, **36**, 45–55.

Murphy D. V., G. P. Sparling, I. R. P. Fillery, A. M. McNeill and P. Branberger, 1998b, Mineralisation of soil organic nitrogen and microbial respiration after simulated summer rainfall events in an agricultural soil, *Aust. J. Soil Res.*, **36**, 231–246.

Murphy D. V., I. R. P. Fillery and G. P. Sparling, 1998c, Seasonal fluctuations in gross N mineralisation, ammonium consumption, and microbial biomass in a Western Australian soil under different land uses, *Aust. J. Agric. Res.*, **49**, 523–535.

Németh K., 1979, The availability of nutrients in the soil as determined by electro-ultrafiltration (EUF), *Adv. Agron.*, **31**, 155–188.

Németh K., 1985, Recent advances in EUF research (1980–1983), *Plant Soil*, **83**, 1–19.

Németh K., H. Bartels, M. Vogel and K. Mengel, 1988, Organic nitrogen compounds extracted from arable and forest soils by electro-ultrafiltration and recovery rates of amino acids, *Biol. Fertil. Soil.*, **5**, 271–275.

Novak J. M., P. M. Bertch and G. L. Mills, 1992, Carbon-13 nuclear magnetic resonance spectra of soil water-soluble organic carbon, *J. Environ. Qual.*, **21**, 537–539.

Paul E. A. and N. G. Juma, 1981, 'Terrestrial Nitrogen Cycles. Processes, Ecosystems, Strategies and Management Inputs', F. E. Clark and T. Rosswall (eds.), Stockholm, Ecological Bulletin 33, pp. 179–204.

Paul E. A. and R. P. Voroney, 1980, 'Contemporary Microbial Ecology', D. C. Ellwood, J. N. Heldger, M. J. Latham and J. M. Slater (eds.), Academic Press, New York, p. 215.

Poss R., A. D. Noble, F. X. Dunin and W. Reyenga, 1995, Evaluation of ceramic cup samplers to measure nitrate leaching in the field, *Eur. J. Soil Sci.*, **46**, 667–674.

Qualls R. G. and B. L. Haines, 1991, Geochemistry of dissolved organic nutrients in water percolating through a forest ecosystem, *Soil Sci. Soc. Am. J.*, **55**, 1112–1123.

Qualls R. G., B. L. Haines and W. T. Swank, 1991, Fluxes of dissolved organic nutrients and humic substances in a deciduous forest, *Ecology*, **72**, 254–266.

Recke H. and K. Németh, 1985, Relationships between EUF-N fractions, N uptake and quality of sugar beet in deep loess soils of Southern Lower Saxony, *Plant Soil*, **83**, 133–141.

Recous S., J. M. Machet and B. Mary, 1988, The fate of ^{15}N urea and ammonium nitrate applied to a winter wheat crop. II. Plant uptake and N efficiency, *Plant Soil*, **112**, 215–224.

Rex M., K. Németh and T. Harrach, 1985, The EUF-N contents of arable soils as an indicator of their yield potential, *Plant Soil*, **83**, 117–125.

Ross D. J., 1992, Influence of sieve mesh size on estimates of microbial carbon and nitrogen by fumigation-extraction procedures in soils under pasture, *Soil Biol. Biochem.*, **24**, 343–350.

Saint Fort R., K. D. Frank and J. S. Schepers, 1990, Role of nitrogen mineralization in fertilizer recommendations, *Commun. Soil Sci. Plant. Anal.*, **21**, 1945–1958.

Scarsbrook C. E., 1965, 'Soil Nitrogen', W. V. Bartholomew and F. E. Clark (eds.), ASA, Madison, Agronomy 10, p. 481.

Schmidt S. and G. R. Stewart, 1997, Waterlogging and fire impacts on nitrogen availability and utilization in a subtropical wet heathland (wallum), *Plant Cell Environ.*, **20**, 1231–1241.

Sheehan M. P., 1985, Experiments on the reproducibility of results from EUF soil extracts with possible improvements resulting from these experiments, *Plant Soil*, **83**, 85–92.

Shen S. M., P. B. S. Hart, D. S. Powlson and D. S. Jenkinson, 1989, The nitrogen cycle in the Broadbalk wheat experiment: ^{15}N-labelled fertilizer residues in the soil and in the soil microbial biomass, *Soil Biol. Biochem.*, **21**, 529–533.

Shepherd M. A., E. A. Stockdale, D. S. Powlson and S. C. Jarvis, 1996, The influence of organic nitrogen mineralization on the management of agricultural systems in the UK, *Soil Use Manag.*, **12**, 76–85.

Smart M. M., F. A. Reid and J. R. Jones, 1981, A comparison of a persulfate digestion and the kjeldahl procedure for determination of total nitrogen in freshwater samples, *Water Res.*, **15**, 919–921.

Smith J. L., R. R. Schnabel, B. L. McNeal and G. S. Campbell, 1980, Potential errors in the first-order model for estimating soil nitrogen mineralization potentials, *Soil Sci. Soc. Am. J.*, **44**, 996–1000.

Smith S. J., 1987, Soluble organic nitrogen losses associated with recovery of mineralized nitrogen, *Soil Sci. Soc. Am. J.*, **51**, 1191–1194.

Smolander A., V. Kitunen, O. Priha and E. Malkonen, 1995, Nitrogen transformations in limed and nitrogen-fertilized soil in Norway Spruce stands, *Plant Soil*, **172**, 107–115.

Solórzano L. and J. H. Sharp, 1980, Determination of total dissolved nitrogen in natural waters, *Limnol. Oceanogr.*, **25**, 751–754.

Sparling G. P. and C. Zhu, 1993, Evaluation and calibration of methods to measure microbial biomass C and N in soils from Western Australia, *Soil Biol. Biochem.*, **25**, 1793–1801.

Sparling G. P., C. Zhu and I. R. P. Fillery, 1996, Microbial immobilization of ^{15}N from legume residues in soils of differing textures: Measurement by persulfate oxidation and ammonia diffusion methods, *Soil Biol. Biochem.*, **28**, 1707–1715.

Stanford G. and S. J. Smith, 1972, Nitrogen mineralization potentials of soils, *Soil Sci. Soc. Am. Proc.*, **36**, 465–472.

Stevens P. A. and C. P. Wannop, 1987, Dissolved organic nitrogen and nitrate in an acid forest soil, *Plant Soil*, **102**, 137–139.

Stevenson F. J., 1982, 'Nitrogen in Agricultural Soils', F. J. Stevenson (ed.), ASA, CSSA, SSSA, Madison, pp. 1–42.

Stockdale E. A., J. L. Gaunt and J. Vos, 1997, Soil-plant nitrogen dynamics: What concepts are required?, *Eur. J. Agron.*, **7**, 145–159.

Streeter J. G. and A. L. Barta, 1988, 'Physiological Basis of Crop Growth and Development', M. B. Tesar (ed.), ASA, CSSA, Madison, p. 176.

Tam S. C. and G. Sposito, 1993, Fluorescence spectroscopy of aqueous pine litter extracts: Effects of humification and aluminium complexation, *J. Soil Sci.*, **44**, 513–524.

Titus B. D. and M. K. Mahendrappa, 1996, 'Lysimeter Systems Designs Used In Soil Research: A Review', Canadian Forest Service, Canada, p. 14.

Tyson K. C., D. Scholefield, S. C. Jarvis and A. C. Stone, 1997, A comparison of animal output and nitrogen leaching losses recorded from drained fertilized grass and grass/clover pasture, *J. Agric. Sci.*, **129**, 315–323.

Vinten A. J. A. and M. H. Redman, 1990, Calibration and validation of a model of non-interactive solute leaching in a clay-loam arable soil, *J. Soil Sci.*, **41**, 199–214.

Warman P. R. and R. A. Isnor, 1989, Evidence of peptides in low-molecular-weight fractions of soil organic matter, *Biol. Fertil. Soil*, **8**, 25–28.

Webster, C. P., M. A. Shepherd, K. W. T. Goulding and E. Lord, 1993, Comparisons of methods for measuring the leaching of mineral nitrogen from arable land, *J. Soil Sci.*, **44**, 49–62.

Wiklicky L., 1982, Application of the EUF procedure in sugar beet cultivation, *Plant Soil*, **64**, 115–127.

Williams B. L., C. A. Shand, M. Hill, C. Ohara, S. Smith and M. E. Young, 1995, A procedure for the simultaneous oxidation of total soluble nitrogen and phosphorus in extracts of fresh and fumigated soils and litters, *Comm. Soil Sci. Plant Anal.*, **26**, 91–106.

Wilson W. S., K. L. Moore, A. D. Rochford and L. V. Vaidyanathan, 1996, Fertilizer nitrogen addition to winter wheat crops in England: Comparison of farm practices with recommendations allowing for soil nitrogen supply, *J. Agric. Sci.*, **127**, 11–22.

Wissmar R. C., 1991, Forest detritus and cycling of nitrogen in a mountain lake, *Can. J. For. Res.*, **21**, 990–998.

Yavitt J. B. and T. J. Fahey, 1986, Litter decay and leaching from the forest floor in *Pinus contorta* (Lodgeploe pine) ecosystems, *J. Ecology*, **74**, 525–545.

Young J. L. and R. W. Aldag, 1982, 'Nitrogen in Agricultural Soils', ASA, CSSA, SSSA, Madison, Agronomy 22, p. 43.

Yu S., R. R. Northup and R. A. Dahlgren, 1994, Determination of dissolved organic nitrogen using persulfate oxidation and conductimetric quantification of nitrate-nitrogen, *Comm. Soil Sci. Plant Anal.*, **25**, 3161–3169.

Zimmermann C. F., M. T. Price and J. R. Montgomery, 1978, A comparison of ceramic and teflon *in situ* samplers for nutrient pore water determinations, *Estuarine Coastal Marine Sci.*, **7**, 93–97.

6

Average Fertiliser Nitrogen Use and Estimates of Nitrogen Uptake by Major Arable Crops in England and Wales

W. S. Wilson[1], M. H. B. Hayes[2*] and L. V. Vaidyanathan[2]

[1] DEPARTMENT OF BIOLOGICAL SCIENCES, UNIVERSITY OF ESSEX, WIVENHOE PARK, COLCHESTER CO4 3SQ, UK
[2] SCHOOL OF CHEMISTRY, THE UNIVERSITY OF BIRMINGHAM, EDGBASTON, BIRMINGHAM B15 2TT, UK

Abstract

Average amounts of fertiliser nitrogen applied are compared with estimates of crop uptake calculated using average yields and typical average nitrogen content in whole crops prior to harvest for wheat, barley, oilseed rape, main crop potatoes and sugar beet grown in England and Wales during the period 1974 to 1995. Soil nitrogen is not included in these estimates. The records show that there is some scope for reducing fertiliser nitrogen application to wheat and main crop potatoes thereby limiting the accumulation of nitrate in soils. The nitrogen status of soils with barley and oilseed rape was not increased by general fertiliser usage while fertiliser treated sugar beet soils had soil nitrogen levels lower than untreated soils.

1 Introduction

Nitrogen (N) is a key nutrient element in crop productivity. It is derived by plants, in part, from the soil and supplemented by fertiliser N when the soil supply is inadequate for economic crop production. Amounts of N to be applied are determined in early Spring and are based on soil type, field cropping history and yield expectations. The economic (break-even) cost of a kilogram (kg) of fertiliser N during the period of the investigation was equivalent to: three kg of wheat or barley, two kg of oilseed rape, five to eight

---

* New address: Department of Chemical and Environmental Sciences, Foundation Building, University of Limerick, Limerick, Ireland.

kg of ware potatoes and about 15 kg of clean sugar beet – the five major arable crops under investigation.

Inherent variability of field factors and unpredictable weather conditions after deciding the amount of fertiliser N to apply in the Spring make calculated, experimental optima imprecise. Present attempts at recommendations, taking account of soil and field factors, through measured or modelled soil nitrogen supply (SNS), are improving precision mostly in fields with large SNS (Vaidyananthan and Leitch[1]). However, these methods are as critically dependent on realistic expectation by early Spring as are other methods requiring timely soil analysis and computer application. Experimental results of applying such predetermined amounts of nitrogen fertiliser are subject to +12 to − 15% variability (Wilson *et al.*[2]). Growers have to follow the guidance of advisers, if not their own judgement, about the contribution of SNS. Crops invariably show a response to N addition by becoming greener and luxuriant. This creates an optimism for higher yields not always realised; on the contrary, lodging, increased disease and delayed maturity and loss of quality can occur.

Until the implementation of the European Union 50 mg NO_3^- l^{-1} standard, water authorities tolerated natural leakage of nitrate from farmland. Criticism of high application of fertiliser N during the 1980s brought about restraint. Consequently, fertiliser N applications per hectare stayed nearly constant or showed a modest decline (Addiscott[3]).

2   Method

Average amounts of nutrients applied as fertilisers are determined through a statistically robust average yield of crops, also by collating returns of harvest information.[4,5] Analysis of experimental crops at or prior to harvest provide typical average nutrient concentrations in the whole crop. Amounts of nutrients (N in the present context) accumulated by plants close to maturity, considered as justifiable estimates of nutrients are calculated by multiplying yield (whole crop) by its typical nutrient concentration called Uptake N. Residues returned to the soil decompose and contribute to SNS used by succeeding crops. The ensuing text provides a comparison of fertiliser use and crop N uptake for wheat, barley, oilseed rape, main crop potatoes and sugar beet grown in England and Wales during the period 1974 to 1995[4,5].

3   Results and Discussion

3.1   Wheat

Data for winter wheat are presented (Table 1a). The area growing wheat nearly doubled at the expense of barley from one million hectares at 1978 to over 1.9 million hectares by the late 1980s. Average fertiliser N additions increased steadily at 8 kg ha^{-1} from under 100 kg N ha^{-1} in 1974 to around 185 kg N ha^{-1}. Characteristically, average yields also increased at an annual rate of 0.2 t ha^{-1} up to 1986 from under 5.0 t ha^{-1} to nearly 7.0 t ha^{-1} in

Table 1 *Average yield, fertiliser N applications and estimates of crop uptake of N at harvest, 1974–1996*

	a) Winter wheat			b) Winter barley			c) Spring barley		
	Yield (t ha⁻¹) 85% D.M.	Fertiliser (kg N ha⁻¹)	Crop uptake (kg N ha⁻¹)	Yield (t ha⁻¹) 85% D.M.	Fertiliser (kg N ha⁻¹)	Crop uptake (kg N ha⁻¹)	Yield (t ha⁻¹) 85% D.M.	Fertiliser (kg N ha⁻¹)	Crop uptake (kg N ha⁻¹)
1974	5.00	98	115	4.00	77	84			
1975	4.30	101	99	3.40	81	71		↓	
1976	3.80	106	87	3.40	89	71			
1977	4.90	116	113	4.20	86	88			
1978	5.25	128	121	4.19	89	80			
1979	5.22	137	120	4.10	97	86			
1980	5.88	146	135	4.43	105	93			
1981	5.84	153	134	4.39	118	92			
1982	6.20	167	143	4.93	120	104			
1983	6.37	183	147	4.66	130	98			
1984	7.72	189	178	6.15	150	141	4.93	100	99
1985	6.37	192	147	5.54	151	127	4.59	103	92
1986	6.96	187	160	5.69	149	131	4.75	106	95
1987	5.99	193	138	5.46	154	126	4.32	101	86
1988	6.23	189	143	5.26	142	121	3.96	91	79
1989	6.74	182	155	5.47	143	126	3.77	94	75
1990	6.97	177	160	5.53	132	127	4.10	86	82
1991	7.25	187	167	5.90	139	136	4.81	90	96
1992	6.82	187	157	6.11	140	141	4.61	90	92
1993	7.33	185	169	5.79	133	133	4.67	94	93
1994	7.35	186	169	5.78	142	133	4.68	103	96
1995	7.70	193	177	6.12	142	141	4.86	101	97
1996	8.12	Not avail	Not avail	6.51	Not avail	150	5.33	Not avail	107

Mixed Winter and Spring Barley data

(a)

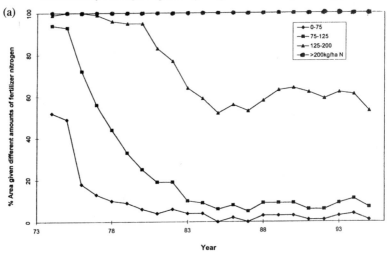

(b)

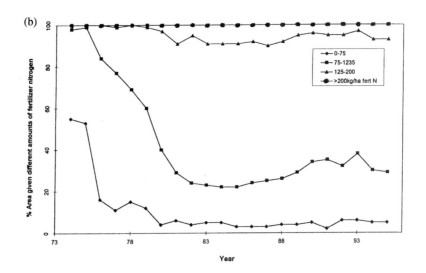

1995. There was a sharp decline during the next two seasons (not due to decrease in fertiliser use) to 6.0 t ha^{-1} but yields increased again at a similar annual rate to nearly 7.5 t ha^{-1} during the mid 1990s. An important trend was that during the last eight years of the study, when yields increased steadily, the average fertiliser N addition declined moderately from the peak of 193 kg N ha^{-1}. This suggests that N supply to crops was not a constraint to yield when seasonal growing conditions were favourable.

The proportion of the wheat area receiving different amounts of fertiliser N is shown in Fig. 1a. The area receiving >125 kg N ha^{-1} increased sharply from 6% in 1974 to 97% in 1987; the area given >200 kg N ha^{-1} increased from none in 1976 to 5% in 1980 and hence to 44–48% during 1985–87 and declining slightly thereafter. Notably from 1989 the area given >200 kg N

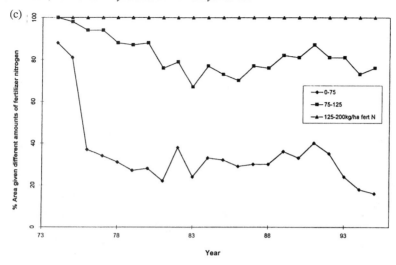

Figure 1   *Percentage area given different amounts of fertilizer nitrogen: (a) wheat; (b) winter barley; (c) spring barley*

ha^{-1} was around 38% but yields were increasing from 6.74 to 7.70 t ha^{-1} emphasising that N supply was not the limiting factor.

The absence of a link between increasing yields and large fertiliser use indicates an accelerating or compounding effect during the 1980s when too many growers became over optimistic expecting large N additions to deliver larger yields without due consideration of soil type, weather, management efficiency and cultural constraints. Awareness of the nitrate pollution problem and a more informed appreciation of the contribution of SNS related to previous cropping in the field halted this trend. Finding simple and grower-friendly ways of identifying fields where the large N additions (>200 kg N ha^{-1}) would not result in a yield response should be the priority for advice on fertiliser N recommendations for wheat.

3.2 Barley

Barley was the major cereal grown up to 1982. The proportion of cereal area under barley declined from over 60% to about 30%, from 1.8 million to around 780,000 hectares between 1974 and 1994. The spring sown barley area declined even faster from over 1.6 million to around 200,000 hectares between 1974 and 1994. However, the area under winter barley increased from around 160,000 in 1974 to nearly 950,000 hectares by 1986, declining to about 570,000 hectares in 1994. Spring barley received only moderate fertiliser N inputs to keep N concentration in the grain suitable for malting. Even so, amounts added increased from 76 kg in 1974 to over 100 kg by 1987 but declined to about 90 kg N ha^{-1} in 1992 again rising to around 100 kg in 1994 (Tables 1b and 1c). Winter barley also started with less than 100 kg ha^{-1} but steadily increased to a peak of over 150 kg ha^{-1} by 1987 declining slightly to around 140 kg ha^{-1}. Separate yields of the two types were available from 1984. Table 1b illustrates the combined data (weighted for area under each type and fertiliser use) up to 1983. Fertiliser additions to both types were in slight excess over uptake from 1979 to 1990 and remained equal to 1994. Excess of addition over uptake from spring barley increased from nothing to under 20 kg by 1989 followed by balance to 1994. Excess fertiliser N added to winter barley increased from under 10 kg to a peak of 26 kg ha^{-1} between 1984 and 1987 then tended to nearly equal crop uptake. Nearly a third of the spring barley crop received less than 75 kg ha^{-1}; over half the area was given up to 125 kg ha^{-1} and the rest received 125–200 kg ha^{-1}: none of the crops was given more than 200 kg fertiliser N ha^{-1} (Fig 1c). Amounts of fertiliser N added to the winter barley crop contrasted sharply. Only more than 50% of the very small area receiving up to 75 kg N ha^{-1}, around 5% of the steeply increasing remained so. Two thirds of the area was given 75–120 kg N ha^{-1} up to 1977 but the proportion declined sharply to under 20% during the mid 1980s and then levelled off around 20–30% to the end of the period. The proportions of the area receiving 125–200 kg N ha^{-1} rose sharply from almost none in 1974 to just under 80% by the mid 1990s and then abated to around 65–70%. Only 10% or less of the area received >200kg fertiliser N ha^{-1} (Fig 1b). Thus the barley crop, unlike the wheat, received fertiliser N close to crop requirements. Risk of lodging caused by liberal fertiliser N additions – short, stiff strawed cultivars were unavailable – despite some protection by timely spraying with stem-shortening chemicals had prevented profligacy in fertiliser use.

3.3   Oilseed Rape

Only winter oilseed rape is considered. This is a comparatively new crop starting with just 13,000 hectares in 1974 rising to 400,000 hectares in 1991. Fertiliser N additions have been generous up to a peak of 280 kg N ha^{-1} in 1984, except during 1974 and 1975. The amount added decreased to around 225 kg N ha^{-1} between 1984 and 1991 declining further to between 180 and 190 kg N ha^{-1} by April (Table 2a).

The proportion of the area given 125 kg N fertiliser ha^{-1} or less was mostly at 10% excepting in 1978 and 1979 and during 1993 to 1995 when it reached about 20%. Only 15% of the area received >200 kg N ha^{-1} during 1974 and 1975; from 1976–89 a large part of the area (80–95%) received 225–280 kg N ha^{-1}. Following a drastic cut in European Commission Subsidy during mid 1980s the proportions of the area given >200 kg N ha^{-1} declined to less than 50% with a concomitant increase in the area given 125 to 200 kg N ha^{-1} *i.e.* from less than 10% to 40% (Fig 2a).

Seed yield was mostly a little under 3.0 t ha^{-1} (91% dry matter) except for a few years with about +0.5 t ha^{-1} fluctuation. This brassica crop grows luxuriantly during the post-rosette pre-flowering stage, using much nitrogen; thus crop N uptake approximates to the amounts of fertiliser N added, except in the four seasons with yield around 2.5 t ha^{-1}. An important feature during

Table 2 Average yield, fertiliser N applications and estimates of crop uptake of N at harvest, 1974–1996

	a) Oilseed rape			b) Main crop potatoes			c) Sugar beet		
	Yield (t ha⁻¹) 91% D.M.	Fertiliser (kg N ha⁻¹)	Crop uptake (kg N ha⁻¹)	Yield (t ha⁻¹)	Fertiliser (kg N ha⁻¹)	Crop uptake (kg N ha⁻¹)	Yield (t ha⁻¹) 16% sugar	Fertiliser (kg N ha⁻¹)	Crop uptake (kg N ha⁻¹)
1974	Not avail	186	Not avail	34.0	175	126	22.86	148	109
1975	Not avail	143	Not avail	28.1	174	82	24.88	150	118
1976	2.3	240	179	20.8	179	77	26.66	156	127
1977	2.6	238	203	29.4	181	109	33.94	145	162
1978	2.4	234	187	37.2	188	138	35.57	149	169
1979	2.7	225	211	34.0	196	126	38.13	157	181
1980	3.3	254	257	27.2	188	138	37.21	150	177
1981	2.6	263	203	35.5	199	131	36.17	162	172
1982	3.3	265	257	38.8	200	144	50.70	145	241
1983	2.5	276	195	31.7	205	117	38.82	158	185
1984	3.4	281	265	39.2	216	145	47.05	150	224
1985	3.0	272	234	36.7	201	136	41.9	127	199
1986	3.24	261	253	37.4	196	138	45.8	130	218
1987	3.49	265	265	39.0	207	144	41.4	128	197
1988	3.00	244	234	39.7	205	147	44.9	121	199
1989	2.94	233	229	37.5	208	139	44.4	121	211
1990	3.23	212	252	37.4	190	138	43.8	127	208
1991	2.90	207	226	37.0	199	137	44.4	127	211
1992	2.88	199	225	45.0	192	167	51.5	122	245
1993	2.73	181	213	43.0	193	160	49.1	115	234
1994	2.62	186	204	41.9	203	155	44.7	127	213
1995	2.91	190	207	39.1	202	145	43.0	122	205
1996	3.49	Not avail	272	41.0	188	152	48.0	Not avail	228

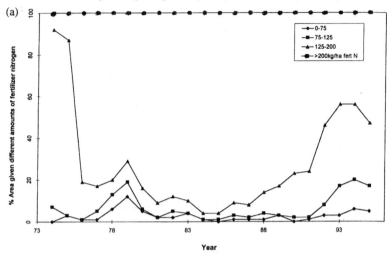

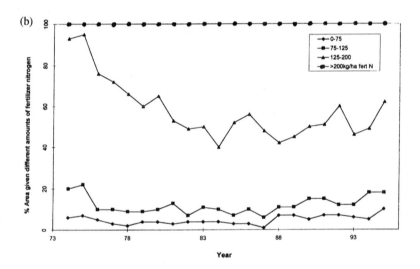

the period since 1990 when the amounts of fertiliser N use decreased to about two thirds of the peak rate, *i.e.* by 20 kg to 180 kg N ha^{-1}, seed yields remained close to or even larger than 3.0 t ha^{-1}. Thus the luxuriant vegetative growth supported by generous fertiliser N additions was not essential for larger seed yields.

It is noteworthy that studies on the effect of removing most of the older expanded leaves during early January, *i.e.* the end of leaf differentiation and the onset of floral initials in the main apex showed some remarkable responses [Agricultural Development and Allied Services (ADAS) Cambridge, Internal Report[6]]. Expansion of young leaves, enclosing the main auxilliary apices, was curtailed. Stem elongation was inhibited resulting in a significant dwarfing of the plants until the flower buds on the main stem apex started to mature. The

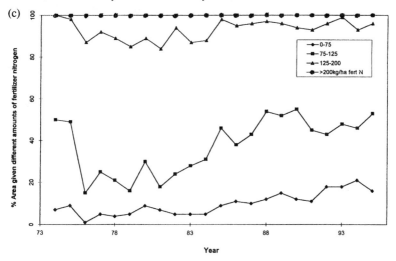

Figure 2   *Percentage area given different amounts of fertilizer nitrogen: (a) oilseed rape; (b) main crop potatoes; (c) sugar beet*

start of flowering was delayed by three to five days but ended synchronously compared with non-defoliated plants; thus causing a more peaked, less extended flowering period. Seed yield was further unaffected or significantly enhanced; thus curbing the post-rosette, pre-flowering vegetative growth seems to be a desirable and beneficial option. This is supported by results of experiments using Paclobutrazol for stem shortening [Agricultural Development and Allied Service (ADAS), Rosemaund Report, Herefordshire[7]].

The very liberal use of fertiliser N during the mid 1970s to late 1980s was therefore unnecessary. That the uptake was large and the apparent unused amount of fertiliser N was small is not relevant because the crop sheds most of the larger expanded leaves as flower development progresses. The harvested seed amounts to a little over 40% of N uptake, returning a large part of the N taken up to the soil (except in a small proportion of fields where the oilseed rape straw is removed for animal feed).

3.4 Main Crop Potatoes

The area under this late spring planted crop diminished from nearly 140,000 hectares in 1976 to less than 100,000 hectares during the 1980s and to even less, 63,000 hectares in 1992 fluctuating up to 100,000 hectares on occasions. The crop received about 200 kg N ha^{-1} except during 1974–78 when it received about 180 kg N ha^{-1} (Table 2b). The proportion of area given >200 kg fertiliser N ha^{-1} increased sharply in 1976 from under 10% to 24% and kept rising steadily to 50% or more going up to 60% in 1984; 80–90% of the area received >125 kg fertiliser N ha^{-1}. More than a third of the area was given large dressings of farmyard manure but growers did not make any significant allowance for the nutrients derived from this source when adding fertiliser nitrogen (Fig 2b).

Average ware yields fluctuated between 30 and 40 t ha^{-1} depending more on weather conditions than on nutrient (N in this context) with a peak yield of 45 t ha^{-1} in 1992. Although a leafy crop, N uptake was 130 to 150 kg N ha^{-1} except for much smaller uptakes during 1974–77 which accounted for about 70% of the added fertiliser N. However, much of the N taken up during growth is returned to the soil at harvest.

3.5   Sugar Beet

This crop was grown in about 200,000 hectares – fluctuating between 173,000 and 227,000 hectares. Sown in the Spring in wide rows, plants take up nutrients (N in this context) quite slowly until early Summer and continue growth and uptake well into the early Autumn months. Fertiliser use had been moderate at around 150 kg N ha^{-1} to the mid 1980s but sharply declined to around 120 kg N ha^{-1} in prompt and sensible response to research findings (Table 2c) involving the penalty for juice impurity associated with large N uptake which also depressed sugar concentration in the roots.

All the area under this crop was given <200 kg fertiliser N ha^{-1} during 1976–82 (around 50% of the area in 1974 and 1975, declining gradually thereafter to 50–60% of the area up to 1986 and decreasing further to around 45% of the area with occasional fluctuations below and above this level). The proportion of the area given 75–125 kg fertiliser N ha^{-1} rose from around 15% (about 40% in 1974 and 1975) to 35–40% by 1990 but declined to 25–35% in the recent years. Less than 10% of the area received <75 kg fertiliser N ha^{-1} until mid 1980s; this proportion increased steadily to reach 21% in 1994. About 30% of the sugar beet area was given farmyard manure. Relatively greater account was given to adjust fertiliser N additions to allow for N available from the manure than in the case of the potato crop (Fig 2c).

Clean root yield increased steadily from about 23 t ha^{-1} in 1974 reaching a peak of about 51 t ha^{-1} in 1982, declining slowly during the next five years but rising again to another peak of 57.5 t ha^{-1} in 1992, falling to around 43–49 t ha^{-1} during the final three years of the period observed. Nitrogen uptake by the crop increased from 109 kg N ha^{-1} in 1974 to a peak of 245 kg N ha^{-1} in 1992 even though fertiliser N additions were much less than the uptake since 1977, more so from 1982. In the 1990s overall average usage has stabilised at about 105 kg N ha^{-1} (Draycott *et al.*[8]). Thus amounts of fertiliser N applied exceeded crop uptake only during 1974–76. Since then the deep rooted crop has been depleting soil N supply at a steadily increasing rate with fertiliser N providing only 50–60% of crop N uptake. However, much of the N taken up by the crop is retained in the soil at harvest except in a small proportion of the area where the tops are grazed by sheep whose droppings return and redistribute part of what they eat.

Conclusions

The survey of fertiliser use over a significant period of time (1974–95) and the uptake by major arable crops puts into perspective the distribution of mineral nitrogen during the growing season of the individual crops. Also revealed is the cropping history, techniques within the growing season and especially the incidence and duration of established rooting systems. The variability and intensity of weather factors have also been related to the uptake or leaching of nitrate. The investigation has undoubtedly signified that the periodic relative abundance of potentially leachable nitrate in the soil is due to the activities of micro-organisms in effecting mineralisation of currently added or permanently resident soil organic matter. The apparent high levels of nitrate consequent upon the application of high amounts of fertiliser N are a secondary rather than a direct effect.

References

1. Vaidyanathan L. V. and Leitch M. H., Use of Fertiliser and Soil Nitrogen by Winter Wheat Established with and without Soil Cultivation Prior to Drilling, *Journal of the Science of Food and Agric.*, Vol. 31, pp. 850–853, Cambridge University Press, 1980.

2. Wilson W. S., Moore K. L., Rockford A. D. and Vaidyanathan L. V., Fertiliser nitrogen addition to winter wheat crops in England: comparison of farm practices with recommendations allowing for soil nitrogen supply, *Journal of Agricultural Science*, Vol 127, pp. 11–32, Cambridge University Press, 1996.

3. Addiscott T. M., Reprinted from *Agricultural Chemicals and the Environment*, Eds. R.E. Hester and R.M. Harrison, *Issues in Environmental Science and Technology* no 5, The Royal Society of Chemistry, Cambridge, 1996.

4. Anon, Survey of Fertiliser Practice. Fertiliser use in farm type regions of England and Wales, Agricultural Development and Advisory Service (ADAS), Rothamsted Experimental Station and Fertiliser Manufacturers Association, 1974–91.

5. Anon, British Survey of Fertiliser Practice. Fertiliser use of farm crops, MAFF, Fertiliser Manufacturers Association and the Scottish Office, 1992–96.

6. Anon, Agricultural Development and Advisory Service (ADAS), Cambridge, Experiments on agronomy of oilseed rape. Internal Report, 1989.

7. Anon, Agricultural Development and Advisory Service (ADAS), Rosemaund, Herefordshire.

8. Draycott A. P., Allison M. F. and Armstrong M. J., 'Changes in Fertiliser Usage in Sugar Beet Production', Theme-Environmental Aspects of Sugar Beet Growing Proc., International Institute of Sugar Beet Research 60th Congress, Cambridge, pp. 39–54, 1997.

7

Nitrate in Ensiled Grass: A Risk or Benefit to Livestock Production?

J. Hill

DEPARTMENT OF AGRICULTURE AND RURAL MANAGEMENT
WRITTLE COLLEGE, CHELMSFORD CM1 3RR, UK

Abstract

Elevated concentrations of nitrate (> 1.5 g NO_3^- kg^{-1} DM) in herbage for ensiling are relatively rare in the UK: 0.37% of grass analysed in 1995 (n = 268), 0.58% in 1996 (n = 346) and 0.39% in 1997 (n = 254) (Collins *pers. comm*). The main reason for this is the practice of leaving at least six weeks between the application of fertiliser and harvesting of grass, a period sufficient for fertiliser – nitrogen (N) assimilation to protein. When nitrate or nitrite occurs in herbage, the formation of the toxic gases NO_x and N_2O during ensiling is likely to increase leading to a potential risk to the health of livestock and to workers during silo filling and emptying. The processes of NO_x and N_2O formation during conservation are the result of the reduction of nitrate and nitrite during the early stages of ensiling (*i.e.* before day 20) by plant and heterofermentative bacterial reductases. Clostridial activity, as a result of soil contamination during harvesting, poses a significant risk to animal health, performance (*i.e.* botulism, enterotoxaemia and reduced voluntary intake of silage) and silage quality (process of secondary fermentation). Nitrite is a powerful inhibitor of clostridial activity during the ensiling process and it has been incorporated into several silage additives used in the European Union.

1 Introduction

Improved grassland and rough grazing accounts for approximately 67% of UK agricultural land (Table 1; MAFF, 1996). Nitrogen (N) is arguably the most important nutrient which influences the productivity of herbage grasses and therefore the production of milk and meat from pasture and conserved grass. The average use of fertiliser N on grassland has increased since 1945

Table 1 *Grassland statistics for the UK (proportion of land in production)*

	Average 1985–1987	Year 1992	1996
Total agricultural area ('000 ha)	18680	18511	18401
Grassland less than 5 years ('000 ha)	1734 (0.093)	1562 (0.084)	1376 (0.075)
Grassland over 5 years ('000 ha)	5066 (0.271)	5289 (0.286)	5213 (0.283)
Sole right rough grazing ('000 ha)	4829 (0.259)	4680 (0.253)	4489 (0.243)
Common right rough grazing ('000 ha)	1216 (0.065)	1230 (0.066)	1237 (0.067)

from about 5 kg N ha^{-1} to 125 kg N ha^{-1} at present (Frame, 1993). The amount of fertiliser applied to grassland is, however, variable dependent on reserves of N in the soil, site and soil conditions, environmental considerations [*e.g.* accumulated degree days (T-sum)] and land use. The highest applications of N fertiliser are likely to occur in intensive dairy farming systems, especially on areas of grassland which are to be used for conservation (Jarvis, 1993).

The development of dry matter (DM) yield of grass shows a linear increase (about 20 kg DM kg N^{-1} added) to an increase in supply up to about 300 kg N ha^{-1} (Unwin and Vellinga, 1994). Above 300 kg N ha^{-1} the rate of response of DM yield declines and N applications in excess of 600 kg N ha^{-1} are uneconomic. Heavy applications of N fertiliser may no longer be acceptable not only from an economic point of view but because of environmental issues. Concern has been focused mainly on the role and effect of elevated concentrations of nitrate (NO_3^-) in both surface water and aquifers as a result of 'leakage' from agricultural systems (McKinney *et al.*, 1999; Jarvis, 1999) and the effect of elevated concentrations of ammonia and nitrous oxide released to the atmosphere (Fowler *et al.*, 1999). From an economic point of view, high rates of application of N to intensive grassland do not always result in the efficient utilisation of applied N by the grass or the cow. The poor efficiency of utilisation of N (*i.e.* the conversion of fertiliser N to product sold) can lead to a reduction in gross margin (Vellinga and Verburg, 1995). This paper assesses the role of nitrate and nitrite (NO_2^-) in grass when the crop is harvested for silage production and there is risk of ingestion of elevated concentrations of nitrate by grazing cattle.

2 Methods

2.1 General Methodology: Occurrence of Nitrate in Grass for Silage

Nitrate and ammonium are the main sources of inorganic N taken up by herbage grass (Marschner, 1995). Ammonium is generally incorporated into N containing organic compounds in the root of the plant whereas nitrate is highly mobile (transported *via* the xylem) and is re-distributed to both root and shoot. The timing of the main application of fertiliser N for intensive grass production for conservation is approximately six to eight weeks before harvesting the grass. This period allows complete assimilation of N from the fertiliser into protein which is about 75–90% of total N in herbage DM; the remaining N being non-protein fractions, for example free amines, ureides, nucleotides, chlorophyll and nitrate (McDonald *et al.*, 1991). If, however, any sustained period of drought stress occurs after the application of fertiliser N, the uptake of nitrate and ammonium by the plant is reduced. High rates of N fertiliser can lead to reduced levels of fermentable carbohydrate, increased concentrations of protein and an increase in buffering capacity of the grass. Hence, samples are taken before cutting to determine the concentrations of DM, sugar and nitrate and the buffering capacity and those data are used to determine the suitability of the crop for ensiling. The data collated from one UK laboratory, presented in Figure 1, shows the distribution of concentration of nitrate in samples of grass cut for first-cut grass silage production in the UK in 1995 to 1997.

2.2 General Methodology: Impact of Nitrate on Silage Fermentation

Elevated concentrations of nitrate in herbage for ensiling are considered as undesirable. Nitrate *per se* is relatively non-toxic to ruminant livestock, but the reduction of nitrate to nitrite and formation of toxic gases (NO_x and N_2O) as a result of ensiling could pose a risk to livestock health and the farmer (Spoelstra, 1985).

The reduction of nitrate, which occurs within hours of ensiling, is a process (*e.g.* by enterobacteria for example *Hafnia alvei*, clostridia and certain lactoba-cilli) producing nitrite, nitric oxide (gas formation) and ammonia (Spoelstra, 1985) but plant reductases also play an important role in reduction during the first five days of ensiling. One of the most important factors that affect the rate of nitrate reduction during these early stages is the rate of acidification of the forage mass. If the supply of readily fermentable carbohydrate in the herbage before ensiling is limited, the rate of acidification may be slowed resulting in an increased rate of conversion of nitrate to ammonium. As a consequence, a poorly fermented silage with high concentrations of butyric acid will result which will have poor intake characteristics and low feeding value for ruminant livestock (Weissbach *et al.*, 1993).

A primary growth of Italian ryegrass was cut by hand on 25 April 1997 and 10 May 1997 and wilted for 24 hours in poor weather. The sward had received a split application of fertiliser on 3 March 1997 (55 kg N ha^{-1}) and 2 April 1997 (69 kg N ha^{-1}). Polythene bag silos (four per treatment; 5 kg fresh weight) were constructed according to Wilson and Wilkins (1972). The silos were opened after 90 days storage and contents were analysed for nutritive and chemical composition. The nutritive and chemical compositional data were subjected to analysis of variance procedures (Genstat 5 Committee, 1993).

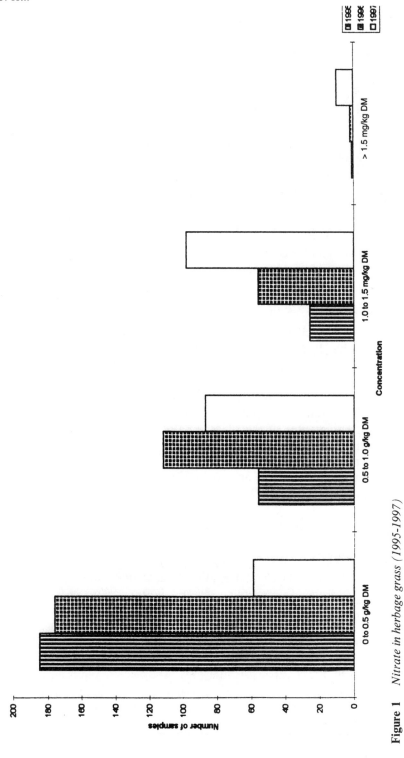

Figure 1 *Nitrate in herbage grass (1995-1997)*

3   Results and Discussion

3.1   Occurrence of Nitrate in Grass for Silage

The range of concentrations of nitrate in herbage cut immediately prior to ensiling in 1995 to 1997 are in Figure 1. There was a very low incidence of excessive levels of nitrate in herbage grass for ensiling in 1995 and 1996 but not in 1997.

Periods of drought *per se* do not affect the concentration of nitrate in the herbage markedly, but rapid uptake of nitrate and ammonium does occur during any period of rainfall after the drought. If the period after a 'drought-rainfall' cycle is too short (*i.e.* less than 4 weeks) before harvesting, high concentrations of nitrate can occur in the herbage (> 1.5 g NO_3^- kg^{-1} DM; ranging from 2 to 8 g NO_3^- kg^{-1} DM; Spoelstra, 1987). Interestingly, extensification of grassland management may well reduce the concentration of nitrate in herbage and hence the effect of the rate of acidification of the forage mass may be of greater importance in ensuring well fermented silage (Weissbach, 1996).

3.2   Impact of Nitrate on Silage Fermentation

The chemical composition of silages made from grass containing low concentrations of nitrate (10 May 1997) or high concentrations of nitrate (25 April 1997) are in Table 2.

The processes of fermentation in silage made from low nitrate grass are dominated by the formation of butyric acid as a result of the fermentation of lactate. The fermentation of lactate as a result of clostridial action leads to a reduction in feeding value of the silage, loss of gross energy during storage (approximately 18% of gross energy; Weissbach, 1996) and, potentially, animal health problems (*e.g.* enterotoxaemia and botulism; Wilkinson, 1988). In the absence of nitrate, butyric acid and hydrogen are formed as a result of the fermentation of lactate (Figure 2). Nitrate does however have a positive role to

Table 2 *Impact of nitrate on chemical composition and fermentation of grass silage*

Date of harvest	Low nitrate grass 10 May 1997		High nitrate grass 25 April 1997		S.E.D (silage)
	Grass	Silage	Grass	Silage	
Corrected dry matter (g kg^{-1} DM)	168	178	165	184	4.62
Nitrate (g kg^{-1} DM)	1.0	0	3.1	0.1	0.25
Sugar (g kg^{-1} DM)	154	28	146	7	11.85
Crude protein (g kg^{-1} DM)	186	176	175	147	10.56
Ammonium N (g kg^{-1} total N)	14.2	103	8.6	193	12.12**
pH	–	4.7	–	4.6	0.85
Lactic acid (g kg^{-1} DM)	–	27.9	–	79.9	11.25**
Acetic acid (g kg^{-1} DM)	–	36.1	–	59.1	8.56*
Butyric acid (g kg^{-1} DM)	–	19.6	–	2.1	4.22***

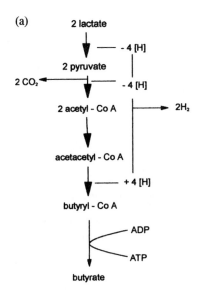

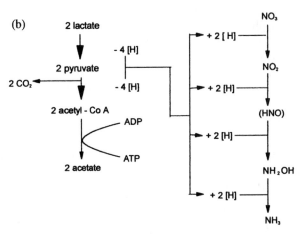

Figure 2   *Fermentation of lactate in the absence of nitrate (from Weissenbach et al., 1993)*

play in the fermentation of grass (Table 2; high nitrate grass). The first product of reduction of nitrate is nitrite. Nitrite has been shown to have a strong inhibitory effect on *Clostridia* and is incorporated into various silage additives (for example kofasil) used in the European Union (Wilkinson *et al.*, 1996; Haigh, 1996; Lattemae and Lingvall, 1996). Therefore in the presence of nitrate and nitrite, the fermentation of lactate yields acetic acid rather than butyric acid (Figure 2).

As nitrate acts as both a hydrogen and an electron acceptor during reduction, this can have an effect on the processes of fermentation (Table 3). The effect of nitrate on the process of fermentation of fermentable carbohy-

Table 3 *Conversion of nitrate to various end products during ensiling (after Spoelstra, 1985)*

Reaction	pH	Organism
1. $NO_3^- + AH_2 + H^+ \rightarrow NO_2^- + A + H_2O$	>5.0	Enterobacteria
	>5.5	Plant reductases
	>4.0	Lactobacilli
2. $2NO_2^- + 2AH_2 + 2H^+ \rightarrow N_2O + 2A + 3H_2O$	>5.0	Enterobacteria
3. $NO_2^- + 3AH_2 + 2H^+ \rightarrow NH_4^+ + 3A + 2H_2O$	>4.5	Enterobacteria
4. $3NO_2^- + 2H^+ \rightarrow NO_3^- + 2NO + H_2O$	<6.0	Chemical conversion

Overall stoichimometry: $NO_3^- + 8H^+ + 8e^- \rightarrow NH_3 + 2H_2O + OH^-$
AH_2 = Electron donor, for instance carbohydrate or NAD

drate by heterolactic bacteria is shown in Figure 3. In the absence of nitrate, glucose is converted to lactic acid and ethanol with the hydrogen (proton) released at the conversion of glucose 6-phosphate to 6-phosphogluconate and the conversion of 6-phosphogluconate to ribulose 5-phosphate. The protons released are utilised during the conversion of acetyl P and acetaldehyde to ethanol. If, however, nitrate is present, the hydrogen (proton) released during the fermentation of glucose is utilised during the decomposition of nitrate to ammonium. Therefore the conversion of acetyl P to ethanol does not occur and acetate is formed (Figure 3). The formation of acetate allows the formation of an additional mole of ATP and therefore may be a preferred pathway for heterolactic bacteria (Hein, 1970; Weissbach *et al.*, 1993).

Finally if there is a rapid decline in the pH of the forage mass during the initial stages of fermentation, the risk of fermentation of lactate (secondary fermentation) and hence the formation of butyric acid is minimal. For rapid acidification to occur efficiently, the level and availability of fermentable sugar in the grass at harvest has to be adequate ($c.$ 170 g kg^{-1} DM) and the buffering capacity of the grass has to be moderate ($c.$ 270 mE kg^{-1}).

3.3 Formation of Toxic Gases and Volatile Nitrosoamines

Velthof and Oenema (1997) reported on the rates of emission of nitrous oxide from dairy farming systems in the Netherlands. Their calculations derived an emission factor for N_2O release from grass silage of 15 ± 10 g N_2O-N kg^{-1} NO_3^--N in the silage dry matter assuming the concentration of nitrate (NO_3^-) in grass ensiled was 2 g N kg^{-1} DM.

The potential formation of volatile nitrosoamines (N-nitrosodimethylamine, N-nitrosodiethylamine and N-nitrosodipropylamine; potent carcinogen and mutogens) as a result of free secondary amines reacting with nitrite in herbage and silage has been reported (van Broekhoven and Davies, 1981 and 1985). The typical concentrations of volatile N-nitrosoamines reported in grass silage are approximately 2 µg kg^{-1} freshweight, a concentration which is unlikely to have any significant effect on the animal. Furthermore, studies performed by van Broekhoven and Stephany (1978) investigated the formation of N-

(a)

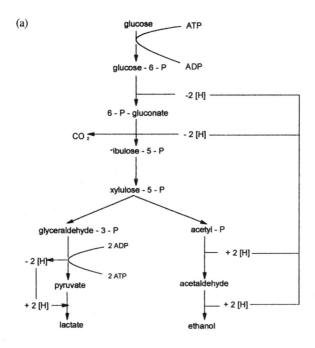

(b)

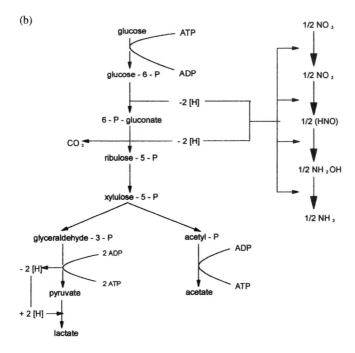

Figure 3 *Heterolactic fermentation in the absence of nitrate (from Weissenbach* et al.*, 1993)*

nitrosoamines in rumen fluid as a result of a reduction of nitrate to nitrite. Only very limited amounts of N-nitrosoamines (less than 1 µg kg^{-1}) were formed even when the supply of nitrate to the rumen was high (120 mg kg live-weight^{-1}).

3.4 Risk of Ingestion of Elevated Concentrations of Nitrate by Ruminant Livestock

Ingestion of nitrate from grass or fertiliser can lead to gastroenteritis (Jones, 1988) but nitrate is relatively non-toxic even at elevated concentrations (minimum lethal dose of 410 mg NO_3^- kg^{-1} live-weight). Ruminants effect the conversion of nitrate to nitrite in the rumen in a similar fashion to the conversion effected by enterobacteria in silage and there is an enhancement in the capacity of conversion as a result of continued feeding of nitrate. This enhancement in reductive capacity might reflect a shift in the population of micro-organisms in the rumen, especially those related to cellulytic degradation as high levels of fermentable metabolisable energy in the diet decrease the formation of nitrite (Geurink *et al.*, 1979).

The minimal lethal doses of nitrite ranges for cattle from 88 to 110 mg kg^{-1} live-weight and from 40 to 50 mg kg^{-1} live-weight for sheep (Radostits *et al.*, 1988). The main risk associated with diets containing high concentrations of nitrate or nitrite is the formation of methaemoglobin and hence anaemic anoxia (Vertregt, 1977). Three basic factors effect the rate of formation of methaemoglobin:

a) the amount of nitrate consumed, *i.e.* the more nitrate consumed per meal the greater the formation of methaemoglobin from haemoglobin (Vertregt, 1977; Kemp *et al.*, 1977);

b) the rate of ingestion of nitrate, *i.e.* the rate of intake of nitrate is dependent on the rate of intake of feed and the concentration of nitrate in the feed. If the rate of intake of nitrate is increased, the rate of formation of methaemoglobin is increased (Geurink *et al.*, 1979);

c) the rate of release of nitrate from the feed to the rumen fluid matrix. The effect of rate of release to the rumen fluid and thus transfer to the blood is important as it controls the rate of formation of methaemoglobin (Vertregt, 1977; Geurink *et al.*, 1979).

In its simplest form, the conversion of haemoglobin to methaemoglobin is shown in Table 4.

The maximum formation of methaemoglobin occurs within five hours after ingestion and death may occur if a lethal dose is presented within 12 to 24 hours, *i.e.* when the concentration of methaemoglobin in blood exceeds 9 g methaemoglobin 100 ml^{-1}. The diagnosis of toxicity is based on the presence of cyanosis and browning of mucosae, dysponoea with rapid, gasping respiration, muscle tremor and weakness leading to unsteady gait (Jones, 1988). The incidence of nitrate toxicity in the UK is relatively uncommon compared to the

Table 4 *The formation of methaemoglobin as a result of nitrate ingestion*

Reaction
$Hb + NO_2^- + 2H^+ \rightarrow HbNO^+ + H_2 + O$
$HbNO^+ + OH^- \rightarrow HiOH + NO$
$2Hb + NO_2^- + H_2O + O_2 \rightarrow 2HiOH + NO_3^-$

Hb = haemoglobin; Hi = methaemoglobin

Netherlands as the rate of application of fertiliser N to pasture is moderate in comparison.

4   Conclusions

The application of fertiliser containing nitrate or nitrate-yielding components will remain integral to the UK dairy sector. Only under conditions of drought, when the utilisation of fertiliser N by the herbage is low, will nitrate become an issue. Correct methods of husbandry minimise the risk associated with nitrate ingestion. Nitrate and nitrite do present a potential problem during silage fermentation. However, in certain cases nitrite can play an important role in the prevention of silage fermentations dominated by *Clostridia.*

Acknowledgements

The author wishes to thank Mrs. C.A. Collins (Natural Resource Management Ltd., Bracknell, UK) for the surveillance data concerning the concentration of nitrate in herbage and The Farmers Club Charitable Trust for sponsorship of part of this work.

References

Fowler, D., Smith, R.I., Skiba, U.M. and M.A. Sutton, 1999. The atmospheric nitrogen cycle and the role of anthropogenic activity. In, *Managing risks of nitrate to humans and the environment.* (Eds. W.S. Wilson, A.S. Ball and R.H. Hinton). Royal Society of Chemistry, Cambridge, pp. 121–138.

Frame, J. 1993. *Improved Grassland Management.* Farming Press.

Genstat 5 Committee, 1993. *Genstat 5 Release 3 Reference Manual.* Clarendon Press, Oxford.

Geurink, J.H., Malestein, A., Kemp, A. and A. Th. Van t'Klooster, 1979. Nitrate poisoning in cattle. 3. The relationship between nitrate intake with hay or fresh roughage and the speed of intake on the formation of methemoglobin. *Neth. J. Agric. Sci.,* **27**: 268–276.

Haigh, P.M. 1996 The effect of dry matter content and silage additives on the fermentation of bunker-made grass silage on commercial farms in England 1984–91. *J. Agric Engng. Res.,* **64**: 249–259.

Hein, E. 1970 *Die Beeinflussung des Garungsverlanfes bei der Grunfuttersilierung durch den Nitratgehalt des Ausgangsmaterials.* Diss., Deutshe Akad. Landwirtschaffswisseuschaften, Berlin.

Jarvis, S.C. 1993 Nitrogen cycling and losses from dairy farms. *Soil Use Manage.* **9:** 99–105.

Jarvis, S.C., 1999 Comparisons between nitrogen dynamics in natural and agricultural ecosystems. In, *Managing risks of nitrate to humans and the environment.* (Eds. W.S. Wilson, A.S. Ball and R.H. Hinton). Royal Society of Chemistry, Cambridge, pp. 2–20.

Jones, T.O, 1988 Nitrate/nitrite poisoning in cattle. *In Practice*, September 1988, pp. 199–204.

Kemp, A., Geurink, J.H., Haalstra, R.T. and A. Malestein, 1977 Nitrate poisoning in cattle. 2. Changes in nitrite in rumen fluid and methemoglobin formation in blood after high nitrate intake. *Neth. J. Agric. Sci.,* **25:** 51–62.

Lattemae, P. and P. Lingvall, 1996 Influence of hexamethylenetetraamine and sodium nitrite in combination with sodium benzoate and sodium propionate on quality of unchopped grass silage. In *Proceedings 11th Int. Silage Conf.*, Aberystwyth, UK, pp. 238–239.

McDonald, P., Henderson, A.R. and S. J. E. Heron, 1991 *The Biochemistry of Silage.* Chalcombe Publications,

McKinney, P.A., Parslow, R. and H.J. Bodansky, 1999 Nitrate exposure and childhood diabetes. In, *Managing risks of nitrate to humans and the environment.* (Eds. W.S. Wilson, A.S. Ball and R.H. Hinton). Royal Society of Chemistry, Cambridge, pp. 327–339.

MAFF, 1996 *Agriculture in the United Kingdom.* HMSO, London.

Marschner, H., 1995 *Mineral nutrition of higher plants.* Academic Press, London.

Radostits, O.M., Blood, D.C. and C.C. Gay, 1988 *Veterinary Medicine – 8th Edition.* Bailliere Tindall, London.

Spoelstra, S.F. 1985 Nitrate in silage. *Grass Forage Sci.,* **40:** 1–11.

Spoelstra, S.F. 1987 Degradation of nitrate by enterobacteria during silage fermentation of grass. *Neth. J. Agric. Sci.* **35:** 43–54.

Unwin, R.J. and Th. V. Vellinga, 1994 Fertiliser recommendations for intensively managed grassland. In, *Grassland and Society.* Proceedings 15th Gen. Meeting of European. Grass. Fed., Wageningen, The Netherlands, pp. 590–602.

van Broekhoven, L.W. and J. A. R. Davies, 1981 The analysis of volatile N-nitrosamines in the rumen fluid of cows. *Neth. J. Agric. Sci.* **29:** 173–177.

van Broekhoven, L.W. and J. A. R. Davies, 1985 The formation of volatile N-nitrosamines in laboratory-scale grass and maize silages. *Neth. J. Agric. Sci.* **33:** 17–22.

van Broekhoven, L.W. and R.W. Stephany, 1978 Formation in vivo of traces of volatile n-nitrosamines in the cow after direct administration of nitrate into the rumen. In, *Environmental aspects of N-nitroso compounds.* IARC Sci. Publ., Lyon, No. 19, pp. 461–463.

Vellinga, Th. V. and S. G. M. Verburg, 1995 *Beheersovereenkomsten op grasland van melkveebedrijven.* Rapport nr. 158, Proefstation voor de Rundveehouderij, Schapenhouderij en Paardenhouderij, Lelystad, The Netherlands.

Velthof, G.L. and Oenema, O., 1997 Nitrous oxide emissions from dairy farm systems in the Netherlands. *Neth. J. Agric. Sci.* **45:** 347–360.

Vertregt, N. 1977 The formation of methemoglobin by the action of nitrite in bovine blood. *Neth. J. Agric. Sci.* **25:** 243–254.

Weissbach, F. 1996 New developments in crop conservation. *Proceedings 11th Int. Silage Conf.*, Aberystwyth, United Kingdom, pp. 11–25.

Weissbach, F., Honig, H. and E. Kaiser, 1993 The effect of nitrate on the silage fermentation. *Proceedings. 10th Int. Conf. Silage. Res,* Dublin, Ireland, pp. 122–123

Wilkinson, J.M. 1988 'Silage and Health', Chalcombe Publ., Marlow, United Kingdom.

Wilkinson, J.M., Wadephul, F. and J. Hill, 1996 *Silage in Europe: A survey of 31 countries.* Chalcombe Publ., Lincoln, United Kingdom.

Wilson, R.F. and R.J. Wilkins. 1972. An evaluation of laboratory ensiling techniques. *J. Science Food Agric.,* **23:** 377–385.

8

The Effects of Elevated Atmospheric Carbon Dioxide Concentrations on Nitrogen Cycling in a Grassland Ecosystem

A. S. Ball and S. Pocock

DEPARTMENT OF BIOLOGICAL SCIENCES, JOHN TABOR LABORATORIES, UNIVERSITY OF ESSEX, WIVENHOE PARK, COLCHESTER CO4 3SQ, UK

Abstract

Grasslands comprise 70% of all agricultural land and cover nearly one fifth of the world's land surface. Grasslands play an important role in N cycling, accounting for approximately 20% of terrestrial N fluxes and may be amongst the earliest systems to exhibit the effects of elevated atmospheric CO_2. In this study the effects of elevated atmospheric CO_2 (700 μmol mol^{-1}) and nitrogen additions (140 and 560 kg ha^{-1} y^{-1}) on the decomposition of ryegrass (*Lolium perenne*) and C and N dynamics were examined.

Mineralisation rates of naturally senescent aerial plant litter was increased by 50% when litter was derived from plants grown in high N and ambient CO_2 (340 μmol mol^{-1}), compared with the degradation of litter from low N treatments. A reduction in this increase to only 25% was observed in the degradation rates of litter from plants grown in high N and elevated CO_2. A general correlation between the C:N ratio and the lignin:N ratio of plant litter and decomposition rates was observed with high nutrient ratios, detected in plants grown in low N and elevated CO_2 respectively, correlating with lower rates of decomposition. The results suggest that increased atmospheric CO_2 will affect N dynamics in managed grassland systems, leading to changes in the cycling of N in the environment.

1 Introduction

Greenhouse gases have been increasing due to anthropogenic activities (Ramanathan, 1988) and currently atmospheric CO_2 is at its highest concentration at around 356 ppm (Schimel, 1996), and is thought to be increasing by 0.5% every year (Watson *et al.*, 1990). Soils represent important sources and sinks of carbon and nitrogen and play an important role in carbon and nitrogen cycling (Van Breeman and Feijtel, 1990). Recent studies have highlighted the importance of climate change on nutrient cycling and soil processes (Ball and Drake, 1997; Schapendonk and Goudriaan, 1995; Curtis *et al.*, 1994). Elevated CO_2 has been found to have a fertilising effect on plant growth (Schimel, 1996), and it has been suggested that this response may lead to 'secondary consequences' in the soil system by altering root growth, fine root turnover, microbial dynamics in the rhizosphere, litter quality, decomposition and nutrient cycling (Wullschenger *et al.*, 1994).

Grasslands comprise of 70% of all agricultural land and cover nearly one fifth of the world's land surface (Parton *et al.*, 1995). It has been suggested that grassland ecosystems will be amongst the earliest systems to exhibit the effects of climate change (Ojima *et al.*, 1993). Experimental work has shown that under elevated CO_2, an increase in grassland soil respiration occurs (Ross *et al.*, 1995; Newton *et al.*, 1995). A recent long term study found soil respiration often to surpass net carbon accumulation in grassland soils exposed to elevated CO_2 (Schapendonk and Goudriaan, 1995). It was concluded by Ross *et al.* (1995) that the cycling of C and N under elevated CO_2 would be more rapid. The application of N fertiliser may therefore be very important under conditions of elevated atmospheric CO_2, as studies have indicated that the nitrogen content of plant material is reduced when plants are grown in elevated CO_2. (Akin *et al.*, 1995). Studies have suggested that changes in soil respiration will be dependent on the form of nitrogen that is applied (Cheshire and Chapman, 1994), the regularity of application (Leuken *et al.*, 1962) and the nutrient status of the soil. Lutze and Gifford (1995) examined the effects of both CO_2 and nitrogen enrichment of grass sward growth over one year and observed that the C:N ratios of grasses increased at elevated CO_2. It was concluded that if high C:N ratios occurred as a response to elevated CO_2, this could reduce productivity by decreasing nitrogen mineralisation rates which could lead to grasslands becoming part of an unidentified sink of carbon.

Decomposition represents one of the most important processes which determine nutrient cycling in a grassland ecosystem. The main influences on decomposition are the physico-chemical environment, the microbial community present and the quality of litter (Swift *et al.*, 1979). Studies on the effect of elevated CO_2 on plant litter decomposition have concentrated on the effect of changes in litter quality. Litter quality refers to the chemical and physical composition of plant litter and how suitable the litter is as a substrate for micro-organisms. The most widely accepted parameter used as a determinant in predicting decomposition rates is the C:N ratio of plant litter (Reinertsien *et al.*, 1983). Enquirez *et al.* (1993) considered that a C:N ratio of 25:1 or lower within plant litter is most ideal for decomposition while above this the nitrogen will be mineralised and converted to microbial protein, while the carbon is respired.

To date very few studies have examined the effects of both elevated CO_2 and nitrogen on plant residue decomposition. However, both nitrogen and CO_2

concentration are clearly important determinants of litter quality and will therefore affect the decomposability of plant litter which in turn affects C and N cycling. We examine in this paper the indirect effects of elevated atmospheric CO_2 and nitrogen concentration on the activity of soil microflora using litter from *Lolium perenne* grown under various nitrogen and atmospheric CO_2 concentrations. Exposure of the plants used in this study to elevated CO_2 led to an increased photosynthetic rate and an increase in carbon accumulation.

2 Materials and Methods

2.1 Plant Material and Soil Samples

Freshly fallen aerial plant materials from all the plants used in this study were gathered and cut into uniform size (approx 2 cm in length) pieces before use. Plant material was grown within the Free Air CO_2 Enrichment (FACE) system and was kindly donated by Dr H Blum (Swiss Federal Institute of Technology, Zurich). *Lolium perenne* was grown either at ambient (354 μmol mol^{-1}) or elevated CO_2 (700 μmol mol^{-1}) and high (560 kg N ha^{-1} y^{-1}) or low (140 kg N ha^{-1} y^{-1}) nitrogen regimes. The soil used for this study was a sandy loam with a C:N ratio of 14:1 and 10% moisture (w/w) content which is a water potential of approximately -0.8 MPa (Taylor and Ball, 1994). The soil was obtained from an undisturbed parkland site at Wivenhoe Park (The University of Essex) and was taken from the top 10–15 cm. The soil was ground through a 2 mm mesh sieve in order to remove plant material and provide a relatively homogeneous soil for measurement of respiration.

2.2 Experimental Details

Soil respiratory activity was measured using an infrared gas analyser (IRGA, ADC 225) attached to a twelve point multi-channel selector (ADC WA161). The degradation studies were carried out using water-jacketed respiration chambers through which sterile humidified air (containing ambient concentrations of CO_2) was passed. Carbon dioxide flux was recorded for a dwell time of five minutes. The chambers were kept in the dark at 20°C, and the moisture content was maintained at 10% (w/w) by the daily addition of sterile water. Each chamber contained 80 g of soil. Following the stabilization of respiratory activity in unamended soils, triplicate chambers were amended with 2.0 g (representing approximately 50% of the initial soil C) of aerial material from *Lolium perenne* grown under either ambient (354 μmol mol^{-1}) or elevated (700 μmol mol^{-1}) CO_2 and low (140 kg ha^{-1} y^{-1}) or high (560 kg ha^{-1} y^{-1}) N and then incubated for 40 days. Soil respiratory measurements were recorded constantly using a chart recorder (Kipp and Zonen BD111).

The carbon:nitrogen ratio of soil and leaf material used in the study was measured on an automated Perkin-Elmer CHNS/O analyser (Series ii, 2400). Soil (20 mg) and plant (5 mg) samples were ground to small particle sizes and placed in foil capsules for analysis by combustion and spectral characterization.

The carbon and nitrogen content of the samples (expressed as percentage) was calculated from standards containing known amounts of carbon and nitrogen.

The analyses of the polymeric components of plant material used in this study were determined by sequential extraction of triplicate samples of plant material followed by gravimetric analysis as previously described (Harper and Lynch, 1981).

3   Results

3.1   Soil Respiratory Activity

Soil respiration measurements were taken daily over a period of forty days. Fig 1 shows the cumulative soil respiration rates (less the background respiration level in unamended soil control) for soils amended with plant litter from various treatments. At the time of addition, the soil respiration rate in the chambers was found to vary from 60–75 μg CO_2-C g^{-1} oven dry soil day^{-1}.

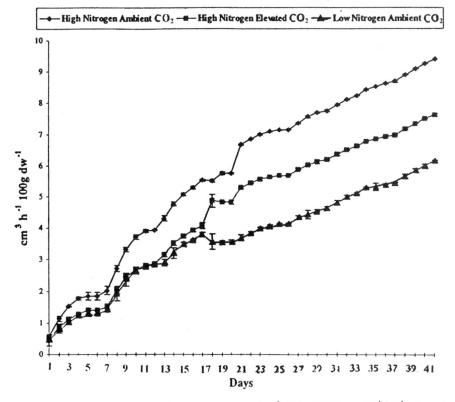

Figure 1   *Cumulative recorded respiration rates (cm^3 CO_2-C 100 g soil^{-1} h^{-1}) for soil containing plant litter from L. perenne grown at ambient and elevated CO_2 and high and low nitrogen concentrations. Values represent the mean of three replicates, with standard errors within 10% of the mean values in all cases*

The addition of plant material led to a three to four-fold increase in soil respiratory activity over the length of the experiment. Soil respiration did not vary in the unamended control chambers throughout the incubation (data not shown). The maximum cumulative respiration rates were found to occur in soils amended with plant material from *Lolium perenne* grown at high N and ambient CO_2, with total CO_2-C release significantly greater (26.5%, $p < 0.05$) than in soils amended with litter from *L. perenne* grown at high N and elevated CO_2. Soils amended with plant material from *L. perenne* grown in low N and ambient CO_2 showed the lowest cumulative production of CO_2-C, with total production some 40% lower than that from soils amended with plant material from *L. perenne* grown at high N and elevated CO_2.

3.2   Analysis of Plant Litter

Analysis of the C:N ratio of plant litter used in this study indicated that the C:N ratio of litter from *L. perenne* grown in elevated CO_2 was significantly higher (unpaired T-test, $p < 0.05$) than litter from *L. perenne* grown in ambient CO_2 (Table 1). Although the C:N ratio was significantly reduced in litter from plants grown in high N conditions the difference between the two CO_2 treatment remained. Gravimetric analysis of the main polymeric components of plant litter revealed no significant differences in the polymeric composition of plant material grown under elevated and ambient atmospheric CO_2 or high and low nitrogen (Table 1). Gravimetric analyses of the polymeric components of plant litter from *Lolium perenne* grown under different environmental conditions showed that the lignin composition remained constant at around 25% (w/w), with hemicellulose and cellullose making up approx 30 and 23% respectively (Table 1).

Elemental analysis of the soil used in this study showed that the C:N ratio of

Table 1   *Elemental and gravimetric analysis of plant litter from Lolium perenne grown in elevated (700 mol mol^{-1}) and ambient (354 mol mol^{-1}) CO_2 concentrations and high (560 kg N ha^{-1}) and low (140 kg N ha^{-1}) nitrogen. Results are the means of three replicates with all standard deviations within 5% of the mean values shown*

| | Plant litter from | | | |
| | C:N Ratio | Lignin | Cellulose | Hemicellulose |
	In oven-dry plant material (%)			
ambient CO_2-high N	18:1[1,2]	24	27	32
elevated CO_2-high N	22:1[1,2]	24	27	32
ambient CO_2-low N	27:1[1,2]	26	23	30
elevated CO_2-low N	42:1[1,2]	25	23	35

[1]   Significant differences (T-test, $p<0.05$) between the C:N ratio of litter from plants grown in ambient and elevated CO_2.
[2]   Significant differences (T-test, $p<0.05$) between the C:N ratio of litter from plants grown in high and low nitrogen.

soils remained unchanged throughout the experiment at 14:1, irrespective of the plant treatment.

4 Discussion

4.1 Soil Respiratory Activity in Chambers

Soil respiratory activity has often been used as an indicator of biodegradation of plant litter (Ball and Drake, 1997). The addition of plant litter to soils led to an immediate three to four-fold increase in soil respiratory activity, presumably caused by the microbial degradation of readily labile materials in the litter, *e.g.* simple sugars, amino acids (Taylor and Ball, 1994). Differences in soil respiratory activity between chambers inoculated with litter from *L. perenne* grown in different conditions were detected during the 41 day incubation. Despite experimental conditions that were far from natural (constant moisture and temperature, relatively large additions of substrate), the results indicate a significant difference between the respiratory rates of soils amended with plant litter from *L. perenne* grown ambient and elevated CO_2 and at high and low nitrogen concentrations, with greatest respiratory activity detected in soils amended with litter from plants grown in high N and elevated CO_2. This study necessarily examined the effects of elevated CO_2 on the initial stages of biodegradation. It is likely that the soil respiration rate would decrease once the labile components of the fresh litter had all decomposed. Previous studies have found that growth in elevated atmospheric CO_2 decreases decomposition rates of plant litter (Ball and Drake, 1997), and that growth in high nitrogen increases plant litter decomposability (Enquirez *et al.*, 1993). The results from this study confirm these observations.

4.2 Analysis of Plant Litter

The C:N ratio of plant litter has been shown to be an important determinant of decomposition rates. Analysis of the naturally senescent litter from *L. perenne* used in this study shows a significant difference between the C:N ratio of litter, according to the plant treatments. The lowest C:N ratio was detected in litter from plants grown in ambient CO_2 and high N. The highest C:N ratio was detected in litter from plants grown in low N and elevated CO_2. Other studies have also found that plants grown in elevated CO_2 show increased C:N ratios (Ball and Drake, 1997; Lutze and Gifford, 1995; Aerts *et al.*, 1995). Elevated CO_2 has been found to dilute the N content of plant litter (Akin *et al.*, 1995; O'Neill and Norby, 1996). These changes in N content of plants have been found to be reflected in the C:N ratios of plant litter. Many studies have linked the rate of decomposition with litter quality, particularly focusing on the C:N ratio as a representation of the degradability of plant material (Reienertsen *et al.*, 1983; Taylor *et al.*, 1989). Plant materials with C:N ratios of 25:1 and above have decreased decomposition rates, as above this level microbial decomposition is limited by N availability and net immobilisation occurs (Haynes, 1986). The N content of plant litter is an important rate-regulating factor in the preliminary stages of decomposition (Berg, 1984). Lignin concentration has also been shown to be an important determinant in the decomposition of plant litter (Melillo, 1982; Taylor *et al.*, 1989) and increased lignin concentrations have been reported in plants exposed to elevated CO_2 (Cipollini *et al.*, 1993), although these results show that lignin concentrations remained unaffected by the atmospheric CO_2 concentrations. A correlation was clearly evident between the soil respiration rates and the litter C:N ratio. Currently there is not sufficient evidence for the changes in C:N ratios found in plant litter grown in elevated CO_2 to be attributed to any particular plant component. It is possible that changes in C:N ratio reflect changes in the composition of plant structural components, which would make them less susceptible to enzyme attack and therefore decrease decomposition rates (Trigo and Ball, 1994).

No change in the C:N ratio of soils used in the litter decomposition experiments could be detected. This is not surprising as changes in the carbon and nitrogen content of soils occur very slowly taking hundreds if not thousands of years (Parton *et al.*, 1995).

5   Conclusions

In conclusion the results indicate that in a climate of elevated atmospheric CO_2 the maintenance of rates of litter decomposition in a grassland ecosystem depend upon increased levels of nitrogen fertilisation in order to maintain the C:N ratio of naturally senescent litter. If current levels of N-fertilisation are maintained, this will lead to reduced litter decomposition and a reduction in the availability of soil organic nitrogen. If increased levels of N-fertilization are used, this will lead to deeper rooting systems which will then compound the problem.

References

Aerts, R., R. Van Logtesijn, M. Van Staaldunen and S. Toet 1995. Nitrogen supply effects on productivity and potential leaf litter decay of Carex species from peatlands differing in nutrient limitations. *Oecologia* **104**, 447–453.

Akin, D.E., L.L. Rigsby, G.R. Gamble, W.H. Morrison, B.A. Kimball, P.J. Pinter, G.W. Wall, R.L. Garcia and R.L. LaMorte 1995. Biodegradation of plant cell walls, wall carbohydrates and wall aromatics in wheat grown in ambient or enriched CO_2 concentrations. *Journal of Food Science and Agriculture* **67**, 399–406

Ball, A.S. and B.G. Drake 1997. Short-term decomposition of litter produced by plants grown in ambient and elevated atmospheric CO_2 concentrations. *Global Change Biology* **3**, 29–35.

Berg, B. 1984. Decomposition of root litter and some factors regulating the process; long-term root litter decomposition in a Scots pine forest. *Soil Biology and Biochemistry* **16**, 609–617.

Cheshire M.V. and S.J. Chapman 1994. Influence of the N and P status of plant material and of added N and P on the mineralisation of C from ^{14}C-labelled ryegrass in soil. *Biology and Fertility of Soils* **21**, 166–170.

Cipollini, M.L., B.G. Drake and D. Whigham 1993. Effects of elevated CO_2 on growth and carbon/nutrient balance in the deciduous woody shrub *Lindera benzoin* (L.) Blume (Lauracea). *Oecologia* **96**, 339–346.

Curtis, P.S., E.G. O'Neill, J.A. Terri, D.R. Zak and K.S. Pregitzer 1994. Below ground responses to rising atmospheric CO_2: Implications for plants, soil biota and ecosystem processes. *Plant and Soil* **165**, 1–6.

Enquirez, S., C.M. Dunrte and K. Sand Jensen 1993. Patterns in decomposition rates amongst photosynthetic organisms: the importance of detritus C:N:P content. *Oecologia* **94**, 457–471.

Harper, H.T. and J.M. Lynch 1981. The chemical components and decomposition of wheat straw, leaves, internodes and stems. *Journal of Science of Food and Agriculture* **32**, 1057–1062.

Haynes, R.J. 1986. The nitrogen cycle. *In* Mineral nitrogen in the plant-soil system. Academic Press, New York.

Leuken, H., W.L. Hutchinson and E.A. Paul 1962. The influence of nitrogen on the decomposition of crop residues in the soil. *Canadian Journal of Soil Science* **42**, 276–288.

Lutze, J.L. and R.M. Gifford 1995. Carbon storage and productivity of a carbon dioxide enriched grass sward after one years growth. *Journal of Biogeography* **22**, 227–233.

Melillo, J.M. 1982. Nitrogen and lignin control of hardwood leaf litter decomposition dynamics. *Ecology* **63**, 621–626.

Newton, P.C.D., H. Clark, C.C. Bell, E.M Glasgow, K.R. Tate, D.J. Ross, G.W. Yeates and S. Saggar 1995. Plant growth and soil processes in temperate grassland communities at elevated CO_2. *New Zealand Journal of Soil Science* **3**, 66–72.

Ojima, D.S., W.J. Parton, D.S. Schimel, J.M.O. Scurlock and T.G.F. Kittel 1993. Modelling the effects of climatic and CO_2 changes on grassland storage of soil C. *Water, Air and Soil Pollution* **70**, 643–657.

O'Neill, E.G. and R.J. Norby 1996. Litter quality and decomposition rates of foliar litter produced under CO_2 enrichment. *In* G.W. Koch and H.A. Mooney (eds), *Carbon dioxide and terrestrial ecosystems* 87–103. Academic Press, London.

Parton, W.J., J.M.O. Scurlock, D.S. Ojima, D.S. Schimel, D.O. Hall and SCOPE-GRAM group members 1995. Impact of climate change on grassland production and soil carbon worldwide. *Global Change Biology* **1**, 13–22.

Ramanathan, V. 1988. The greenhouse theory of climate change: A test by an inadvertant global experiment. *Science* **240**, 293–299.

Reinertsen, S.A., L.F. Elliot, V.L. Cochran and G.S. Campbell 1983. Role of available carbon and nitrogen in determining the rate of wheat straw decomposition. *Soil Biology and Biochemistry* **16**, 459–464.

Ross, D.J., K.R. Tate and P.C.D. Newton 1995. Elevated CO_2 and temperature effects on soil carbon and nitrogen cycling in ryegrass/white clover turves of an Endoaquept soil. *Soil Biology and Biochemistry* **27**, 240–250.

Schapendonk, A.H.C.M. and J. Goudriaan 1995. The effects of elevated CO_2 on the annual carbon balance of a *Lolium perenne* sward. *Photosynthesis* **5**, 779–784.

Schimel, D.S. 1996. Terrestrial ecosystems and the carbon cycle. *Global Change Biology* **1**, 77–91.

Swift, M.J., O.W. Heal and J.M. Anderson 1979. *In*: Decomposition in terrestrial ecosystems. Blackwell Science, Oxford.

Taylor, B.R., D. Parkinson and W.F.J. Parsons 1989. Nitrogen and lignin content as predictors of litter decay rates: a microcosm test. *Ecology* **70**, 97–104.

Taylor, J. and A.S. Ball 1994. The effect of plant material grown under elevated CO_2 on soil respiratory activity. *Plant and Soil* **126**, 315–318.

Trigo, C. and A.S. Ball 1994. Is the solubilized product from the degradation of lignocellulose by actinomycetes a precursor of humic substances? *Microbiology* **140**, 3145–3152.

Van Breeman, N. and T.C.T. Feijtel 1990. Soil processes and properties involved in the production of greenhouse gases. *In:* A.F. Bouwman (Ed), *Soils and the greenhouse effect*. Wiley, New York.

Watson, R.T., H. Rodhe, H. Oeschger and U. Siegenthaler 1990. Greenhouse gases and aerosols. *In:* J.T. Houghton, G.J. Jenkins and J.J. Ephramus (Eds), *Climate Change: The IPCC Scientific Assessment*, pp 1–40. Cambridge University Press, Cambridge.

Wullschenger, S.D., J.P. Lynch and M.B. Glenn 1994. Modelling the below ground response of plants and soil biota to edaphic and climatic change-What can we expect to gain? *Plant and Soil* **165**, 149–160.

Section 2

Environmental Aspects of Nitrates Introductory Comments

By P. B. Tinker, University of Oxford

Nitrate in the environment is important in various ways. There is the behaviour of nitrate and its duration in the environment; the way in which environmental nitrogen becomes part of water supplies; the impact on the biota in the waters and the possible damage from this. The distinction between agricultural and environmental nitrate is often unclear: the nitrogen cycle of the world operates in both natural and agricultural landscapes, and both contribute to the total water supply.

The atmospheric part of the nitrogen cycle (Fowler *et al.*) is greatly altered by anthropogenic nitrogen, from combustion and from agricultural operations, especially intensive stockfarming. Human activities in countries like United Kingdom now dominate the N cycle, particularly NH_3 deposition. Nitrogen is now as important as sulfate in forming acid rain, and the deposition of fixed nitrogen per hectare is so large in some areas that natural vegetation growsfaster, and species may alter.

The addition of nitrogen to waters can also occur from human sewage, directly or after application of sewage wastes to land (Clapp *et al*). Large quantities of total nitrogen were applied in sludge over 20 years, but by careful management no pollutants escaped from the watershed. Land application will become more important as sea dumping is forbidden, and allows the beneficial recycling of nitrogen.

Quite apart from the human health issue, nitrate (and other soluble nitrogen) can damage surface waters by eutrophication if nitrogen is the nutrient limiting algal growth – though most United Kingdom waters are phosphorus limited – or by acidification (Hornung). Nitrogen levels in waters in uncultivated uplands and with high rainfall are usually very low, but as rivers reach the lowlands agriculture and sewage increase the load they are carrying. A further mechanism of water damage arises because nitrite and nitrogen dioxide can damage fish, invertebrates and hence the rest of the food chain (Kelso *et al.*).

Nevertheless, the input into surface waters in a natural landscape is never as concentrated as that from agricultural parts of a catchment, and there is a major task in preventing such nitrate in soil reaching surface and groundwater (Cook). The methods of minimizing this problem within agriculture are discussed in another section. However, many preventive actions of an agri-environmental type have been initiated, such as buffer strips of vegetation along water courses, and the designation of sensitive areas and vulnerable zones, from which aquifers used for public supplies are fed. Within these there are controls on the agricultural use of nitrogen. All these measures aim to maintain sustainable water resources, but the sheer complexity of the environment can defeat such schemes. It may be difficult to predict either the degree of control required in a particular situation, or the reliability of actions intended to provide it.

A large fraction of the N in the environment is present as humic substances. The understanding of their behaviour, in both soil and water, is important, and Hayes *et al.* discuss the use of Nuclear Magnetic Resonance (NMR) to help determine the very complex chemistry of these compounds.

Two papers (Pathak, and Gupta & Chandrashekharan) describe the situation in India. Many factors threaten nitrate pollution there, with rapidly increasing fertilizer use in some areas, large numbers of cattle and areas of low rainfall, and some groundwaters already have high concentrations. Reported cases include some with heavy nitrate pollution, well above permitted limits, but villagers have little option but to use them. As India continues to develop, careful management of land and water will be needed to prevent major nitrate pollution.

9

The Atmospheric Nitrogen Cycle and the Role of Anthropogenic Activity

D. Fowler, R. I. Smith, U. M. Skiba, M. Coyle,
C. Flechard, C. E. R. Pitcairn and M. A. Sutton

INSTITUTE OF TERRESTRIAL ECOLOGY, BUSH ESTATE,
PENICUIK, MIDLOTHIAN EH26 0QB, SCOTLAND

1 Introduction

The emission of gaseous nitrogen from terrestrial (and marine) ecosystems forms part of the natural biogeochemical cycling of this vital nutrient. The form of the emission ranges from molecular nitrogen (N_2) to the radiatively active gas N_2O and the much more reactive NO and NH_3. These gases are therefore part of entirely natural processes and are represented on simple schemes depicting the annual cycling of nitrogen between the atmosphere and biosphere in standard texts. Anthropogenic activities, especially those following the industrial and agricultural revolutions beginning in the 19th century, have greatly increased the emissions of the reactive fixed nitrogen as NO from combustion processes and NH_3 from agricultural activity. The anthropogenic contribution to the oxidized nitrogen emissions to the atmosphere now represent the dominant component of the total and effects of these emissions are detectable throughout the global atmosphere. The effects include modification of the oxidizing capacity of the atmosphere, acidic deposition, the formation of photochemical oxidants (and ozone in particular), radiative forcing of climate and eutrophication of terrestrial and marine ecosystems. Thus the emissions of oxidized nitrogen contribute directly to many of the regional and global environmental problems of the late 20th century.

Similarly, the emission of reduced nitrogen as NH_3, from intensive animal production and also from fertilized cropland and industrial processes, is now a major contribution to the atmospheric transport and deposition of fixed nitrogen globally (Dentener, 1993). The current scale of anthropogenic fluxes of fixed nitrogen to the atmosphere now exceed the natural fluxes at a global scale and taken together with the industrial fixation of nitrogen for agriculture, there is justification for the view that human activities have taken over the global nitrogen cycle.

This paper summarizes the current understanding of the global nitrogen cycle and considers in more detail the current atmospheric budgets of oxidized and reduced nitrogen over the UK. The fate of UK emissions and the consequences of deposited nitrogen of terrestrial ecosystems are also briefly reviewed.

1.1 Emissions

The emissions of NO_x, largely as NO from the combustion of fossil fuel, occurs both from oxidation of the N in the fuel and also the combination of atmospheric oxygen and nitrogen in the combustion processes. The global emissions have increased from approximately 2 Tg NO_x-N y^{-1} at the turn of the century to 20 Tg NO_x-N y^{-1} by 1980 (Dignon & Hamseed, 1989). The emissions have increased very rapidly since 1950 (Figure 1) and the rapid global industrial development, especially in S.E. Asia and S. America, are projected to continue the increases in NO_x emission well into the 21st century.

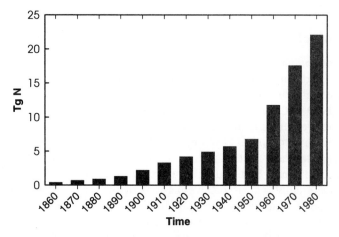

Figure 1 *Global emissions of NOx from combustion*

The estimates of future industrial NO_x emission are uncertain but projections by the IPCC indicate a steady growth of emissions to 46 Tg NO-N y^{-1} by the year 2025 (IPCC 1997). The emissions are currently concentrated in the mid-northern latitudes between 20°N and 60°N, but increasingly the new emissions will be concentrated at lower latitudes, well into tropical regions.

Emissions of reduced nitrogen are much less certain than those of oxidized nitrogen, but are dominated by livestock production in regions with intensive agriculture. The current emissions of oxidized nitrogen have been estimated at 55 Tg N^{-1} of which 70% is of anthropogenic origin while emissions of reduced nitrogen have been estimated at 46 Tg N y^{-1} of which 66% are anthropogenic (Table 1). The total emission of fixed reactive nitrogen (*i.e.* ignoring N_2) to the atmosphere is therefore approximately 111 Tg N year^{-1} of which 62% is anthropogenic and 39% of natural origin (Figure 2).

Table 1   *Global terrestrial emissions of N*

	Emissions (Tg N y^{-1})	*Anthropogenic* % (Tg N y^{-1})	*Natural* % (Tg N y^{-1})
NOx	55	70% (38.5)	30% (16.5)
NH$_3$	46	66% (30.4)	34% (15.6)
Total	111	62% (62)	38% (38)

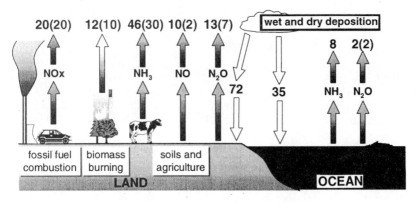

Figure 2   *Atmospheric surface exchange of fixed nitrogen (Tg N y^{-1})*

These global emission inventory values are subject to considerable uncertainty. The fuel use statistics are considered reasonably reliable for the major emission regions of Europe, North America and Japan where the development of control measures for air pollutants remains a high priority for the governments concerned and emissions data are scrutinized in considerable detail. In other parts of the world where environmental regulations are either in development or have not yet been considered, few reliable emission statistics are available.

The emissions of nitric oxide from soils are less easily quantified at national or regional scales. The magnitude of the flux of NO is generally small, in the range 1 to 20 ng NO-N m^{-2} s^{-1}, but over regional scales these small emissions make an important contribution to the total. The NO is formed as an intermediate during the microbial reduction of NO$_3^-$ to N$_2$ (denitrification) or the oxidation of NH$_4^+$ to NO$_3^-$ (nitrification). The simple 'hole in the pipe' model introduced by Firestone and Tiedje (1979) has proved a valuable concept. In this model, illustrated in Figure 3, the major transformations, reactants and pathways are illustrated and the relative magnitudes of the component fluxes are indicated by the scale of the anions. The bulk of the NO formed is believed to result from nitrification and being a microbial process it is very sensitive to temperature and substrate supply. The soil to atmosphere emission flux increases with available soil nitrogen, soil temperature and is at an optimum with intermediate soil moisture content. The loss of soil NH$_4^+$ as NO during

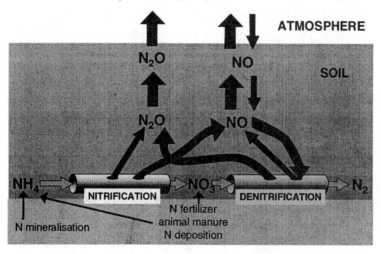

Figure 3   *Transformations of N compounds in soil*

nitrification is generally in the range 1% to 4% of NH_4^+ oxidized. However, as indicated in Fig. 3, some NO is also produced as an intermediate in the de-nitrification pathway and this NO production may exceed that of nitrification (Remde, *et al.*, 1989). The fraction of the total NO emission from soil resulting from denitrification is limited, however, by the soil conditions which favour denitrification, *i.e.* wet, anaerobic conditions in which gaseous diffusion within the soil profile is limited by the water filled pore space. In these conditions the majority of the NO produced is reduced further to N_2 before is reaches the atmosphere (Skiba *et al.*, 1997).

The temperature response of soil NO emission has been characterized by activation energies in the range 40 to 100 kJ mol^{-1} by Skiba *et al.* (1993) and by Johansson and Granat (1984). In the field, the determination of activation energies are complicated by the temperature profile with depth and its daily cycle, since the average depth of NO production is uncertain and likely to be spatially variable due to spatial variability in the thermal properties of soil. The strong relationship between NO emissions and soil temperature (generally in the upper horizons of soil between the surface and a depth of 5 cm) have been used to develop inventories of soil NO emission (Williams *et al.*, 1992; Stohl *et al.*, 1996). The approach by both was to use soil temperature (T) and an empirical factor for each land use (A) to calculate soil NO emissions for the United States (Williams *et al.*, 1992) and Europe (Stohl *et al.*, 1996) according to

$$\text{NO flux (ng NO-N m}^{-2}\text{ s}^{-1}) = A \exp{[(0.071 \text{ Å } 0.007)T]}$$

For arable land Stohl *et al.* (1996) replaced the uniform land ice factor (A) during the growing season with a fertilizer induced emission factor assuming that during this period 5% of the mineral N applied as a fertilizer is emitted as NO. This is a rather large fraction of the N applied when compared with field measurements from which Skiba *et al.* (1997) calculated an equivalent factor of 0.3%. In multiple regression analyses of soil NO emissions from a range of agricultural, woodland and moorland sites in southern Scotland, the emissions of NO were controlled mainly by soil NO_3^- concentrations and soil temperature (Skiba *et al.*, 1993).

1.2 Soil Water and Gaseous Diffusivity

The emissions of NO from soil is limited in very dry soils by sufficient moisture for microbial activity, and at very high soil water content by rates of gaseous diffusion in soil and pathways to the free atmosphere. Thus NO emission is optimal at water filled pore space (WFPS) percentages in the range 30% to 60%, and vary little in response to changes in soil water contents in this range (Davidson, 1991).

1.3 Soil Nitrogen

The supply of mineral nitrogen has been shown in a range of laboratory and field studies to stimulate NO emission from soil (Ludwig, 1994; Skiba, Smith & Fowler, 1993). The fraction of applied nitrogen released as NO varies from 0.003% to 11% with a geometric mean of published values close to 0.3%. Considering both soil temperature and soil nitrogen, Skiba *et al.* (1993) showed that approximately 60% of the variability in observed NO emission fluxes from a range of arable, grassland, forest and moorland soils could be explained by a regression model of the form

$$\text{NO flux (ng NO-N m}^{-2}\text{ s}^{-1}) = -3.23 + 1.01 \log[\text{soil NO}_3^-] + 0.165\,(T_5)$$

in which T_5 is the soil temperature at a depth of 5 cm.

1.4 Compensation Points

The emission flux of NO from soil occurs along a concentration gradient from the sites of production to the atmosphere or a sink within the soil. Ambient concentrations of NO are also influenced by combustion sources and the NO concentrations in the air above soil may therefore vary and influence the soil surface – atmosphere gradient in NO concentration. At large ambient NO concentrations the soil acts as a sink for ambient NO. Thus the exchange of NO must be considered bi-directional with the existence of a compensation point representing the ambient concentration at which no net soil-atmosphere exchange occurs. The process is represented schematically in Figure 4 with the broad range of fluxes indicating typical fluxes, their decline with ambient NO concentration and the range of compensation points from 1 to 10 ppb NO. In practice, heavily fertilized soils may show substantially larger compensation points.

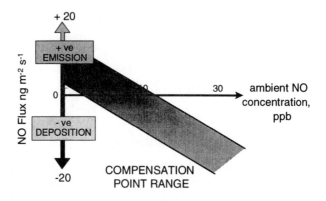

Figure 4   *Compensation point for NO emissions from soil*

1.5   Emission of Reduced Nitrogen

The emissions of NH_3 occur following the hydrolysis of urea from livestock waste which liberates ammonia into solution.

$$CO(NH_2)_2 + H_2 \xrightarrow{\text{Urease}} CO_2 + 2NH_3$$

The high solubility of NH_3 relative to CO_2 increases the solution pH and restricts NH_4^+ which promotes the release of NH_3 from solution by volatilization. The release occurs from livestock buildings, land spreading of manure and slurry. In the case of slurry spreading between 20% and 70% of the slurry nitrogen is volatilized. On a global scale the NH_3 emission from farm animals represents approximately 22 Tg NH_3-N annually or about half of the total. The remaining sources include fertilizer volatilization (20%), emissions from the ocean (18%), biomass burning (12%) and much smaller contributors from industry and humans (Bouwman 1995). While the NH_3 emission from farm waste is always a source, cropland may be both a source or a sink for NH_3 depending on the nitrogen status of the vegetation, the wetness of the crop canopy and ambient concentrations of NH_3. The NH_3 exchange is bi-directional with a compensation point determined by apoplast NH_4^+ and pH, the sink strength of external foliar surfaces and ambient NH_3 concentration, as illustrated in Figure 5a. In field conditions the compensation point is most usefully considered as a canopy compensation point (Figure 5b) in which the numerical representation of surface-atmosphere fluxes is considered as a resistance analogue (Sutton *et al.*, 1995).

The reactive nature of atmospheric ammonia leads to very rapid deposition close to the source areas, and since the sources are very numerous and variable in size the ambient NH_3 concentrations show great spatial variability.

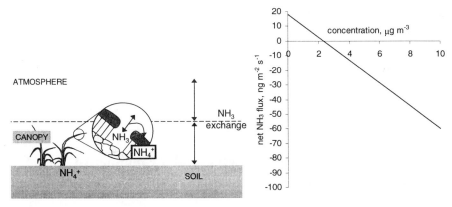

Figure 5a *NH₃ exchange between vegetation,
soils and the atmosphere*

Figure 5b *NH₃ compensation point*

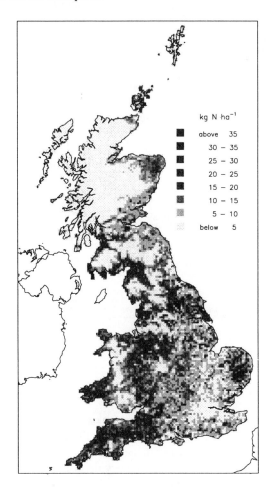

Figure 6 *1988 UK NH₃ Emissions (kg N ha⁻¹)*

The NH_3 emissions in the UK have been estimated at 290 kt NH_3-N y^{-1} of which approximately 80% are of agricultural origin (Sutton *et al.*, 1997), and cattle contribute in excess of 50%. The very patchy nature of UK NH_3 emissions is illustrated in Figure 6 which clearly identifies the large emission areas including Cheshire, Devon and Somerset, East Anglia and the Vale of York, while the remote highlands of West Scotland and the uplands of Wales and the Scotland/England border counties are the smallest source areas.

2 The Atmospheric Chemistry and Transport of Reactive Nitrogen

2.1 Oxidized Nitrogen

The emitted NO is readily oxidized to NO_2 by ambient ozone so that in urban areas and close to roads the local ambient ozone concentrations are depleted by this simple reaction. However, further chemistry of oxidized nitrogen differs considerably between night and day because key intermediates are photolysed by solar radiation.

The NO_2, produced by oxidation of NO, may be further oxidized to NO_3 by O_3 (Figure 7a) and the NH_3 radical produced may react further with NO_2 to N_2O_5 which reacts with water on aerosol, fog or cloud droplets to form HNO_3.

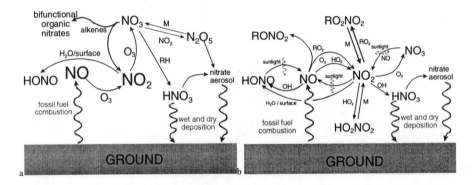

Figure 7   *(a) Nocturnal interconversions of oxidised nitrogen compounds in the atmosphere. (b) Day time interconversions of oxidised nitrogen compounds in the atmosphere.*

The day time chemistry is rather more complex because photolysis of NO_2 by solar radiation leads to a photostationary equilibrium between NO, NO_2 and O_3 (Figure 7b). The NO_2 may be further oxidized to HNO_3 directly by the OH radical or deposited onto vegetation by dry deposition. The HNO_3 may be deposited directly at the ground by dry deposition or incorporated in aerosol and cloud.

2.2 NH$_3$

The emitted NH$_3$ is readily re-deposited onto natural surfaces, especially if they are wet, or is incorporated into acid aerosols (HNO$_3$ or H$_2$SO$_4$) forming (NH$_4$)$_2$SO$_4$ and NH$_4$NO$_3$ aerosols. The gas to particle conversion (Figure 8) represents an important source of aerosols and the particles are readily incorporated into cloud and rain and removed from the atmosphere by rain. The rain and aerosol chemistry over N. Europe (and N. America) is dominated by the ions NO$_3^-$, NH$_4^+$, SO$_4^{2-}$ and H$^+$. Increasingly as SO$_2$ emissions are declining the reduced and oxidized nitrogen are becoming the dominant ions in aerosol and precipitation and the major source of acidic deposition. The chemistry of reduced nitrogen is relatively simple and is restricted mainly to aerosol formation processes. However the presence of so much NH$_3$ and NH$_4^+$ is very important as a regulator of cloud and aerosol pH and thus strongly influences the chemistry of other reactive pollutants including SO$_2$ in particular. In the case of SO$_2$ the supply of NH$_3$ maintains cloud droplet pH at sufficiently high values to promote oxidation of SO$_2$ to SO$_4^{2-}$ by ozone.

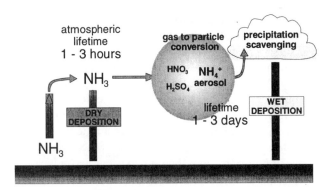

Figure 8   *The deposition and transformation of NH$_3$ emissions*

2.3 Ozone Formation

The detailed atmospheric processing of oxidized nitrogen lies outside the scope of this paper. However, as the role of oxidized nitrogen is central to photochemical oxidant formation and currently ground level ozone represents the major regional gaseous pollutant in Europe and North America, a brief outline of the process leading to ozone formation is necessary.

The oxidation of NO to NO$_2$ by O$_3$ and photolysis of NO$_2$ by solar radiation does not lead to net O$_3$ formation. However, in the presence of reactive volatile organic compounds, their oxidation to peroxy radical provides an alternative oxidant to convert NO to NO$_2$. The NO$_2$ produced may then undergo photolysis and net O$_3$ production results. The process is illustrated at its simplest in Figure 9. The number of O$_3$ molecules produced per NO$_2$ molecule

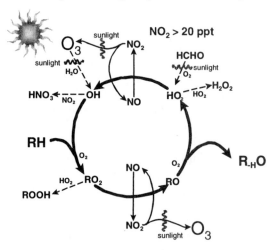

Figure 9   *Ozone production and loss in the oxidation of saturated hydrocarbons*

oxidized to NO_3 varies in the range 2 to 12 (PORG, 1997) and is complicated by the very large range of VOCs and their different photochemical oxidant creation potentials (Derwent and Davies, 1994).

The background ozone concentrations in Europe have doubled since the latter decades of the 19th century and this increase is a consequence of both NO_x and VOC emissions. During summer, anticyclonic weather and gentle airflow over source regions for NO_x and VOC compounds in bright sunshine provide ideal conditions for the production of large ambient concentrations of O_3 and its transport to almost anywhere in Europe. In the UK, peak hourly O_3 concentrations reach 100 ppb to 150 ppb and persistent warm weather allows a series of days with these large maxima. The consequences of elevated ozone concentration, those in excess of thresholds for effects on vegetation and human health, are currently producing the main political interest and the development of control measures.

2.4   Deposition of Oxidized and Reduced Nitrogen

The emitted NO may be deposited if ambient concentration exceeds the compensation point. In practice however, most soils are net sources of NO, as ambient NO concentrations in rural areas are below the compensation point most of the time. The exceptions are soils in polluted urban areas.

Following emission of NO from the soil, reaction of NO with O_3 within or just above the crop canopy converts NO to NO_2 (Figure 10) and the NO_2 is readily absorbed by stomata (Thoene *et al.*, 1991). The effect of this process is to reduce the net loss of soil emitted NO before it is released from the upper layers of the crop canopy as discussed by Yienger and Levy (1995). The importance of soil emissions of NO on the net land-atmosphere exchange of oxidized nitrogen is both as a large area source of atmospheric NO_x in areas

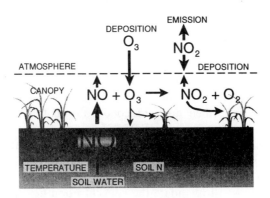

Figure 10 *NO emission from soil*

remote from combustion sources, and as a mechanism reducing the canopy-atmosphere gradient in NO_2 concentration and hence reducing the effective sink strength of vegetation for ambient NO_2.

For NO_2 the removal by terrestrial surfaces is almost entirely due to stomatal uptake (Hargreaves *et al.*, 1992; Thoene *et al.*, 1991). The absence of a significant cuticular uptake pathway leads to negligible deposition at night (vg œ 1 mm s^{-1}). There also appears to be no significant mesophyll resistance for stomatal NO_2 uptake (Thoene *et al.*, 1991) as the flux of NO_2 to crop canopies can be estimated reasonably well from canopy conductances for water vapour (in the absence of surface water on the canopy) making allowance for differences in molecular diffusivity ties of the two gases. In this case the maximum rates of NO_2 deposition would be of the order 5 to 6 mm s^{-1} for typical summer daytime conditions over grassland.

To calculate NO_2 dry deposition over regional scales requires both land-use and meteorological information to calculate rates of deposition and the ambient NO_2 concentration to calculate the flux. The latter is provided either by long-range transport models (EMEP, 1997) or by direct measurements. For the UK a monitoring network provides the measured NO_2 concentrations (RGAR, 1997) and dry deposition is simulated using a resistance model (Smith *et al.*, 1998). The dry deposition of NO_2 to the UK is approximately 40 kt NO_2-N y^{-1} and is primarily uptake by stomata, and therefore is deposited during the warm spring and summer conditions.

The pronounced seasonal dependence of NO_2 deposition is one of the causes (along with the presence of low level temperature inversions) of the enhanced winter concentration of NO_2. The other significant dry deposition input of oxidized nitrogen is that of HNO_3, which being a very reactive gas is deposited at the maximum rate provided by the turbulent deposition process (Muller *et al.*, 1993). The potential terrestrial inputs through this pathway lie in the range 0.1 to 5 kg HNO_3-N ha^{-1} annually (Erisman *et al.*, 1996). It is not currently possible to map the dry deposition of HNO_3 throughout the UK because ambient HNO_3 concentration measurements are not made routinely in a UK network and are only available for a few selected sites and for rather short periods.

2.5   Dry Deposition of Ammonia

The fluxes of NH_3 to vegetation are bidirectional, with emission to the atmosphere occurring when ambient concentration falls below the canopy compensation point (Sutton *et al.*, 1997). However, for a substantial fraction of the UK land area the vegetation is an efficient sink for ammonia and rates of deposition are large. The ambient concentration of NH_3 is routinely monitored by a national 70-site network (Sutton *et al.*, 1997) and these measurements are used to validate the concentration field produced using an atmospheric dispersion model with the spatially disaggregated UK NH_3 emission data at a spatial scale of 5 km $\times$ 5 km (Sutton *et al.*, 1997). The concentration data are then used within a canopy compensation point model to calculate the net flux of NH_3 throughout the year at 5 km $\times$ 5 km resolution for the UK. The resulting annual deposition flux of 100 kt NH_3-N represents approximately 40% of annual emission, emphasizing the importance of this deposition pathway. The land-use specific dry deposition values are generally in the range from 0.1 to 30 kg NH_3-N ha^{-1} y^{-1} and are largest for woodland close to large sources of NH_3.

2.6   Wet Deposition

The deposition of both oxidized and reduced nitrogen in precipitation is monitored throughout the UK by a network of precipitation chemistry collectors containing 5 daily wet-only collectors (the primary sites) and 33 weekly bulk collectors. These measurements provide the concentrations of all major ions, of which NO_3^- and NH_4^+ are amongst the dominant (non-marine) components along with SO_4^{2-} and H^+.

The annual mean concentrations of NO_3^- and NH_4^+ in rain range from 5 µmol l^{-1} to 70 µmol l^{-1} with the largest concentrations in the East Midlands and the smallest concentrations at remote sites in the NW Highlands of Scotland. The wet deposition of reduced and oxidized nitrogen is obtained as the product of the annual precipitation, obtained from the very large network of rain collectors and the precipitation weighted mean concentrations of NH_4^+ and NO_3^-. In addition, an orographic enhancement is applied to quantify the additional wet deposition which occurs in the uplands of the UK as a consequence of seeder-feeder scavenging of polluted hill cloud by falling rain (Fowler *et al.*, 1988).

The wet deposition of nitrogen ranges from 3 to 30 kg ha^{-1} throughout the country with a wet deposition of 120 kt NH_4^+-N and 110 kt NO_3^--N for the UK (Figure 11 a&b). The total wet and dry deposition of nitrogen (oxidized and reduced) amounts to 380 kt N y^{-1} which is equivalent to about 14 kg N ha^{-1}, with maximum 20 km $\times$ 20 km averaged deposition of 35 kg N ha^{-1} and minimum values of about 4 kg N ha^{-1} y^{-1}. These 400 km^2 average deposition

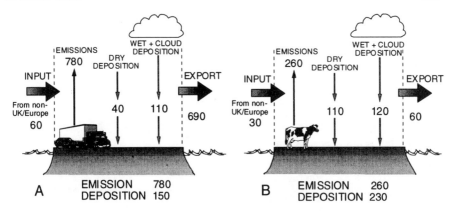

Figure 11 *UK Budgets of (a) oxidised and (b) reduced nitrogen for 1992–94*

values conceal considerable within grid square variability, so that at sites close to large NH_3 sources, or in the very high rainfall uplands in polluted parts of the country, the peak values are larger by at least a factor of two.

The national atmospheric budgets of oxidized and reduced nitrogen reveal quite different characteristics for the two pollutants. In the case of oxidized nitrogen (Figure 11a), of the annual emission of 780 kt N only 150 kt N are deposited within the UK before the wind advects the pollutants out of the country, so that only 19% of UK emissions are deposited within the country, the remaining 81% being exported as NO_2, HNO_3 and aerosol NO_3^-. In the case of reduced nitrogen, 230 kt of the 290 kt emitted are deposited within the UK so that 79% of emissions are deposited within the country. Thus the mean transport distance before deposition for reduced nitrogen is much shorter than that of oxidized nitrogen. These data can be represented by a footprint in which 80% of the UK emissions are depicted (Fig. 12). While only a schematic representation obtained from the long range transport and deposition model used to compute inter-country exchange of pollutants in Europe (EMEP, 1997), it illustrates the very different transport distances of the two pollutants.

3 Terrestrial Effects of Deposited Nitrogen

The data in Fig. 11 also shows that even though reduced nitrogen emissions contribute only 25% of the emissions of fixed reactive nitrogen to the atmosphere in the UK, they represent 60% of the terrestrial inputs. Terrestrial effects of deposited nitrogen in the UK are therefore dominated by reduced nitrogen.

The inputs of atmospheric nitrogen to UK terrestrial surfaces may be represented by a frequency distribution from the 20 km × 20 km mapped data (Fig. 13). These data show a broad peak between 15 kg N ha^{-1} and 30 kg N ha^{-1} and an upper tail approaching 50 kg N ha^{-1}. For comparison the early nitrogen deposition measurements by Lawes and Gilbert (1851) at Rothamsted have been used to estimate a probable distribution of nitrogen deposition in

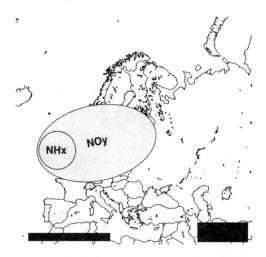

Figure 12   *The footprint of UK NH_x and Noy emissions*

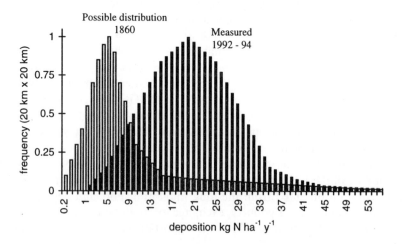

Figure 13   *Frequency distribution of atmospheric nitrogen deposition*

the mid 19th century (Fig. 13). While the peak of the distribution seems reasonable based on the Rothamsted data the upper tail is entirely speculative but has been included because urban areas were subject to large NH_3 concentrations from houses and domestic coal fires, and the acidic atmosphere from SO_2 emissions would have promoted rapid deposition. The peaks in two data sets are sufficiently different to conclude that atmosphere inputs of nitrogen have increased by approximately an order of magnitude since the mid 19th century.

The rapid deposition rates of NH_3, especially on to moorland and forests, lead to much larger nitrogen inputs to these land use categories. The dry deposition modelling used land use dependent deposition velocities with the

Table 2 *Partitioning of N deposition to land cover types[a]*

	Forest	Moorland	Grassland	Arable
Total N deposition, kt	68	124	98	124
Area, $\times 10^6$ ha	2	7.9	6.5	7.9
Mean deposition, kg N ha^{-1}	33	16	15	16
% Reduced N	78%	65%	56%	65%

[a] Total deposition, 380 kt N; reduced, 230 kt N; oxidised, 150 kt N

concentration fields and therefore provides the inputs within each grid square to each of the main land uses (forest, moorland, grassland, arable and urban). The deposited nitrogen may therefore be partitioned by land use as well as oxidized and reduced form as in Table 2.

The relatively small area for forest ($\sim$ 10% of UK) receives a mean deposition of 33 kg N ha^{-1} annually of which 78% is $NH_4^+ + NH_3$, while the moorland receives 16 kg N ha^{-1} annually of which 65% is in reduced form. The most nitrogen sensitive land uses are the semi-natural woodland and wetland areas, both of which receive the bulk of their inputs as reduced nitrogen. This exercise can be taken one further step by introducing the concept of critical levels for nutrient nitrogen. In this case the critical load is taken as the maximum long-term annual input of nitrogen to the ecosystem that can occur without causing change to species composition of the flora. A ratio of 15–20 kg N ha^{-1} year^{-1} has been set for nitrogen for deciduous forest (Grennfelt and Thornelof, 1992). Taking the value of 20 kg N ha^{-1} year^{-1} as the critical load and assuming it applies to all forest in the UK, the deposition maps can be compared with the areal distribution of forest for both oxidized and reduced nitrogen. The results (Fig. 14) show that a substantial fraction of UK forest exceeds the critical loads for nitrogen deposition due to reduced nitrogen deposition alone, where for oxidized nitrogen there are no exceedances.

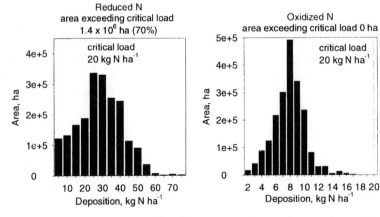

Figure 14 *Frequency distribution of N deposition on forests showing the areal extent of exceedance*

The data used for this exercise are purely illustrative and cannot be used in the absence of empirical measurements to quantify damage. However, the analysis does identify reduced nitrogen as the major contributor to atmospheric inputs of nitrogen.

The major terrestrial effects of atmospheric nitrogen inputs in the UK are believed to be the changes in species composition in sensitive plant communities (Pitcairn *et al.* 1991), although the magnitude of the changes and their geographical distribution are uncertain. In the proximity of large sources of NH_3 clear changes in species competition have been observed (Pitcairn *et al.* 1998).

The other terrestrial effects of emitted nitrogen include:

(i) Exceedance of human health effect thresholds for ozone as a consequence of NO_x and VOC, and exceedances of critical levels for damage to crops and semi-natural vegetation (PORG, 1997).

(ii) Production of aerosol (NH_4^+ NO_3^-, $(NH_4)_2SO_4$) from both oxidized and reduced nitrogen. The aerosol contributes to radiative forcing of climate by increasing the effective albedo of the earth both directly and indirectly by the modification of optical properties of clouds (IPCC, 1997).

(iii) Deposited nitrogen contributes to ecosystem acidification. The number of hydrogen ions produced per deposited NH_4^+, NH_3, NO_2 or HNO_3 varies depending on the fate of the deposited nitrogen (Sutton *et al.* 1993). In general it is assumed that one H^+ is produced per deposited N atom.

The nitrogen related component of acidic deposition was until 1990 the minor factor of the total, with sulfur deposition contributing the majority of the acidification. However, sulfur deposition has declined quite rapidly during the last decade and the amounts of acidity contributed by N and S are now similar. As the sulfur emissions and deposition continue to decline the nitrogen will become the dominant acidifying input.

It is clear therefore that the environmental effects of emissions of fixed nitrogen both oxidized and reduced are among the most important of the regional and global pollution issues. The local problems in urban areas of vehicle NO_x and in rural areas of agricultural NH_3 emissions show that nitrogen compounds are also important at local scale, pollution problems.

Acknowledgements

The authors gratefully acknowledge the Department of the Environment Transport and the Regions, DETR (Air Quality) and the Ministry of Agriculture, Fisheries and Food (MAFF) for supporting the field studies from which this paper has been produced.

References

Bouwman, A.F. 1995 Compilation of a global inventory of environs of nitrous oxide. Ph.D. Thesis, Dutch Agricultural University, Wageningen, The Netherlands.

Davidson, E.A. 1991. Fluxes of nitrous oxide and nitric oxide from terrestrial ecosystems. In: *Microbial Production and Consumption of Greenhouse Gases: Methane, Nitrogen Oxides and Halomethanes.* (Ed. J.E. Rogers and W.B. Whitman) American Society for Microbiology, Washington, pp. 219–235.

Dentener, F. 1993. Heterogeneous Chemistry in the Troposphere. Ph.D. Thesis CIP-Gegevens Koninkluke Bibliotheek, Den Haag, The Netherlands.

Derwent, R.G., Davies, T.J. 1994. Modelling the impact of NO_x or hydrocarbon control on photochemical ozone in Europe. *Atmospheric Environment,* **28**: 2039–2052

Dignon, J., Hamseed, S. 1989. Global emissions of nitrogen and sulfur-oxides from 1860 to 1980. *The Journal of the Air & Waste Management Association,* **39**: 180–186.

EMEP, 1997. Transboundary Air Pollution in Europe. MSC-W Status Report 1997. Part 1: Emissions, dispersion and trends of acidifying and eutrophying agents. (Ed. E. Berge). Co-operative programme for monitoring and evaluation of the long range transmission of air pollutants in Europe. Meteorological Synthesising Centre-West, The Norwegian Meteorological Institute, Oslo.

Erisman, J.W., Mennen, M.G., Fowler, D., Flechard, C.R., Spindler, G., Gruner, A., Duyzer, J.H., Ruigrok, W., Wyers, G.P. 1996. Towards development of a deposition monitoring network for air pollution in Europe. Report No. 722108015, RIVM Bilthoven, The Netherlands.

Firestone, M.K., Tiedje, J.M. 1979. Temporal changes in nitrous oxide and dinitrogen from denitrification following onset on anaerobiosis. *Applied and Environmental Microbiology,* **38**: 673–679.

Fowler, D., Cape, J.N., Leith, I.D., Choularton, T.W., Gay, M.J. & Jones, A. 1988. The influence of altitude on rainfall composition at Great Dun Fell. *Atmos. Environ.* **22**, 1355–1362.

Grennfelt, P. & Thörnelöf, E. (Eds.) 1992. *Critical loads for nitrogen.* Report of the Lökeberg Workshop. Nord (Miljörapport) 41. Nordic Council of Ministers, Copenhagen.

Hargreaves, K.J., Fowler, D., Storeton-West, R.L. Duyzer, J.H. 1992. The exchange of nitric oxide, nitrogen dioxide and ozone between pasture and the atmosphere. *Environmental Pollution,* **75**: 53–60.

IPCC 1997. International Panel on Climate Change Guidlines for National Greenhouse Gas Inventories, OECD, Paris.

Johansson, C., Granat, L. 1984. Emission of nitric oxide from arable land. *Tellus,* **36**: 25–37.

Lawes, J.B. and Gilbert, J.H. 1851. On agricultural chemistry. *J. R. Agric. Soc.* **12**, 1–40.

Ludwig, J. 1994. *Untersuchungen zum Austausch von Stickoxiden zwischen Biosphre und Atmosphäre,* Ph.D. thesis, University Bayreuth, Bayreuth, Germany.

Muller, K., Kramm, G., Meixner, F., Dollard, G.J., Folwer, D., & Possanzini, M. 1993. Determination of HNO3 dry deposition by modified Bowen ratio and aerodynamic profile techniques. *Tellus,* **45B**, 346–367.

Pitcairn, C.E.R., Fowler, D. & Grace, J. 1991. Changes in species composition of semi-natural vegetation associated with the increase in atmospheric inputs of nitrogen.

Final report. Published by NCC as CSD report no.1246. Nature Conservancy Council.

Pitcairn, C.E.R., Leith, I.D., Sheppard, L.J., Sutton, M.A., Fowler, D., Munro, R.C., Tang, S. & Wilson, D. 1998. The relationship between nitrogen deposition species composition and foliar nitrogen concentrations in woodland flora in the vicinity of livestock farms. *Environ. Pollut. (In press)*.

PORG, 1997. *Ozone in the UK*. Fourth report of the Photochemical Oxidants Review Group. Department of the Environment, Transport and the Regions, London. (ITE Edinburgh).

Remde, A., Slemr, F., Conrad, R. 1989. Microbial production and uptake of nitric oxide in soil. *FEMS Microbial Ecology*, **62**: 221–230.

RGAR, 1997. *Acid deposition in the United Kingdom*. Fourth report of the Review Group on Acid Rain. London: Department of the Environment, Transport and the Regions.

Skiba, U., Smith, K.A. and Fowler, D. 1993. Nitrification and denitrification as sources of nitric oxide and nitrous oxide in a sandy loam soil. *Soil Biology & Biochemistry* **25**: 1527– 1536.

Skiba, U., Fowler, D., Smith, K.A. 1997. Nitric oxide emissions from agricultural soils in temperate and tropical climates: sources, controls and mitigation options. *Nutrient Cycling in Agroecosystems*, **48**: 139–153.

Stohl, A., Williams, E., Wotawa, G. & Kromp-Kolb, H. 1996. An European inventory of soil nitric oxide emissions and the effect of these emissions on the photochemical formation of ozone. *Atmospheric Environment*, **30**, 3741–3755.

Sutton, M.A., Pitcairn, C.E.R. & Fowler, D. 1993. The exchange of ammonia between the atmosphere and plant communities. *Adv. Ecol. Research*. **24**, 301–393.

Sutton M.A., Place C.J., Eager M., Fowler D. and Smith R.I. 1995. Assessment of the magnitude of ammonia emissions in the United Kingdom. *Atmos. Environ.* **29**, 1393–1411.

Sutton M.A., Miners B.P., Wyers G.P., Duyzer J.H., Milford C., Cape J.N. and Fowler D. (1997) *National ammonia concentration monitoring in the United Kingdom: sampling intercomparison, network structure and initial network results*. Interim Report to the Department of the Environment (EPG 1/3/58), Insitute of Terrestrial Ecology, Edinburgh.

Thoene, B., Schröder, P., Papen, H., Egger, A., Rennenberg, H. 1991. Absorption of atmospheric NO_2 by spruce (*Picea abies* L. Harts.) trees 1. NO_2 influx and its correlation with nitrate reduction. *New Phytologist*, **117**: 575–585.

Williams, E.J., Guenther, A., Fehsenfeld, F.C. 1992. An inventory of nitric oxide emissions from soils in the United States. *Journal of Geophysical Research*, **97**: 7511–7520.

Yienger, J.J., Levy II, H. 1995. Empirical model of the global soil-biogenic NO_x emissions. *J. Geophys. Res.*, **100**: 11447–11464.

10

Nitrates in Soils and Waters from Sewage Wastes on Land

C.E. Clapp, R. Liu, D.R. Linden, W.E. Larson, and R.H. Dowdy

USDA-ARS AND DEPARTMENT OF SOIL, WATER & CLIMATE,
UNIVERSITY OF MINNESOTA, ST. PAUL, MN 55108, USA

Abstract

Research was initiated in 1973 at St. Paul, Minnesota, with objectives to develop efficient, practical, and environmentally safe methods for utilizing sewage wastes on land in harmony with agricultural usage. Applications of municipal sewage sludge and/or wastewater effluent were studied. Liquid digested sewage sludges from several wastewater treatment plants were applied to a 16-ha terraced watershed cropped to maize (*Zea mays* L.) and reed canarygrass (*Phalaris arundinacea* L.). The sludge was transported to the site by tank truck, stored in lagoons, and spread by combinations of traveling gun and subsurface injection. Sludge was applied to the maize areas for 20 years (total of 68 cm, 224 tonnes ha^{-1} solids, and 9460 kg ha^{-1} total N) and to the reed canarygrass areas for 12 years (total of 96 cm, 173 tonnes ha^{-1} solids, and 10 040 kg ha^{-1} total N). Crop yields were high with normal plant tissue concentrations of N, P, and K. Analysis of water samples from runoff, soil, and ground water showed no movement of potentially polluting materials out of the watershed *via* surface runoff or leaching. Wastewater effluent was sprinkled onto a 2-ha area containing maize and eight forage species for a four-year period at about 5 and 10 cm wk^{-1} (total of 120 to 280 cm yr^{-1} and 260 to 680 kg N ha^{-1} yr^{-1}). Tile drainage, ground and soil water, soils, and crops were analyzed for N, P, K, and trace metals. Maize and forage grasses produced high yields (average of 13.4 and 11.1 tonnes ha^{-1} for maize and reed canarygrass, respectively). Removal of N and P from effluent by the combination of crop uptake and soil sorption was satisfactory. No increase in trace metal concentrations was detected in either the water percolate or crops. Special management of the crop was important for maintaining adequate infiltration rates, for producing high yields of dry matter, and for maximum removal of N.

1   Introduction

Disposal of the products of municipal wastewater treatment is a major environmental problem. Finding environmentally acceptable, socially responsible, and economically feasible plans for carrying out this task is receiving much attention from both research and regulatory agencies, as well as from the public. Increasingly rigid water and air quality standards have forced municipal officials, environmental engineers, and wastewater treatment plant operators to look beyond conventional systems of treatment and disposal. Land application of sewage sludge and effluent is an alternative that returns materials to a natural cycle, which is agriculturally beneficial. In turn, renewed interest in treatment on land has stimulated scientists to examine in detail the physical, chemical, and biological processes in soil that influence waste renovation. Historically, land treatment methods have been based on sewage disposal, whereas we now emphasize a more modern approach to wastewater treatment followed by recycling on agricultural land.

Increasing nitrate levels in surface and ground water can be of concern when caused by high applications of sewage sludge and wastewater. The toxic effects of nitrates on human and animal health are well documented elsewhere in this publication. Excessive amounts of nitrates in drinking water can cause severe health problems such as methemoglobinemia for infants and adults, hypo and hypertension, and Balkan nephropathy (Bruning-Fann and Kaneene, 1993; Prasad and Power, 1995). Methemoglobinemia or 'blue baby' syndrome in human infants is the most widely known toxic effect of nitrate ingestion (Schmidt, 1956). Nitrites produced from nitrates are also reported to react in the stomach with secondary amines resulting from the breakdown of meat and fish, forming N-nitroso compounds, which in turn might cause stomach cancer (Prasad and Power, 1995).

The Rosemount watershed study was initiated in 1973 on land of the University of Minnesota's Agricultural Experiment Station near Rosemount, MN (Clapp *et al.*, 1977, 1983, 1994). The primary goal of the research was to develop efficient, practical, and environmentally safe methods for utilizing municipal sewage sludge on land in harmony with agricultural usage. The long-term study at the Rosemount watershed is one of the best examples of detailed analyses of the environmental and agricultural impacts of sludge application to land. After 20 years of research, results have shown that there are many benefits from using sludge as a plant nutrient source for maize (*Zea mays* L.) and reed canarygrass (*Phalaris arundinacea* L.). Also, this research showed that the concentrations of Cd, Cr, Cu, Ni, and Pb in plant tissue were not increased by sludge applications. Sludge-borne Zn did increase slightly the Zn levels in maize stover (Dowdy *et al.*, 1994). Historically, yields on land where sludge was applied have been slightly better than for the fertilized control areas within the same watershed (Linden *et al.*, 1995).

The Apple Valley wastewater experiment was centred on renovation of municipal wastewater effluent by a crop irrigation system (Clapp *et al.*, 1977; Clapp *et al.*, 1978; Linden *et al.*, 1984). The study involved applications of large amounts of effluent with its associated nutrients, and this in turn required the development of crop and soil management practices to provide the most efficient treatment in balance with high agricultural production. Under a high loading of effluent, however, there are questions with regard to the best soil and crop management systems. The practices examined included the selection of crop species and populations for the removal of the greatest amounts of nitrogen (N), the tillage and residue management for row crops, and the production of two crops during one season.

Associated experiments identifying nitrogen (nitrate-N) interactions have been carried out at several locations in the US (Melsted, 1973; Sopper and Kardos, 1973; Bouwer and Chaney, 1974; Palazzo and McKim, 1978; Spaulding *et al.*, 1993; Frink and Sawhney, 1994). In general, these experiments indicated that if the rates of N application were limited to no greater than 20% above the amounts taken up by the plants, then nitrate-N was not an environmental problem.

The overall objective of the studies reported below was to develop agronomic practices to give maximum nutrient utilization by crops treated with municipal sludges and wastewater effluent. The N (especially nitrates) in the soil-water-crop systems was monitored by measuring the characteristic components in field experiments.

2 Experimental

2.1 Sewage Sludge

Research was conducted on a 16-ha watershed at the Rosemount Agricultural Experiment Station of the University of Minnesota near Rosemount, Minnesota. Table 1 summarizes the site description and experimental set-up. Additional details of sludge storage, application methods, water monitoring, and crop and soil management are given by Duncomb *et al.* (1982), Clapp *et al.* (1994), and Linden *et al.* (1995).

2.1.1 Materials. During the 20-yr study, municipal sewage sludge was applied to the maize areas at an average rate of 4 cm liquid yr^{-1}, 11 tonnes solids ha^{-1} yr^{-1}, and 475 kg N ha^{-1} yr^{-1}. There were 28 applications totaling 68 cm of sludge, 224 tonnes ha^{-1} solids, and 9460 kg ha^{-1} total N. On grass areas, sludge was applied for 12 years at an average rate of 8 cm liquid yr^{-1}, 15 tonnes solids ha^{-1} yr^{-1}, and 830 kg N ha^{-1} yr^{-1}. There were 61 applications totaling 96 cm of sludge, 173 tonnes ha^{-1} solids, and 10 040 kg ha^{-1} total N. Total sludge and nutrients applied per cropping season are shown in Table 2. Sludge analyses are given in Table 3.

2.1.2 Methods. The watershed had surface tile inlets that directed runoff water underground to an outlet at a central drainage channel. Runoff water from each area was collected separately by automatic samplers modified to

Table 1 *Site and experimental set-up for Rosemount sewage sludge watershed*

Location: Rosemount, Minnesota, USA (44% 41′ 34″ north latitude and 93% 04′ 45″ west longitude).
Soils: Port Byron silt loam (Typic Hapludoll); Bold silt loam (Typic Udorthent); Tallula silt loam (Typic Hapludoll).
Watershed: Parallel terraces with tile outlets; 10 treatment areas, ~ 1.5 ha each.
Fertilized Control: Optimum N, P, K for maximum production.
Sludge Treatments: 5 treatment plants, digested, liquid, ~ 8 cm yr^{-1}; Applied over 20 yr, 1974–93.
Application: Tankwagon, traveling gun, chisel injector.
Crops and Treatment Areas: Maize & reed canarygrass; 1 control, 4 sludge.
Monitoring: Sludge, crops, soil, water (soil, runoff, wells).

Table 2 *Sludge and nutrients applied by cropping season for Rosemount sewage sludge watershed (1974–93)*

| | | | | | | Nutrients | |
Years	Applications	Sludge (cm)	(%)	Solids (tonnes ha^{-1})	N	P (kg ha^{-1})	K
Maize							
1974–77	11	20.3	1.40	39.1	2220	1180	190
1978–81	7	24.0	2.75	67.1	3450	1680	191
1982–85	3	10.8	2.08	30.3	1440	671	62
1986–89	3	6.6	2.25	23.4	936	607	72
1990–93	4	6.7	9.50	64.0	1410	973	89
Total	28	68.4	–	224	9460	5100	604
Grass[†]							
1974–77	18	24.3	1.76	43.2	2530	1110	254
1978–81	33	50.0	1.95	101	5570	2420	370
1982–85	10	21.9	1.31	28.6	1940	600	79
Total	61	96.2	–	173	10040	4130	703

[†] Grass area planted to maize since 1986.

Table 3 *Composition of sludge applied to maize and grass treatment areas for Rosemount sewage sludge watershed (1974–1992)*

| | Maize | | Grass |
Constituent[†]	1974–1989 (%)	1990–1992 (%)	1974–1985 (%)
Total solids	2.96	10.0	1.94
Total C	23.6	15.0	25.7
Total N	6.92	1.83	5.93
NH$_4$N	1.77	0.63	2.58
P	2.51	1.75	2.46
K	0.33	0.17	0.43
EC (dS m^{-1})	4.0	3.3	4.4
pH	7.5	7.5	7.8

[†] Total solids based on 105°C dry weight. Other constituents based on percentage of total solids.

begin collection when flow started and at 1-hr intervals during runoff. Flow rates were measured by a water stage recorder with a slotted tube and a stilling well (Linden *et al.*, 1983). Soil water was sampled at 3- to 4-week intervals by porous ceramic samplers installed at 60- and 150-cm depths at 24 sites on the sludge and control areas during the first 5 years of the study. Samples from 14 shallow ground water monitoring wells within the watershed were collected monthly (see Linden *et al.*, 1995 for site diagram). Background samples from various water sources, both within and around the watershed, were taken at regular intervals for the 1973 season before any sludge was applied on the project site, and data for these were compared with those for water samples taken during the following growing seasons.

Water and sludge samples were refrigerated immediately after collection and analyzed for organic components within one week, or the samples were acidified. Soil samples were dried at 35°C; plant samples were dried at 65°C. Sludge, plant, soil, and water analytical methods are described in Linden *et al.* (1995).

2.2 Wastewater Effluent

The experimental site selected was adjacent to the Apple Valley, Minnesota, wastewater treatment plant which had effluent polishing filters (capacity 5700 m^3 day^{-1}). The site was chosen because of the uniformity of the soil, and because it was isolated from population centres, yet easily accessible (see Clapp *et al.*, 1977 for site diagram). The Waukegan silt loam (Typic Hapludoll) soil is dark-coloured, well-drained, and has a water table at about the 140- to 150-cm depth (see Table 4). Details of drainage, irrigation, cropping and sample collection are summarized by Clapp *et al.* (1977) and by Linden *et al.* (1984).

2.2.1 Materials. Applications of effluent on monthly and on total season rates are given for one year for each treatment in Table 5. Irrigation began about May 15 and stopped at the end of October. Irrigation rates averaged 1.6 cm hr^{-1} on the forage and 1.3 cm hr^{-1} on the maize, with durations of 2

Table 4 *Site and experimental set-up for Apple Valley wastewater effluent project*

Location: Apple Valley, Minnesota, USA (44% 41' 36" north latitude and 93% 10' 08" west longitude.
Soil: Waukegan silt loam (Typic Hapludoll).
Effluent: Secondary from activated sludge process, with polishing filter; applied 4 years, 1974–77.
Treatments: Fertilized control, optimum N, P, K; irrigation as required. Low effluent, 5 cm wk^{-1}; 320 kg N ha yr^{-1}; High effluent, 10 cm wk^{-1}; 580 kg N ha yr^{-1}.
Plots: 18 x 46 m, 2 replications.
Application: Solid set sprinkler system.
Crops: Maize, with and without residue return; Forages, 8 species; cut 2x, 3x, and 4x.

Table 5 *Effluent applications for Apple Valley wastewater effluent project (1975)*

Treatment	Monthly application amounts (cm)							Total (cm)	1975 season totals[†] (kg ha^{-1})	
	May	June	July	Aug.	Sept.	Oct.			Total N	$NH_4 + NO_3$-N
Maize										
Low	0	21.6	32.8	25.1	12.8	17.3	109	186	165
High	0	36.1	56.4	47.5	24.4	32.5	197	337	297
Forage										
Low	7.0	20.6	37.5	32.8	15.2	22.9	137	234	20
High	11.4	42.4	63.1	59.9	21.3	41.5	240	410	362

† Based on 1975 mean effluent nitrogen composition.

hours (low) and 4 hours (high) twice a week. Thus about 6.5 cm wk^{-1} (low) and 13 cm wk^{-1} (high) were applied on the forage and 5 cm wk^{-1} (low) and 10 cm wk^{-1} (high) were applied on the maize. The 10-cm wk^{-1} application was about the maximum amount of effluent that could be applied on the maize without considerable ponding and runoff. Effluent samples were collected in plastic pans at ground level on each block and composited immediately after each irrigation for analysis. Total N applied during the season in individual applications equaled that calculated from effluent composition and effluent applied. Effluent characteristics are given in Table 6; however, only the N components are of interest here.

Table 6 *Wastewater effluent composition for Apple Valley wastewater effluent project*

Characteristic	Mean (mg L^{-1})[†]	$\pm$ S.D.
Total N	20.8	6.3
NH$_4$-N	16.5	5.7
NO$_3$-N	1.5	1.5
Total P	8.4	3.1
PO$_4$-P	7.6	2.6
COD	49	19
Suspended solids	20	13
Volatile solids	13	5
EC (dS m^{-1})	2.1	0.3
pH	8.2	0.2

[†] Mean $\pm$ pooled standard deviation (S.D.) for ~66 irrigations yr^{-1}. Samples collected at several plot locations and composited.

2.2.2 Methods. Soil water was sampled weekly by using porous ceramic samplers installed at soil depths of about 45 to 60 cm (at the silt loam/gravel interface) and 120 to 130 cm (15 cm above the water table) at 36 sites. Duplicate samplers were installed at each depth. Two sampling sites were located within each maize-residue-effluent treatment combination (four per block), and two within each reed canarygrass-effluent treatment combination (two per block). Rain water was collected in plastic measuring gauges and composited for analysis after each rainfall event.

3 Results

3.1 Sewage Sludge

Runoff and snowmelt water samples collected from 1976 to 1981 (Table 7) showed concentrations of NO$_3^-$-N and PO$_4^{3-}$-P that averaged up to three times higher for sludge treatments compared to the control areas. Soil water data indicated significant movement of NO$_3^-$-N through the soil as the result

Table 7   *Analyses of runoff and snowmelt water and average annual losses for Rosemount sewage sludge watershed (1976–1981)*

Treatment	Amount (cm)	Concentration[†] NO$_3$-N (mg L^{-1})	PO$_4$-P	Loss NO$_3$-N (kg ha^{-1} yr^{-1})	PO$_4$-P
Maize runoff					
Control	8.6	3.9	0.4	3.3	0.3
Sludge	9.2	9.7	0.8	8.9	0.8
Grass runoff					
Control	6.0	6.8	1.1	4.1	0.7
Sludge	7.3	19.7	5.0	14.4	3.6
Grass snowmelt					
Control	3.2	8.0	0.8	2.6	0.2
Sludge	3.7	23.9	3.1	8.8	1.1

[†] Means of four replicated samples per year on the maize and grass sludge-treated areas and one replicated sample per year on the control areas.

of sludge treatment (Table 8). Soil water sampled at 60- and 150-cm depths showed that the sludge-treated areas in maize had higher NO$_3$-N than the control, but not as high in the grass areas. Water was sampled from sampling wells that reached temporarily-saturated soil water below the root zone compact glacial till layer in the watershed. The NO$_3$-N levels in the 12 wells over the 1974 to 1993 period (Figure 1) showed a general increase in NO$_3$-N concentration over time, until N application rates were decreased in 1981. However, during 1990 to 1993, when commercial fertilizers were added in amounts equal to the N levels in the sludge, the control areas had considerably higher NO$_3$-N levels. Water analyses over the period of the study, both within

Table 8   *Analysis of soil water for Rosemount sewage sludge watershed (1976–1981)*

Treatment	Depth (cm)	Concentration[†] NO$_3$-N (mg L^{-1})	PO$_4$-P
Maize	60		
Control		116	0.03
Sludge		173	0.04
Grass	60		
Control		20	0.01
Sludge		90	0.13
Maize	150		
Control		97	0.02
Sludge		160	0.02
Grass	150		
Control		22	0.00
Sludge		52	0.05

[†] Means of eight replicated samples from the control areas and 16 replicated samples from the sludge-treated areas taken at 4-week intervals between May and October.

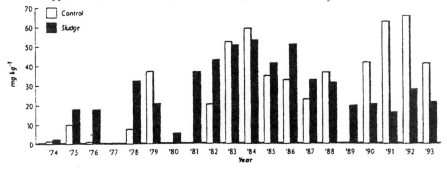

Figure 1   *Nitrate (NO₃-N) concentrations in water samples from shallow wells at the Rosemount sewage sludge watershed (1974–1993)*

and outside the watershed, summarized in Table 9, show that water quality has remained relatively the same during 20 years of sewage sludge application.

3.2 Wastewater Effluent

Monthly and seasonal means of nitrate-N concentrations in soil water at the Apple Valley site for the 1975 year are shown in Table 10 for water collected in porous ceramic samplers at two soil depths. Nitrogen concentrations in soil water measured for three years (Figure 2) were lower under forages than under maize. The maize control (no effluent) treatment had the highest N concentrations throughout the experiment, but decreased with time. Monthly mean nitrate-N concentrations at the 125-cm depth under high effluent treatments of maize and forage crops for two years are shown in Figure 3. In general, concentrations were higher in the spring and autumn than through the summer when high crop uptake demands depleted the soil mineral-N.

4 Discussion

4.1 Sewage Sludge

Protection of water quality is an important consideration when applying fertilizers and sludge as soil amendments. Erosion, runoff, and leaching can be problems on sloping land used for row crops. Sediment and nutrients sorbed on soil particles can cause environmental problems in surface waters. Controlling erosion, as was done on this Rosemount watershed, prevented excessive sediment loss in runoff waters. Incorporation of fertilizers and injection of sludge into the soil was also important in reducing loss of nutrients in runoff (Table 7). Crop type also had an effect on nutrient movement. Runoff waters from reed canarygrass treatment areas had higher NO_3^--N and PO_4^{3-}-P than those from maize areas, in spite of the fact that the maize areas had much less surface protection during the winter. Levels of NO_3-N loss were even higher

Table 9 *Summary of water analyses from various sources at the Rosemount sewage sludge watershed*

Sample location[†]	Year	pH	EC (dS m^{-1})	NO$_3$-N (mg L^{-1})
Shallow well # 1[‡]	1974–93	7.9	0.80	11.3
2	1974–93	8.0	0.87	15.9
3	1974–93	7.9	0.83	13.0
4	1974–93	8.0	0.92	21.4
5	1974–93	8.0	1.06	27.8
6	1974–93	8.0	1.05	19.1
7	1974–93	8.0	0.81	28.5
8	1974–93	7.8	1.18	51.5
9	1974–93	8.0	0.99	32.2
10	1974–93	7.8	1.47	74.6
11	1974–93	7.7	1.11	54.5
12	1974–93	7.9	1.08	31.1
13	1974–93	7.9	1.03	59.5
14	1974–93	8.1	1.17	59.1
Lagoon porous cup (15)	1991–93	8.0	1.22	25.8
Lagoon porous cup (16)	1990–93	8.0	1.20	18.9
Lagoon deep well (18)	1990–93	8.1	0.68	0.7
Farm well (19)	1973	8.3	0.56	0.8
	1973–93	8.0	0.59	0.6
	1993	8.1	0.65	1.05
Watershed reservoir (20)	1973	7.8	0.18	2.1
	1973–93	8.0	0.45	2.0
	1993	8.1	0.56.	2.0
Dam tile (21)	1973	8.0	0.68	6.2
	1973–93	8.0	0.76	11.4
	1993	8.2	0.80	12.0

[†] See Linden *et al.* (1995) for location of sampling wells.
[‡] Shallow wells sample soil water in saturated zone at 50–150 cm deep.

when sludge was surface-applied to grass in winter. Overall, surface water quality was very good at the study site.

High levels of nitrates reduced the quality of soil water (Table 8), but not of deep ground water. Other water quality parameters were not affected by activities at the site. Excessive nitrates occurred due to applications of N which were in excess of crop needs in the early years (Figure 1). Nitrate levels in the shallow ground water decreased in recent years in response to decreased application rates. Lower application rates, however, did not reduce yields. At this site, nitrates did not enter deep ground water, but were contained in the perched water held by the dense glacial till overlaying the deep aquifer. Periodic monitoring of a small stream below the watershed showed no degradation of water quality over the study period.

The long-term study of the Rosemount sewage sludge watershed is an excellent example of environmental and agronomic analysis of sludge application to land. The value of sludge as a nutrient source was established for maize

Table 10 *Total inorganic nitrogen ($NO_3^--N + NH_4^+-N$) concentrations in soil water for Apple Valley wastewater effluent project*

Treatment Month	60-cm depth			125-cm depth		
	Control	Low	High $(mg\ L^{-1})^\dagger$	Control	Low	High
Maize						
June	107	26.5	23.8	23.4	22.6	20.5
July	141	19.6	11.4	33.0	21.2	20.4
August	131	4.0	0.4	36.2	12.0	7.4
September	128	0.8	1.0	31.0	4.7	2.4
October	119	1.6	8.0	33.0	6.4	3.9
Mean	125	10.5	8.9	31.3	13.4	10.9
Forage						
June	1.0	0.3	3.2	5.8	0.1	3.8
July	12.0	0.4	2.8	6.0	0.4	1.8
August	16.8	0.1	1.0	7.3	0.2	1.4
September	–	0.0	1.0	10.5	0.1	0.2
October	13.0	0.2	6.2	11.0	0.1	1.6
Mean	10.7	0.2	2.8	8.1	0.2	1.8

† Means of eight replicated samples on maize and four replicated samples on forage blocks taken by porous ceramic samplers at weekly intervals between June and October 1975.

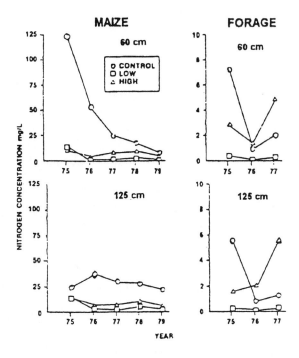

Figure 2 *Season mean NO_3^--N concentrations at two sampling depths under maize and forages for control, low, and high effluent treatments*

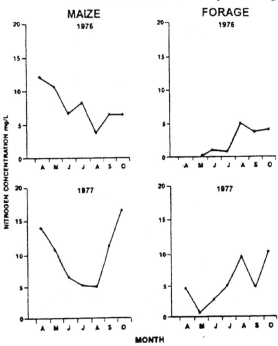

Figure 3   *Monthly mean NO_3^--N concentrations at the 125-cm depth under high effluent treatment of maize and forages for two years*

and grass crops. Grass was more efficient than maize in removing N, P, and K supplied in the sludge. Choice of crop is important if land application of sludge is used primarily as a disposal method. From a water quality viewpoint, the Rosemount watershed study showed that sludge application can be conducted in an environmentally safe fashion. Surface water quality was protected by adequate soil erosion and runoff control. Monitoring shallow ground water allowed adjustments to management practices that protected the deep aquifer. In conclusion, this study is one of a very few to address both the agronomic and environmental issues concerning sludge application to land at the watershed scale over an extended period of time.

4.2 Wastewater Effluent

Nitrogen transformations, availability, and movement under maize and forage cropping for different effluent treatments can be followed by measuring the N concentrations in soil water. Of the total inorganic N concentration, nitrate-N represented 95 to 100 percent; ammonium-N values never exceeded 0.1 mg L^{-1}. Soil water N at the 60-cm depth in the maize blocks showed an order of magnitude difference between fertilizer control and effluent treatments (Table 10). Nitrogen values for the control blocks remained high throughout the season, while those of the effluent treatments decreased markedly following the period of combined higher effluent application and soil nitrification. Increased N concentrations at the end of the season coincided with a decrease in uptake by maize before the rye cover crop was established. In the maize control blocks, N content remained essentially constant at the 125-cm depth during the season. In the maize effluent treatment blocks, N again decreased in midseason and increased at the end of the growing season. Forage crops removed much more N than did maize, as seen by the lower monthly and seasonal mean N concentrations in soil water. Higher values on the forage control blocks at both depths reflect periodic fertilizer applications, while forage on the effluent treatment blocks removed N continuously as applied, allowing very little to move below the root zone.

The high N concentrations under the control treatments were believed to be due to the high application rates of fertilizer N, the mineralization of organic N present in the soil at the initiation of the experiment, and low leaching volumes since irrigations were applied only as crop demand dictated. The organic N reservoir of the soil was apparently being depleted and slowly leached during the course of this experiment. The forage control treatment had higher N concentrations than the forage high and low effluent treatments during 1975 and had intermediate concentrations between the low and high effluent treatments during 1976 and 1977.

Nitrogen concentrations were somewhat higher under the high effluent treatments than under the low effluent treatments for both crops (Figure 2), although these differences were not always significant. There were no N concentration differences between the years of the experiment (except in the case of the control treatment), or between residue or no-residue treatments under the maize crop. Soil water N concentrations under the high effluent forage treatment were higher in 1977 than for the previous 2 years, which reflects the high N application amounts in 1977. The forage control treatment had higher N concentrations than the effluent treatments in 1975. There were no significant differences between treatments in 1976.

The low (< 6 mg N L^{-1}) N concentrations under forage effluent treatments indicate that the crop had a high capacity for removing N from wastewater. Forage crops removed N more effectively than maize because of the high yields and high protein content of the forages (Marten *et al.*, 1980). At slightly lower application rates, however, maize was capable of renovating N from water to an acceptable level because the maize crop also yielded well. It should be emphasized, however, that these conclusions are based upon the fact that the Apple Valley wastewater treatment plant produced a predominantly NH_4-N form of effluent. Higher N concentrations would be expected from a NO_3-N form of effluent (Kardos and Sopper, 1973). Even high application amounts (Table 5) in the summer could not match the crop demands and so the soil solution N was being depleted then. Soil water N concentrations were at a minimum during July and August under maize, and during May and June under forage crops. That corresponded to the maximum growth and N uptake patterns for these two crops.

Concern about ground water contamination by NO_3^--N from application of excessive amounts of effluents has led to extensive monitoring of experimental sites using shallow wells and porous ceramic cup water samplers. The Apple Valley wastewater effluent experiment showed that NO_3^--N levels in soil water at 125-cm depths increased significantly with increased effluent application rates, especially under maize, but only occasionally during the early and late season (Figures 2 and 3), when applications surpassed crop uptake demands. The data from this study suggest that the ratio of total N application to crop removal should not exceed approximately 1.2:1 to prevent NO_3^--N buildup at depths below the rooting zone of crops.

5 Summary

5.1 Sewage Sludge

The long-term study of the Rosemount watershed is a landmark example of a detailed environmental and agronomic analysis of sludge application to land. The value of sludge as a fertilizer substitute was established for maize and grass crops. Yields on sludge-treated areas were on the average slightly better than those on fertilized control areas within the same watershed. Grass was more efficient than maize in removing N, P, and K supplied in the sludge. It was also found that reed canarygrass thrived on sludge amendments, whereas some other forages performed less well when drainage was restricted.

From a water quality viewpoint, the Rosemount watershed study showed that sludge application can be conducted in an environmentally safe fashion. Surface water quality was protected when adequate soil erosion measures were taken at the site. Terraces installed on sloping ground and soil conservation practices were very important for the prevention of deterioration of water quality. Nutrient losses due to runoff and erosion were quite low in comparison to the amount of nutrients applied to the soil.

It was shown that fertilizer and sludge application rates, rates that supplied N in excess of crop uptake, affected the NO_3^--N concentration of near-surface water. At this study site, high NO_3^--N concentration in the near-surface, perched ground water did not affect the quality of the deep aquifer. In other areas, excess NO_3^--N may directly affect ground water quality. This issue emphasizes the need for proper management of all nutrient sources in agriculture to protect ground water resources and to prevent health risks to humans and animals.

5.2 Wastewater Effluent

The Apple Valley wastewater effluent study can be summarized by the following general statements:

Crop yields were good for maize and excellent for grass with respect to dry matter production and quality. Removal of N by reed canarygrass was significantly higher compared to maize. Soil water quality was good, with both N and P below regulatory standards.

Annual NO_3^--N concentrations were generally below 10 mg L^{-1} under both maize and forages at all effluent irrigation rates. This value corresponds to the U.S. Public Health Service drinking water standards limit of 10 mg N L^{-1} (45 mg NO_3 L^{-1}). Data indicated the potential to exceed concentrations of 10 mg N L^{-1} early and late in the season. Nitrate-N concentrations were lower under forages than under maize, indicating the higher uptake of N by the forage grasses. Increases in N applications from effluent irrigation produced larger increases in soil water NO_3-N concentrations under maize than under forages. This result probably denotes the ability of the forage grasses to take up greater proportions of the increased N applications than the maize crop. Nitrogen concentrations in soil water were considerably higher under the fertilized control than for the effluent irrigation treatments under both crops. The high concentrations were the result of high fertilizer application rates and mineralization of soil organic N. The high NO_3^--N values from fertilizer draw attention to potential health risks from N applications which exceed crop requirements.

Municipal wastewater effluent can be successfully renovated by slow-rate irrigation land treatment systems *if* proper soil, crop, and water management procedures are followed.

References

Bouwer, H., and R.L. Chaney. 1974. Land treatment of wastewater. Adv. Agron. 26:133–176.

Bruning-Fann, C.S., and J.B. Kaneene, 1993. The effects of nitrate, nitrite and N-nitroso compounds on human health: a review. Vet. Hum. Toxicol. 35:521–538.

Clapp, C.E., D.R. Linden, W.E. Larson, G.C. Marten, and J.R. Nylund. 1977. Nitrogen removal from municipal wastewater effluent by a crop irrigation system. p. 139-150. *In* R.C. Loehr (ed.) Proc. 1976 Cornell Agric. Waste Manage. Conf., Ann Arbor Science Publ., Ann Arbor, MI.

Clapp, C.E., A.J. Palazzo, W.E. Larson, G.C. Marten, and D.R. Linden. 1978. Uptake of nutrients by plants irrigated with municipal wastewater effluent. p. 395–404. Proc. State of Knowledge in Land Treatment of Wastewater, Int. Symp. U.S.Army Corps of Engin., CRREL, Hanover, NH.

Clapp, C.E., W.E. Larson, R.H. Dowdy, D.R. Linden, G.C. Marten, and D.R. Duncomb. 1983. Utilization of municipal sewage sludge and wastewater effluent on agricultural land in Minnesota. p. 259-292. *In* K. Schallinger (ed.) Proc. 2nd int. symp. on peat and organic matter in agriculture and horticulture. Inst. Soil and Water, Volcani Center, Bet Dagan, Israel.

Clapp, C.E., R.H. Dowdy, D.R. Linden, W.E. Larson, C.M. Hormann, K.E. Smith, T.R. Halbach, H.H. Cheng, and R.C. Polta. 1994. Crop yields, nutrient uptake, soil and water quality during 20 years on the Rosemount sewage sludge watershed. p. 137–148. *In* C.E. Clapp *et al.* (eds.) Sewage sludge: Land utilization and the environment. SSSA Misc. Publ., Soil Sci. Soc. Am., Madison, WI.

Dowdy, R.H., C.E. Clapp, D.R. Linden, W.E. Larson, T.R. Halbach, and R.C. Polta. 1994. Twenty years of trace metals partitioning on the Rosemount sewage sludge watershed. p. 149-155. *In* C.E. Clapp *et al.* (eds.) Sewage sludge: Land utilization and the environment. SSSA Misc. Publ., Soil Sci. Soc. Am. Madison, WI.

Duncomb, D.R., W.E. Larson, C.E. Clapp, R.H. Dowdy, D.R. Linden, and W.K. Johnson. 1982. Effect of liquid wastewater sludge application on crop yield and water quality. J. Water Pollut. Control Fed. 54:1185–1193.

Frink, C.R., and B.L. Sawhney. 1994. Leaching of metals and nitrate from composted sewage sludge. Bull. Conn. Agric. Exp. Stn., New Haven, CT.

Kardos, L.T., and W.E. Sopper. 1973. Renovation of municipal wastewater through land disposal by spray irrigation. p. 148–163. *In* W.E. Sopper and L.T. Kardos (eds.) Recycling treated municipal wastewater and sludge through forest and cropland. The Pennsylvania State Univ. Press, University Park, PA.

Linden, D.R., C.E. Clapp, and R.H. Dowdy. 1983. Hydrologic and nutrient management aspects of municipal wastewater and sludge utilization on land. p. 79–101. *In* A.L. Page *et al.* (eds.) Utilization of municipal wastewater and sludge on land. Univ. of California, Riverside, CA.

Linden, D.R., C.E. Clapp, and W.E. Larson. 1984. Quality of percolate water after treatment of a municipal wastewater effluent by a crop irrigation system. J. Environ. Qual. 13:256–264.

Linden, D.R., W.E. Larson, R.H. Dowdy, and C.E. Clapp. 1995. Agricultural utilization of sewage sludge. Minn. Agric. Exp. Sta. Bull. 606–1995, St. Paul, MN.

Marten, G.C., W.E. Larson, and C.E. Clapp. 1980. Effects of municipal wastewater effluent on performance and feed quality of maize vs. reed canarygrass. J. Environ. Qual. 9:137–141.

Melsted, S.W., 1973. Soil-plant relationships (some practical considerations in waste management). p. 121–128. *In* Proc. of the Joint Conf. on Recycling Municipal Sludges and Effluents on Land. EPA, USDA, and NASULGC, Washington, DC.

Palazzo, A.J., and H.L. McKim. 1978. The growth and nutrient uptake of forage grasses when receiving various application rates of wastewater. p. 157–163. Proc. State of Knowledge in Land Treatment of Wastewater, Int. Symp. U.S.Army Corps of Engin., CRREL, Hanover, NH.

Prasad, R., and J.F. Power. 1995. Nitrification inhibitors for agriculture, health and the environment. Adv. Agron. 54:233–281.

Schmidt, E.L. 1956. Soil nitrification and nitrates in water. Publ. Health Rep. 71:497–503.

Sopper, W.E., and L.T. Kardos (eds.). 1973. Recycling treated municipal wastewater and sludge through forest and cropland. The Pennsylvania State University Press, University Park, PA.

Spaulding, R.F., M.E. Exner, G.E. Martin, and D.D. Snow. 1993. Effects of sludge disposal on groundwater nitrate concentrations. J. Hydrology 142:213–228.

11

The Role of Nitrates in the Eutrophication and Acidification of Surface Waters

M. Hornung

INSTITUTE OF TERRESTRIAL ECOLOGY, MERLEWOOD
RESEARCH STATION, GRANGE OVER SANDS, CUMBRIA
LA11 6JU, UK

Abstract

Nitrogen is an essential nutrient in aquatic ecosystems but when nutrient availability increases eutrophication can result. Chemical changes are paralleled by changes in biological productivity, the composition and diversity of biota and consequently in physical conditions of water bodies. The main concerns about eutrophication are limitations on the water use and increased costs of treatment but health risks can also be associated with algal blooms. Nitrogen is rarely the limiting nutrient in aquatic systems; in most temperate regions waters are P limited but N can be limiting in estuarine and marine waters. Atmospheric inputs of pollutant S and N have lead to the acidification of aquatic ecosystems in acid sensitive areas. The chemical changes lead to changes in biota and a reduction in diversity. Sulfate deposition has been the primary driver of acidification but the importance of N is increasing as S deposition decreases. Both acidification and eutrophication can take place naturally; recent concerns relate to acceleration of the processes as a result of 'human' activities.

Regional patterns of nitrate concentrations in surface waters in the UK show a trend from lower concentrations in the extensively used uplands of the north and west to higher levels in the more densely populated and intensively farmed south and east. Atmospheric deposition can be the main nitrate input to upland waters but inputs from agricultural land and from water treatment are the main sources in the lowlands. A few areas of enhanced concentrations have been identified in the uplands; atmospheric inputs have probably saturated the retention capacity of the catchments in these areas. Areas of acidified waters have been identified in the UK. There is no single data source for the definition of UK waters which have been affected by eutrophication but limited surveys have provided 'snapshots' of the occurrence of waters with algal bloom problems.

The critical load concept is being applied to calculate the maximum nitrate plus sulfate inputs from the atmosphere that will not produce acidification. Guidelines and models have also been developed to set or determine the maximum nutrient inputs to lakes to prevent eutrophication.

1   Introduction

Nitrogen is an essential element in both terrestrial and aquatic ecosystems, being a component of essential amino acids. In terrestrial ecosystems N is often the limiting nutrient while in surface waters P is more commonly limiting. N is cycled efficiently in natural and semi-natural ecosystems, such that in relatively undisturbed terrestrial systems the N cycle is essentially closed and there is little leakage of nitrate to surface waters (Clarke & Rosswall 1981, Melillo 1981). The main input of N to these systems is atmospheric deposition of NO_x and NH_y. These inputs are generally small, a few kg N ha^{-1} yr^{-1}, although pollution resulting from the burning of fossil fuels has resulted in large increases in atmospheric inputs even in relatively remote areas. Soils generally contain large pools of N, associated with soil organic matter, but only a very small fraction is in the ammonium or nitrate form at any one time, with rapid utilisation by the biota. The processes of mineralisation and nitrification of organic forms of N largely control the availability of nitrate in soils (Sprent 1987).

Atmospheric deposition and the small amount of leaching from the catchment dominated N inputs to pristine surface waters. Disturbance of natural terrestrial ecosystems results in increased leaching of nitrate (*e.g.* Gosz 1981). In managed agricultural ecosystems, fertilisers (organic and inorganic) become the main N input and nitrate losses to waters increase further. In these situations, catchment-derived inputs become the dominant input to surface waters.

When N is available in excess of the requirements of the biological system enhanced nitrate leaching can occur with resulting acidification of sensitive soils and eventually linked surface water systems. Additionally the nitrification of NH_y inputs to soils, from atmospheric deposition, animal wastes or in fertilisers is also an acidifying process, which can eventually impact on soils and waters. Nitrogen is one of the major nutrients required to support production in aquatic ecosystems (Reynolds 1997). In natural waters concentrations of the nutrient elements are generally low and the microflora have adapted to rapidly sequester and utilise available supplies. Increases in nutrient availability can therefore produce increases in growth and production, which are associated with eutrophication.

The processes of acidification and eutrophication can both take place naturally. The natural acidification of soils and surface waters is driven by internally generated acidity and generally takes place over hundreds or thousands of years. In the current context the term is used to refer to the recent, relatively rapid acidification of soils and surface waters resulting primarily from the atmospheric inputs of acidifying pollutants, sulfur and nitrogen oxides and ammonia (Chadwick & Hutton 1991). This acidification has involved changes in river and lake chemistry, which have produced secondary changes in the aquatic biota.

Similarly, although an increase in nutrient status of waters, the underlying driver of eutrophication, can take place naturally, the term eutrophication is generally used to refer to the biological effects of increases in concentrations of available nutrients resulting from human activities, sometimes referred to as 'cultural', 'artificial', or 'anthropogenic' eutrophication. Harper (1992) defines it as 'the biological effects of an increase in concentration of nutrients, usually N and P but sometimes Si, K, Ca, Fe or Mn, on aquatic ecosystems.' The publication 'The UK Environment' (Anon 1992) refers to eutrophic waters as those in which 'the level of nutrients is such that the growth of algae is no longer limited.' The OECD definition makes a specific reference to water use: 'The nutrient enrichment of waters which results in the stimulation of an array of symptomatic changes, among which increased production of algae and macrophytes, deterioration of water quality and other symptomatic changes, are found to be undesirable and interfere with water use' (OECD 1982). The potential impacts on water use are key factors, be it for water supply, recreation or irrigation.

Both acidification and eutrophication result in major changes in the species composition of the biotic communities, with generally a reduction in diversity, and changes in productivity. The main concern in the case of acidification has been the loss of fish stocks and of diversity while in the case of eutrophication it is the impact on water treatment costs and possible health impacts. The paper summarises the chemical and biological changes associated with eutrophication and acidification; reviews the current patterns of nitrate concentrations in the surface waters of UK and the factors controlling spatial and temporal variations; then considers approaches to setting limits for nitrate concentration and/or loadings to prevent acidification or eutrophication.

2 Acidification

2.1 Background

Regional surveys of water chemistry (SFT 1987) and paleolimnological studies (*e.g.* Battarbee *et al.*, 1990) from lakes have shown a marked acidification of surface waters in parts of north west Europe, including the UK, and North America within the last 150 years. Acidic surface waters occur naturally, primarily in areas dominated by acidic, organic soils or where bedrock contains significant concentrations of sulfides. Natural acidification of surface waters in the UK since the end of the last glaciation can be traced in sediment cores from lakes. Data of this type show a gradual acidification in the early part of the post glacial period but at the majority of sites there is a stabilisation at several thousand years before present with little change between then and

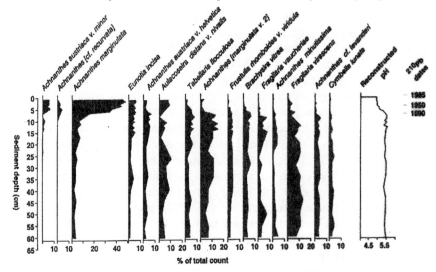

Figure 1   *Variation in diatom community over time and reconstructed pH from a core from Lochnagar*
(From Battarbee *et al.*, 1996)

the last 150 years or less. Thus, a sediment core from Lochnagar indicates that the lake pH was relatively stable at around 5.5 for many hundreds of years and that recent acidification began in about 1890 (Battarbee *et al.*, 1996, Jones *et al.*, 1993) (Figure 1). The timing of the onset of this recent acidification has varied. It dates from about the time of the industrial revolution in parts of northern England and southern Scotland while in Scandinavia and the more remote parts of Scotland the increase in acidity has taken place during the last 50 years (Battarbee 1989).

The recent acidification has been linked to the increased burning of fossil fuels, with emissions of sulfur and nitrogen oxides and subsequent increased atmospheric inputs of sulfate, nitrate and ammonium to terrestrial and aquatic systems (Chadwick & Hutton 1991). The acidification has taken place in areas with poorly buffered soils which have been depleted of base cations, primarily calcium and magnesium, from the soil cation exchange complex and which contain few weatherable minerals containing base cations.

Sulfate has generally been the prime driver of the recent acidification, with nitrate playing a minor role. However, sulfur emissions and sulfate deposition have decreased over the last 10 to 20 years in north west Europe as a result of emission controls and changes in the balance of fossil fuels burnt. In contrast emissions of nitrogen compounds continued to increase over most of this period before stabilising in the last few years. A number of studies in North America and Scandinavia have shown an increasing contribution of nitrate to the acidification of lakes. Thus regional surveys of lakes in Norway in 1974 and 1986 have shown that nitrate concentrations in lakes in southern Norway had almost doubled between the two surveys and that in 1986 the largest nitrate concentrations were in the most acid lakes (Henriksen and Brakke 1988). The increased nitrate concentrations were also associated with increased levels of aluminium. In North America, Sullivan *et al.* (1997) concluded that nitrate made a significant contribution to acidification of lakes in the Adirondacks. The increases in the relative contribution of nitrate to the acidification over the last 20 years shown by such studies may limit further recovery as sulfate emissions continue to decline.

In terms of acidification, sulfur dioxide emissions are the only form of sulfur that need to be considered but, in the case of nitrogen, emissions of both NO_x and NH_y have to be taken into account. The former are derived from the burning of fossil fuels in power stations, industry and motor vehicles while the latter is derived primarily from agriculture with small amounts emitted by industry. While there is more NO_x than NH_y emitted to the atmosphere from UK sources, there is more NH_y deposited to the UK land surface than NO_x (INDITE 1994, Review Group on Acid Rain, 1997). Nitrification of ammonia deposited to soils is a powerful acidifying process and can lead to rapid acidification of sensitive soils.

2.2 Chemistry of Acidified Waters

Acidified waters are generally characterised by low pH, high SO_4^{2-} and /or NO_3^- concentrations, low dissolved organic carbon (DOC) and relatively high concentration of aluminium. The acidity is linked to the high concentrations of strong acid anions. In extreme cases there may also be higher than background concentrations of heavy metals. In contrast, naturally acidic waters in areas of acid organic soils have low pH, low sulfate and aluminium concentrations, and high DOC; the acidity is associated with organic anions. Acidification of surface waters is generally associated with reductions in pH, Ca and Mg concentration, and in acid neutralising capacity (ANC) but increases in strong acid anions (SO_4^{2-} and NO_3^-), inorganic forms of aluminium and toxic metals such as Cd, Ni and Zn. There is also a reduction in colour, with a consequent increase in transparency.

Chemical limits have not been set for the definition of acidified waters but they can be identified by the combination of relatively high sulfate, nitrate and inorganic aluminium concentrations. In broad terms the waters have a pH of <5.5 and inorganic aluminium concentrations of > 1.0 mg l^{-1}.

2.3 Biological impacts of acidification

Changes can be seen at all trophic levels of aquatic biota along the gradient from circum-neutral to acidic waters. Thus, there are changes in the phytoplankton and zooplankton species composition and a reduction in the species diversity of phytoplankton; for example Almer *et al.* (1978) reported 70 species in circum-neutral lakes and 10 species in lakes in the pH range 4.1–5.1. The phytoplankton biomass is not, however, affected until there is a significant increase in toxic metals. The dominant species of diatoms vary with pH and there is a general reduction in diversity of the diatom flora at low pH; these changes form the basis of the pH re-constructions in paleolimnological studies using lake sediment cores. There are also reductions in the species diversity and overall biomass of macro-invertebrates with increasing acidity. The floral and faunal changes are accompanied by a reduction in the rates of organic matter decomposition. There is also a change in fish species and, as conditions become more acid, a reduction in survival and biomass (Muniz & Walloe 1990). In the UK reduced densities and survival of brown trout have been shown in both regional surveys and experimental studies (Harriman & Morrison 1982, Milner & Varallo 1990). The changes in invertebrate and fish populations have also been linked to a reduction in the densities and impaired breeding performance of some river birds, especially the dipper, and in the carrying capacity of streams for otters (*e.g.* Ormerod *et al.*, 1996, Mason & Macdonald 1987). There is an increase in filamentous algae, associated with reduced grazing intensity and rates of decomposition and, in extreme cases, colonisation of lake beds by Sphagnum species. The extreme case is exemplified by some highly sensitive water bodies in Scandinavia which are now fishless, with clear water and beds covered in a growth of algae.

A key feature of the chemistry of recently acidified waters, with respect to biological changes, is the raised concentrations of inorganic aluminium in the low pH, low conductivity waters. Thus, laboratory studies, experimental manipulations of stream water pH and aluminium concentrations, and regional surveys have demonstrated close links between water aluminium concentrations and the status of the aquatic biota (*e.g.* Ormerod *et al.*, 1990). Data from a regional survey in Wales have shown a relationship between a reduction in diversity of invertebrate populations in streams and increasing aluminium concentrations. Mayflies, caddis, mollusc and crustacea populations show the most striking changes. An inverse relationship has also been shown between fish numbers and survival and aluminium concentrations (Milner & Varallo 1990).

2.4   Extent of Acidified Waters in the UK

Areas with waters at risk of acidification can be identified using information on geology, soils and land use. In the UK the main areas at risk are in the north and west, which are underlain by thin acidic soils and hard, slowly weathering bedrock containing small amounts of weatherable silicate minerals or carbonates (Hornung *et al.*, 1995). However, small areas of sensitive soils and geology also occur in the south east, mainly on areas of quartz-rich Quaternary sands.

The critical loads approach (see below) provides an alternative method of identifying surface waters sensitive to acidification. Mapping of critical loads of acidity for UK surface waters shows low critical loads, *i.e.* acid sensitive waters, in similar areas to those defined using data on geology, soils and land use (Harriman *et al.*, 1995).

Regional surveys and paleolimnological studies have confirmed the existence of acidified surface waters in the Southern Uplands and parts of the Highlands in Scotland, the Lake District and the Northern Pennines, the Welsh uplands, the mountain areas of Northern Ireland and areas of Quaternary sands in southern England (Harriman 1989, Kreiser *et al.*, 1995) (Figure 2).

3   Eutrophication

3.1   Background

Over the last 20 to 30 years, there has been growing international concern about the impacts of eutrophication of waters, although the problem has been recognised, and the nutrient status of water bodies has been manipulated for a much longer period. Thus, for example, the United States Environmental Protection Agency identifies eutrophication as the critical problem in those US surface waters with impaired water quality (US Environmental Protection Agency 1996), while UNEP states that eutrophication is 'probably the most pervasive water quality problem on a global scale' (UNEP 1991). There is a large literature on the topic and overviews are presented by, for example, Harper (1992), Ryding and Rast (1989) and Sutcliffe and Jones (1992). Eutrophication of water bodies can take place naturally over time as nutrients accumulate in the system and/or the inflow of nutrients from the catchment increases. In Ryding and Rast (1989) this is referred to as the natural ageing process of lakes, a process which takes place over hundreds or thousands of years. However, most cases of eutrophication occur as a result of increased inflows of nutrients resulting from human activities. Thus, the sedimentary record from Blelham Tarn in the Lake District shows a slow but continuous increase in nitrogen from 8000 B.C. until c.1000 A.D followed by a sharp increase. The gradual increase unto A.D 1000 mainly reflects natural processes while the sharp increase results from the first major period of forest disturbance (Pennington 1981). The major increases in nutrient status associated with eutrophication, as currently defined, are much more recent; for example, in the Blelham Tarn core the modern enrichment takes place post 1930. It is seen earlier in lakes with more densely populated catchments and/or dominated by intensive agriculture.

Although eutrophication of waters is generally seen as a deleterious change, it should be remembered that the nutrient status of water bodies has long been manipulated to enhance fish production. In this context, and for similar specific purposes, eutrophication could be seen as beneficial.

3.2   Limiting Nutrients: P versus N

Aquatic biotic systems require a range of essential nutrients. The availability and the balance between the availability of these nutrients controls the productivity of the system (Reynolds 1997). Waters are often classified on the basis of their nutrient status, and associated biology, into oligotrophic

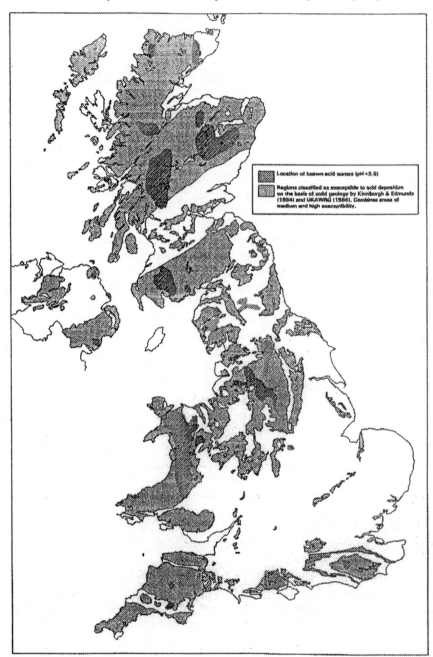

Figure 2   *Distribution of acidified waters in the UK*
(From UK Acid Waters Review Group, 1989)

(nutrient poor), mesotrophic and eutrophic. The key nutrients are carbon, phosphorus, nitrogen, silica, potassium, iron and manganese. As in all biological systems the productivity is limited, within the constraints imposed by the physical environment, by the availability of any nutrient for which demand exceeds supply; this is the 'limiting nutrient'. Phosphorus is most commonly the limiting nutrient in lakes and reservoirs although nitrogen can be limiting in some low latitude lakes and examples of potassium, iron and silica limitation have been reported. In contrast to the situation in lakes and reservoirs, nitrogen is generally thought to be the limiting nutrient in seas and oceans (Nixon 1981). Either nitrogen or phosphorus can be limiting in estuaries and the limitation can vary seasonally, with P limiting in the spring and nitrogen in the summer and autumn (Lee *et al.*, 1996, Fisher *et al.*, 1992).

In most natural water bodies concentrations of nutrients are low and the microflora are adapted to efficiently sequester and rapidly utilise these nutrients (Reynolds 1997). Responses to an increase in supply of the limiting nutrient are therefore generally rapid, resulting in large increases in production in the lower trophic levels of the system. If the initial limiting nutrient is supplied in large enough amounts to satisfy demand, availability of one of the other nutrients may become limiting and it may be possible to control productivity by managing inputs/availability of the other nutrients.

3.3   Eutrophication of UK Waters

There is no data set that can be used to define the extent of the eutrophication problem and distribution of waters in the UK affected by eutrophication. At the broadest level, the content and availability of nutrients of almost all water bodies in the UK has increased compared to their pristine state. River quality in general is reviewed regularly by the water industry and the relevant regulatory organisations, formerly the National Rivers Authority (NRA) and currently the Environment Agency (EA) and the Scottish Environment Protection Agency (SEPA). The 1990 survey of river quality in England and Wales classified 11% of rivers as being of 'poor' or 'bad' quality; the classification being based on fishery status, ammonia concentrations, dissolved oxygen saturation and biochemical oxygen demand (BOD) (National Rivers Authority 1991b). In 'poor' quality rivers fish are 'absent or only sporadically present' and 'bad' quality rivers are 'grossly polluted and likely to form a nuisance'. Various sources of pollution could produce the poor or bad river quality, including eutrophication; however, the main concentrations of the 'poor' and 'bad' quality rivers were in Teeside, Merseyside, South and West Yorkshire and the Midlands, suggesting industrial pollution.

If severe eutrophication is defined by the occurrence of toxic algal blooms (c.f. the definition of eutrophic waters in 'The UK Environment'), a 1989 survey by the National Rivers Authority provides an indication of the scale of the problem. The survey examined a sample of 915 waters and of these 594 had blue-green algae present with 169 containing high enough concentrations to warrant alerts to owners and to Environmental Health Officers (National

Rivers Authority 1991a). The largest numbers of water bodies with blue-green algae were in the Anglian, North West and South West regions.

3.4   Chemical and Physical Changes Associated with Eutrophication

The key chemical change is an increase in the concentrations of available forms of nutrient elements but particular phosphorus and nitrogen. As discussed below, an increase in availability of limiting nutrients leads to changes in the biological status and productivity of the system which feedback to influence the chemical and physical conditions. Thus, decomposition of accumulated phytoplankton and macrophyte debris, resulting from the increased productivity, leads to a reduction in oxygen status and the increased algal biomass leads to a reduction in light penetration. These changes feedback to produce further changes in the biological system.

As noted above, waters are often divided into oligotrophic, mesotrophic and eutrophic on the basis of their nutrient status. A number of schemes have been developed for the definition of these classes using combinations of chemical, physical and biological parameters. Thus, the OECD classification is based on the concentrations of P, N and chlorophyll a, and light penetration (Table 1). In this classification, for example, oligotrophic lakes have upper P and N concentration limits of 8 µg P l^{-1} and 661 µg N l^{-1} (annual geometric mean) and eutrophic lakes lower limits of 84.4 µg P l^{-1} and 1875 µg N l^{-1}. However, in any classification, a water body may fall into one class on the basis of one parameter but another class for other parameters.

3.5   Biological Impacts of Eutrophication

Increases in the availability of limiting nutrients lead to an increase, often very large, in biomass and productivity. There are also changes in species composition with generally a reduction in species diversity. There is a gradual reduction in the number of species of plankton and diatoms, cyanobacter and unicellular green algae become dominant along the trend from oligotrophic to eutrophic. There is an increase in epiphytes. Mycrophyte biomass generally increases initially but then declines again at very high nutrient levels, as a result of

Table 1   *OECD boundary values for open trophic classification system [annual (geometric) mean values]*

Parameter	Oligotrophic	Mesotrophic	Eutrophic
Total P (µg P/l)	8	26.7	84.4
Total N (µg N/l)	661	753	1875
Chlorophyll a (µg/l)	1.7	4.7	14.3
Chlorophyll a peak value (µg/l)	4.2	16.1	42.6
Sechi depth (m)	9.9	4.2	2.45

competition with green algae for light; the number of macrophyte species declines. Thus, Harper (1978a) reports a large difference between the summer macrophyte biomass of an oligotrophic, 44 g dwt m², and a polytrophic lake, 200 g dwt m², with a parallel reduction in species from 15 to 4. As phytoplankton and macrophyte debris accumulates, decomposition leads to changes in oxygen status, mainly in the hypolimnion. Depletion of oxygen can, in extreme circumstances, lead to fish death. Fish biomass increases with increasing nutrient availability and the increased productivity in the lower trophic levels, but there are also changes in the dominant species and diversity. Thus, corigonids are typical of oligotrophic waters, while percids are most common in mesotrophic and cyprids in eutrophic waters. The end point of eutrophication is dense phytoplankton biomass and a large population of stunted cyprinid fish.

3.6 Potential Health Effects Linked to Eutrophication

There has been increasing concern in recent years about the possible effects on animal and human health of accumulations in lakes and reservoirs of potentially toxic blue-green algae. A number of the blue-green algae, Cyanobacteria, produce toxins which are potentially poisonous to fish and mammals (Codd 1995). Three classes of toxins can be produced by Cyanobacteria (Codd op cit): hepatoxic microcystins which cause pneumonia like symptoms and sickness, neurotoxic anatoxins and lipopolysaccharides which produce skin irritations. The algae, and hence the toxins, are normally present at concentrations too low to form a health risk but the situation can change if large quantities of the algae are accumulated as scums. Surface scums of algae form when stable algal blooms break down as a result of a change in weather conditions. The blooms themselves form during periods of calm, sunny weather and require an abundant supply of nutrients. Once formed, the scums can be driven by the wind, accumulating on windward shores of water bodies in concentrations large enough to have an impact on health if ingested. Contact with or ingestion of scums containing blue-green algae has been linked to rashes, eye irritation, vomiting, diarrhoea, fever and pains in muscles and joints. The scums can also be toxic to livestock and in one specific case, a bloom of toxic blue-green algae at Rutland Water in September 1989 was linked to the death of a number of sheep and dogs (Anon 1992). It should be stressed that the formation of blooms and algal scums are not an inevitable result of eutrophication but their formation requires relatively high levels of nutrients.

4 Patterns of Nitrate Concentrations in UK Surface Waters

Nitrate concentrations for surface waters are not in themselves an indicator of acidification or eutrophication. They do show, however, areas in which nitrate concentrations are large enough to warrant consideration of their possible contribution to acidification or eutrophication and where further investigation, in combination with other information, or management intervention may be necessary.

Data from the UK Harmonised Monitoring Scheme show a trend from low mean annual nitrate concentrations in lakes and rivers (<1.0 mg l^{-1}) in the north and west of the UK to higher concentrations (>5.5 mg l^{-1}) in the south and east (Betton *et al.*, 1991) (Figure 3). A similar general trend in concentrations was shown by a national acid waters survey carried out between 1990 and 1992 (Allott *et al.*, 1995) but this also revealed relatively high concentrations in a few locations in the uplands. These trends reflect the broad patterns of variation in intensity of land use and population density in the UK and the link between nitrate concentrations and these variables is discussed further in the following section. Thus, the highest mean annual concentrations reported by Betton *et al.* (op cit) occur in lowland areas dominated by intensive arable and grassland farming; it is in these sorts of areas that nitrate could be contributing to eutrophication. The same authors also examined the trends in concentrations between 1974 and 1986. Between 1980 and 1986, 51% of sites in the Scheme showed increases in nitrate concentrations, 38% no significant change and 38% a decrease. The increases were highest in lowland areas (<100 m) with the largest increases in East Anglia and the Midlands. The authors suggested an association between the increases in concentration and intensification in agriculture. They also used the data to predict that, at the present rate of increase, large areas of East Anglia will have nitrate-N concentrations greater than the WHO limit of 11.3 mg l^{-1} by the year 2000.

As noted above, acidification of surface waters is only a risk in areas with sensitive soils and bedrock. Most of the areas with relatively high concentrations at present are in the lowlands and overlie low sensitivity, well buffered soils. In addition intensively managed agricultural soils are commonly limed, where necessary to maintain a circum-neutral pH. However, the acid waters survey identified a few instances of relatively high nitrate concentrations in acid sensitive areas in the uplands. Allott *et al.* (1995) screened out of the dataset sites with potential within catchment nitrogen sources to leave sites where atmospheric deposition is the dominant N input to catchments (Figure 4). Clusters of sites with relatively high nitrate concentrations (>20 μeq l^{-1}) remained in the Pennines, the Lake District, the Welsh uplands and the Southern Uplands. This suggests that terrestrial systems in these areas may be becoming 'saturated' with N, mainly derived from atmospheric sources, and that significant leakage of nitrate is taking place with the risk of nitrate-driven acidification.

It has been suggested by Henriksen (1988) that the ratio, in equivalents, of NO_3^- to $(NO_3^- + SO_4^{2-})$ can be used as an index of the importance of NO_3^- relative to SO_4^{2-} in determining acidification status. If atmospheric inputs of S and N are equivalent, neither are retained in the catchment and all deposited NH_4^+ is nitrified, the factor would be approximately 0.5. A value greater than 0.5 would indicate that NO_3^- has a greater influence on acidification status than SO_4^{2-}. The ratio for lakes in Scotland, based on mean annual concentra-

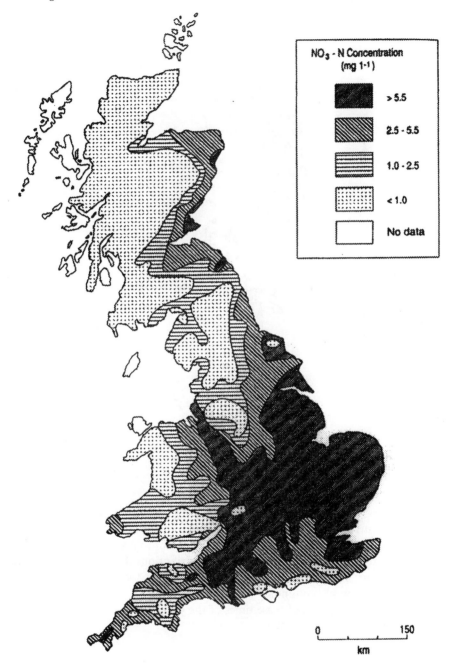

Figure 3   *Nitrate concentrations in UK surface waters*
(From Goudie and Brunsden 1994, after Betton *et al.*, 1991)

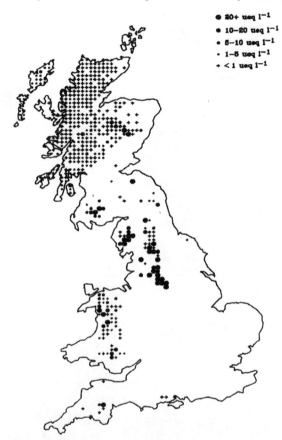

Figure 4   *Nitrate concentrations in surface waters with sites dominated by catchment*
inputs screened out
(From Allot *et al.*, 1995)

tions, is currently less than 0.17 indicating that, at the present, nitrate is making a relatively small contribution to the chronic acidification (Chapman and Edwards, in press). However, the contribution of NO_3^- to acidification can also vary seasonally as concentrations of NO_3^- vary. Marked variations in concentration are commonly found in upland streams, with peaks in the late winter/early spring and at these times of the year nitrate can play an important role in episodic acidification.

5   Controls on Nitrate Concentrations in Surface Waters

Nitrate concentrations in surface waters and their temporal variations are affected by a number of factors; these include atmospheric inputs/pollution climate, the geographical location of the catchment, within catchment factors such as land use, and within water body factors and processes.

5.1 Pollution Cimate

Direct inputs of N from the atmosphere to surface waters play a minor role in determining nitrate concentrations in surface waters in the UK, mainly because of the large ratios of catchment to water area. However, as noted above, atmospheric inputs to the catchment can be an important contribution to the total catchment N load and the dominant N input to catchments in the areas of the UK with extensive land use and low population densities. In the acid waters survey the highest NO_3^- concentrations in upland waters corresponded broadly with areas with the highest atmospheric inputs, but in a statistical analysis only 45% of the variance in NO_3^- concentrations could be explained by N deposition (Allot *et al.*, 1995). Analysis of data from European forest systems indicates that atmospheric inputs >c. 20 kg ha^{-1} yr^{-1} can exceed the N demand of the forest biological system and lead to significant nitrate leaching from soils with the consequent risk of soil and surface water acidification (Dise and Wright 1995). In the UK, atmospheric inputs are largest, and exceed 20 kg ha^{-1} yr^{-1}, in the uplands of Wales, the Pennines, the Southern Uplands and the western Highlands of Scotland, and in an area of lowland England running south eastwards from Worcestershire to Sussex. Deposition reaches a maximum of >50 kg N ha^{-1} yr^{-1} in the southern Pennines, Central Lake District and south east Wales but much larger values of >100 kg ha^{-1} yr^{-1} are known to occur in areas adjacent to intensive agricultural units. In lowland England, inputs of c. 50 kg N ha^{-1} yr^{-1} are much smaller than inputs from agriculture and industry.

5.2 Land Use

Over Great Britain as a whole, land use within the catchment is the main single control on nitrate concentrations in stream and lake waters. The data in Table 2 show rates of loss of nitrate which range from between 1 and 7 kg NO_3^- ha^{-1} yr^{-1} from upland moorland, c. 250 kg NO_3^- ha^{-1} yr^{-1} from intensively managed grassland and over 500 kg NO_3^- ha^{-1} yr^{-1} in an extreme example of an intensively managed agricultural catchment in the Netherlands. In the lowlands nitrate inputs to surface waters are dominant by runoff/drainage from intensive agricultural land and from sewage effluents. In areas of intensive agriculture, the amount of nitrate leached varies considerable between crops, with the amount of crop cover during the winter, when leaching is at a maximum, the frequency and timing of cultivation and the rate of fertiliser addition. It is often thought that the amount of nitrate released can be related simply to fertiliser inputs but studies have shown that this is much too simplistic and that much of the nitrate leached is derived from the soil N pool. In populated areas sewage effluents form major point source inputs and Table 2 shows the magnitude of releases from such sources.

Table 2 *Rates of loss (kg NO_3^- - ha^{-1} yr) from a variety of vegetation/land use sources*

Land use/vegetation type	N loss rate
Upland moorland	1–7 (+ 1–7 NH_4^+-N)
Low intensity grassland	1–9
Swedish forest	9–16
Loch Leven agricultural catchment	33
Swiss lowland catchment	96
River Ouse (agriculture)	117
Swiss lowland agriculture	210
England – intensively managed grassland	250
Perthshire – agriculture	260
Perthshire – urban	352
Perthshire – soft fruit	391
Denmark – intensively managed lake catchment	580–840
Combined storm flow/sewage effluent treatment works	90440

(After INDITE 1994)

Land use is still an important factor in the less intensively managed and more sparsely populated uplands, although overall atmospheric deposition of N assumes a greater significance as a potential source of nitrate to surface waters. As indicated in Table 2, leaching is low from areas dominated by relatively undisturbed semi-natural vegetation although a seasonal pattern is often present with concentrations peaking in late winter or early summer. A change or disturbance in land cover in these areas can, however, lead to large relative increases in nitrate concentrations. Thus, pasture improvement in the uplands can lead to increases in stream nitrate concentrations and these can remain elevated for many years after the initial improvement (Hornung *et al.*, 1985). Afforestation can also lead to an increase in nitrate leaching with further, sometimes large increases at the time of the felling of the crop (*e.g.* Reynolds *et al.*, 1995, Stevens & Hornung 1988). However, the concentrations resulting from such changes in land use/land cover in the uplands are generally small compared with those in streams or rivers receiving drainage from areas of arable or intensive grassland. The increases in concentrations in upland streams can have an impact down stream as the upland waters important role in flushing or diluting the lowland waters.

5.3 Within Water Body Processes

Biological, chemical and physical processes can all influence nitrate concentrations in waters. Thus seasonal patterns of uptake by the aquatic flora can give rise to large, seasonal variations in concentrations, particularly in lakes, with summer minima and winter maxima. Denitrification also takes place in water bodies leading to reductions in nitrate concentrations; again it varies seasonally, peaking in summer. The physical process of mixing, residence time and flow rate can all result in temporal and spatial variations in nitrate concentrations. Thus, concentrations will be lowered by dilution at high flows but may be relative high in areas of stagnant water with weak mixing.

6 Setting of Limits and Targets

6.1 Acidification

The critical loads approach is an effects based approach to setting limits for the emissions of atmospheric pollutants so as to protect ecosystems. The approach has been mainly used in the context of acidfying pollutants and was used in the negotiations for the last sulfur protocol, signed in Oslo in 1996. A critical load is the maximum load of a given pollutant that will not produce damage in specified receptor ecosystems. Methodologies have been developed for the calculation of critical loads of sulfur and these have recently been extended to incorporate the contribution of nitrogen to acidification. The critical load is calculated to ensure that a critical chemical limit for the surface waters is not transgressed as a consequence of the atmospheric inputs. The critical chemical limit is generally set to ensure maintenance of populations of specified, indicator stream biota, *e.g.* brown trout. The methodologies have been combined with the chemical data derived from a survey of a sample of surface waters in the UK, carried out between 1990 and 1992, to produce maps of critical loads for UK freshwaters (United Kingdom Critical Loads Advisory group 1995). These maps have been overlain with deposition data to assess where current sulfur deposition exceeds the critical load and to determine by how much the deposition would have to be reduced to protect the waters from acidification and to allow recovery of already acidified waters. Work is now in progress to provide the input data needed to carry out a similar exercise for total deposited acidity, including nitrogen deposition. A protocol setting limits to national emissions of nitrogen gases to the atmosphere is soon to be negotiated under the auspices of the United Nations Economic Commission for Europe and parallel negotiations under the auspices of the European Union are considering protocols for control of acidity. The data on UK critical loads will form part of the UK input to these negotiations.

6.2 Eutrophication

Nitrate and phosphate concentrations have been used, as noted above, to define the trophic state of water bodies, *i.e.* oligotrophic, mesotrophic and eutrophic, and these could therefore be used to define target concentrations or to identify waters at risk of eutrophication. A number of models have also been developed to aid the understanding and management of eutrophication. In Ryding and Rast (1989) these are grouped as watershed (nutrient load) models, waterbody models and management models. Within each of these categories, empirical and simulation models can be recognised. The empirical models are generally based on statistical relationships derived from existing databases while the simulation models incorporate the mathematical descriptions of a number of the important physical, chemical and biological processes operating in water bodies. The main advantage of the latter models, over the empirically based ones, is that they allow temporal spatial aspects of the eutrophication to be considered. Most of the models are designed to consider P, reflecting the generally dominant role of this element in controlling eutrophication.

As noted above, any increase in nutrient availability in natural aquatic systems could be regarded as eutrophication. The main concerns generally relate to a reduction in water quality and problems with water use. However, increases in nutrient availability in an oligotrophic or mesotrophic water body designated as a conservation site may result in unacceptable changes in the biotic communities long before there was a water use or treatment problem. In other cases nutrient status may be deliberately increased to increase coarse fish production, although sports fisheries generally require oligotrophic or meso-trophic conditions. In a broad sense, therefore, eutrophication and target concentration of nutrients would be best defined with respect to specific end uses.

References

Acid Waters Review group. 1989. Acidity in United Kingdom Freshwaters. HMSO, London.

Allott, T.E.H., Curtis, C.J., Hall, J.R., Harriman, R. 1995. The impacts of nitrogen deposition on upland surface waters in Great Britain: a regional assessment of nitrate leaching. Water, Air and Soil Pollution 85, 297–302.

Almer, D., Dickson, W., Ekstrom, C., & Hornstrom, E., 1978. Sulfur pollution and the aquatic ecosystem. In: Sulfur in the Enviroment. Part II. Ecological Impacts. Edited by J. O. Nriagu. Burlington, Canada; Canada Centre for Inland Waters.

Anon. 1992. The UK Environment. HMSO, London.

Battarbee, R.W. 1989. The acidification of Scottish Lochs. In: Acidification in Scotland, Scottish Development Department, Edinburgh.

Battarbee, R.W., Mason, J. Reinberg, I & Talling, J.F. (Eds). 1990. Palaeolimnology and lake acidification. The Royal Society, London, 219 pp.

Battarbee, R.W., Jones, V. J., Flower, R. J., Appleby, P.G., Rose, N.L. & Rippey, B. 1996. Palaeolimnological evidence for the atmospheric contamination and acidification of high Cairngorm lochs, with special reference to Lochnager. Botanical Journal of Scotland 48, 79–87.

Betton, C., Webb, B. W. & Walling, D. E. 1991. Recent trends in NO_3-N concentration and levels in British Rivers. IAHS Publication 203, 169–80

Chadwick, M. J. & Hutton (editors). 1991. Acid depositions in Europe. Environmental effects, control strategies and policy options ñ Stockholm; Stockholm Environment Institute.

Clarke, F.E. & Rosswall, T. (Eds). 1981. Terrestrial Nitrogen Cycles. Ecol. Bull (Stockholm), 33.

Codd, G.A. 1995. Cyanobacterial toxins: occurrence. Properties and biological signifi-cance. Water Sci Technol 32(4), 149–156.

Dise, N. B. & Wright, R. F. 1995. Nitrogen leaching from European forests in relation to nitrogen deposition. Forest Ecology and Management 71, 153–161

Fisher, T.R., Peele, E.R., Ammerman, J.W., & Harding, L.W. 1992. Nutrient limitation of phytoplankton in Chesapeake Bay. Mar. Ecol., Prog. Ser. 82, 51–63.

Gosz, J.R. 1981. Nitrogen cycling in coniferous ecosystems. In: Terrestrial Nitrogen Cycles. Edited by F.E. Clarke, and T. Rosswall. Ecol. Bull (Stockholm) 33, 402–426.

Goudie, A.S. & Brunsden, D. 1994. The Environment of the British Isles. An Atlas. Clarendon Press, Oxford.

Harper, D.M. 1992. Eutrophication of freshwaters. Chapman & Hall, London.

Harper, D. M. (1979) Limnological Studies on three Scottish Lowland Freshwater Lochs. Ph.D. thesis, University of Dundee.

Harriman, R. 1989. Patterns of surface water acidification in Scotland. In: Acidification in Scotland. Scottish Development Department, Edinburgh.

Harriman, R. & Morrison, B.R.S. 1982. Ecology of streams draining forested and non-forested catchments in an area of Scotland subject to acid precipitation. Hydrobiologia 88, 2510263.

Harriman, R.,Allott, T.E.H., Battarbee, R.W., Curtis, C., Hall, J.R., & Bull, K. 1995. Critical load maps for UK freshwaters. In: Critical Loads of Acid Deposition for United Kingdom Freshwaters. Institute of Terrestrial Ecology, Edinburgh.

Henriksen, A. 1988. Critical Loads of Nitrogen to surface waters. In: Critical Loads for Sulfur and Nitrogen. Edited by J. Nilsson and P. Grennfelt. Copenhagen: Nordic Council of Ministers.

Henriksen, A. & Brakke, D. F. 1988. Increasing contribution of nitrogen to the acidity of surface waters in Norway. Water Air Soil Pollution 42, 183–201.

Hornung, M., Reynolds, B. & Hatton, A.A. 1985. Land management, geological and soil effects on streamwater chemistry in upland mid-Wales. Appl. Geog., 5, 71–80.

Hornung, M., Bull, K.R., Cresser, M., Ullyett, J., Hall, J.R., Langan, S., Loveland, P.J. & Wilson, M.J. 1995. The sensitivity of surface waters of Great Britain to acidification predicted from catchment characteristics. Environmental Pollution 87, 207–214.

INDITE 1994. Impacts of Nitrogen Deposition in Terrestrial Ecosystems, London; United Kingdon Review Group on Impacts of Nitrogen Deposition on Terrestrial Ecosystems

Jones, V.J., Flower, R.J., Appleby, P.G., Natkawski, J., Richardson, N., Rippey, B., Stevenson, A.C. & Battarbee, R.W. 1993. Palaeolimnological evidence for the acidification and atmospheric contamination of lochs in the Cairngorms and Lochnagor areas of Scotland. J. Ecol., 81, 3–24.

Kreiser, A.M., Patrick, S.T., Battarbee, R.W., Hall, J. & Harriman, R. 1995. Mapping water chemistry. In: Critical loads of Acid Deposition for United Kingdom. Freshwaters. Critical Loads Advisory Group. Sub-Group Report on Freshwaters. Institute of Terrestrial Ecology, Edinburgh.

Lee, Y. S, Seiki, T., Mukai, K., & Okada, M. 1996. Limiting nutrients of phytoplanton community in Hiroshima Bay, Japan. Water Research 30, 1490–1494.

Mason, G.F. & Macdonald, S.M. 1987. Acidification and otter Lutra lutra distribution on a British river. Mammalia 51, 62–87.

Melillo, J. M. 1991. Nitrogen cycling in deciduous forests. In: Terrestrial Nitrogen Cycles. Edited by F. E. Clarke and T. Rosswall. Ecol. Bull (Stockholm) 33, 427–442.

Milner, N.J. & Varallo, P.V. 1990. Effects of acidification on fish and fisheries in Wales. In: Acid Waters in Wales, edited by R.W. Edwards, A.S. Gee and J.H. Stoner. Kluwer Academic Publishers, Dordrecht.

Muniz, I. P. & Walloe, L. 1990. The Influence of water quality and catchment characteristics on the survival of fish populations. In: the Surface Waters Acidification Programme. Edited by B. J. Mason. Cambridge: Cambridge University Press.

National Rivers Authority 1991a. Toxic Blue Green Algae. Water Quality Series No. 2. NRA.

National Rivers Authority. 1991b. The Quality of Rivers, Canals and Estuaries in England and Wales. Water Quality Series No. 4. National Rivers Authority, Bristol.

Nixon, S.W. 1981. Remineralisation and nutrient cycling in coastal marine ecosystems. In: Estuaries and nutrients. Edited by B.J. Neilson & L.E. Cronin. Humana Press, Clifton, USA.

OECD 1982. Eutrophication of Waters, Monitoring, Assessment and Control. Final Report. OECD Cooperative Programme on Monitoring of Inland Waters (Eutrophication Control), environmental Directorate. Paris; OECD. 154p.

Ormerod, S.J., Weatherley, N. S. & Gee, A. S. 1990. Modelling the ecological impact of changing acidity in Welsh streams. In: Acid Waters in Wales, Edited by R. W. Edwards, A. S. Gee & J. H. Stoner. Dordrecht: Kluwer Academic Publishers.

Ormerod, S.J., Allinson, N., Hudson, D. & Taylor, S.J. 1996. The distribution of breeding dippers (cinclus cinclus(L.) Aves) in relation to stream acidity in upland Wales. Freshwater Biology 16, 501–507.

Pennington, W. 1981. Records of a lakes life in time ñ the sediments. Hydrobiologio, 79, 197–219.

Review Group on Acid Rain. 1997. Acid Deposition in the United Kingdom 1992–1994. London; DETR

Reynolds, B., Stevens, P.A., Hughes, S., Pokinson, J.A. & Weatherley, N.S. 1995. Stream chemistry impacts of conifer lovestry in velsh catchments. Water, Air and Soil Pollution 79, 147–170.

Reynolds, C.S. 1997. Vegetation processes in the pelagic: A model for Ecosystem Theory. Ecology Institute, Oldendorf/Luhe, Germany.

Ryding, S.O. & Rast, W. 1989. The control of Eutrophication of Lakes and Reservoirs. United Nations Educational Scientific and Cultural Organisation, Paris.

SFT (1987). One Thousand Lake Survey, Norway 1986. SFT Report 283/87. Norwegian state Pollution Control Authority, Oslo.

Sprent, J.I. 1987. The Ecology of the Nitrogen Cycle. Cambridge University Press, Cambridge.

Stevens, P.A. & Hornung, M. 1988. Nitrate leaching from a felled Sitka spruce plantation in Beddgelert Forest, North Wales. Soil Use & Management 4, 3–9.

Sullivan, T. G., Eilers, J. M., Cosby, B. J. & Vache, K. B. 1997. Increasing role of nitrogen in the acidification of surface waters in the Adirondack Mountains, New York. Water Air Soil Pollution, 95, 313–336.

Sutcliffe, D.W. & Jones, J.G. 1992. Eutrophication: research and application to water supply. Freshwater Biological Association, Ambleside.

UNEP, 1991. United Nations Environment Programme Environmental Data Report 1991/92. Blackwell, Oxford.

United Kingdom Critical Loads Advisory Group 1995. Critical loads of acid deposition for United Kingdom freshwaters. Institute of Terrestrial Ecology, Edinburgh

US Environmental Protection Agency. 1996. Environmental indicators of water quality in the United States. USEPA Rep 841-R-96-002. USEPA, Office of Water (4503F). US Gov. Print. Office, Washington D.C.

12
Toxicity of Nitrite to Freshwater Invertebrates

Beverley H. L. Kelso[1], D. Mark Glass[2] and Roger V. Smith[1]

[1] DEPARTMENT OF AGRICULTURAL AND ENVIRONMENTAL SCIENCE, THE QUEEN'S UNIVERSITY OF BELFAST, NEWFORGE LANE, BELFAST, NORTHERN IRELAND, UK
[2] ABC LABORATORIES (EUROPE) LTD., COLERAINE, NORTHERN IRELAND, UK

1 Introduction

Aquatic invertebrates are important components of river biota, providing the vital link in the food chain between their food sources (*e.g.* algae, bacteria and detritus) and fish, to which they themselves become prey. Due to the nature of the aquatic environment, freshwater organisms are particularly vulnerable to the input of pollutants. Northern Ireland rivers are subjected to seasonal nitrite (NO_2^-) fluctuations, which are linked to large inputs of agriculturally derived nitrogenous (N) substrates (Smith *et al.*, 1995, Smith *et al.*, 1997). Greater than 60% of these rivers exceed the E.E.C. drinking water directive guideline of 0.03 mg NO_2^--N L^{-1} (European Economic Community, 1980) and even larger proportions are above the 0.003 and 0.009 mg NO_2^--N L^{-1} recommended for supporting salmonid and cypinid fisheries, respectively (European Economic Community, 1978). The greater proportion of the NO_2^- originates from sediment-water interface transformations of nitrogenous substrates rather than from direct inputs. The impending increase of sewage and sludge applications to agricultural land with a consequent increase in nitrate (NO_3^-) leaching to streams and the transformation of this to NO_2^- could provide added hazards to aquatic biota.

A large part of the aquatic food chain, from micro-organisms and invertebrates, through to fish are adversely affected by elevated NO_2^- concentrations. Animals are exposed directly to NO_2^- in the aquatic environment, which typically causes higher internal NO_2^- concentrations than terrestrial animals (Jensen, 1996). Considerable research has been performed on the toxicity of NO_2^- to fish and crayfish (Arillo *et al.*, 1984; Bath and Eddy, 1980; Eddy *et al.*, 1983; Gutzmer and Tomasso, 1985; Jensen, 1990b; Margiocco *et al.*, 1983;

Perrone and Meade, 1977; Smith and Williams, 1974; Tahon *et al.*, 1988) mainly for commercial and economic reasons, but no information is available on the effects of NO_2^- on other invertebrates. The toxicity of NO_2^- to fish was originally attributed to the conversion of oxyhaemoglobin, the respiratory pigment in red blood cells, to methaemoglobin, a form that is incapable of transporting oxygen around the body (Lewis and Morris, 1986; Eddy and Williams, 1994), leading to physiological hypoxia (Russo, 1985). However, inconsistencies between the degree of methaemoglobinaemia and mortality (Margiocco *et al.*, 1983; Smith and Williams, 1974) suggest that other mechanisms may be involved (Jensen, 1996).

Nitrite may react with nitrogenous compounds *e.g.* amides and amines that are part of the host tissues, forming carcinogenic *N*-nitroso compounds (Wolff and Wasserman, 1972). Jensen *et al.* (1987) reported an elevation of plasma potassium (K^+) levels in NO_2^- exposed carp, which is thought to have intracellular origins (Jensen, 1990a). This K^+ imbalance may disrupt a number of vital physiological functions (Jensen, 1990b) *e.g.* neurotransmission, skeletal muscle contractions and functions of the heart (Jensen, 1996). Structural damage of the liver may ensue from high NO_2^- concentrations and may disrupt the function of many essential hepatic enzymes (Arillo *et al.*, 1984). Reduction in growth rate has been reported by Arillo *et al.* (1984) and Colt and Armstrong (1981), as too has an increase in the susceptibility of fish to disease (Colt and Armstrong, 1981). Hence, the target tissues for NO_2^- toxicity are likely to be wider than the blood pigment haemoglobin, however important its role.

The NO_2^- toxicity issue is further complicated by the influence of environmental factors, especially chloride (Cl^-), affecting the severity of the problem. Chloride has a protective effect against NO_2^- toxicity in fish, thought to arise from competition between Cl^- and NO_2^- for transportation across the gills (Perrone and Meade, 1977; Hilmy *et al.*, 1987; Tomasso *et al.*, 1979). This effect is perceived to be so great, that studies where Cl^- measurements are not documented, are virtually deemed invalid because inter-laboratory comparisons cannot be made.

Because of their wide array of gaseous transport systems, invertebrates present new challenges for studying the toxicity of NO_2^-. In crustaceans, oxygen is transported simply in solution or bound to respiratory pigments *e.g.* haemoglobin, or in larger species, haemocyanin (Barnes, 1987). In contrast to haemoglobin which contain iron atoms that change oxidation state upon binding and release of oxygen, haemocyanin uses copper as the oxygen binding site (Jensen, 1996). However it is postulated that the haemocyanin is oxidized by NO_2^- in a similar manner to haemoglobin, thus preventing oxygen transport (Colt and Armstrong, 1981). *Daphnia*, a crustacean frequently used in toxicity studies, relies on haemoglobin for its respiratory requirements, with the concentration of haemoglobin present directly related to the amount of oxygen available in the surrounding media. Crustaceans containing haemocyanin are thought to be much more sensitive to the effects of NO_2^- than those lacking this pigment (Wickins, 1982).

The present study was initiated to assess the influence of NO_2^- on freshwater invertebrate communities in Northern Ireland through a combination of field studies and laboratory tests. Multivariate statistical methods were employed to analyze ecological data (both chemical and biotic) from streams to 'define an optimal subset of environmental variables which best explains the biotic structure' (Clarke and Ainsworth, 1993). The object of the toxicity testing was to evaluate, in particular, the involvement of the failure of gaseous transport systems in inducing mortality. Acute toxicity was assessed by determination of a 96-h LC_{50} for a variety of invertebrates, whilst sub-lethal effects of NO_2^- were assessed in a lifecycle study of *Daphnia*.

2 Methods

2.1 Study of Native Invertebrate Populations

2.1.1 Chemical Analyses.   The Upper Bann river system, Northern Ireland has been the subject of intensive chemical and biological monitoring on a routine basis as part of a regional water quality survey (Foy and Kirk, 1985). A total of 25 stations were examined on a bi-weekly basis from April – September, 1990–1995. The samples were collected and analyzed for pH, dissolved oxygen (%DO), biological oxygen demand (BOD), nitrite (NO_2^-), ammonium (NH_4^+), free ammonia (NH_3) and nitrate (NO_3^-) as outlined in Smith *et al.* (1997) and soluble reactive phosphate (SRP), total soluble phosphate (TSP) and total phosphate (TP) as reported by Smith (1977). Ion chromatography was used to measure the concentration of chloride (Cl^-) (Weiss, 1986).

2.1.2 Faunal Sampling.   Using a 5 minute kick sampling technique, macroinvertebrates were sampled on one occasion during each summer from 1990–1995. On return to the laboratory, the samples were preserved in a 5% formalin solution. The invertebrates and other organic debris were separated by washing through a series of graded sieves and the samples identified to family level using a range of published keys.

2.1.3 Statistical Analyses.   The invertebrate communitites at the 25 sites examined ranged from as few as 2 species, to a maximum of 25 species. It was necessary to use PRIMER, a computer multivariate statistical package (Clarke and Warwick, 1994), to investigate the relationships between invertebrate community structure and environmental variables. Invertebrate data, grouped by individual stations was transformed on a presence/ absence basis and a Bray-Curtis similarity coefficient calculated between every permutation of sample pairs using the CLUSTER programme, thus producing a triangular similarity matrix. Another data matrix was prepared for the environmental variable data, following the calculation of a normalised Euclidean distance. The BIOENV procedure was employed to maximize a Spearman rank correlation (p_s) between the biotic and abiotic triangular matrices, thus selecting the environmental variables best explaining community structures (Clarke and Ainsworth, 1993).

2.2 Laboratory Toxicity Tests

2.2.1 Acute Toxicity.   Six 96-hour NO_2^- acute toxicity tests were conducted on aquatic invertebrates using static assays. A combination of standard bioassay guidelines given by the American Public Health Association (1992) were used in conjunction with experimental techniques outlined by Parrish (1985).

Invertebrate genera were chosen on the basis of preliminary investigations into the sensitivity of aquatic invertebrates to NO_2^-. The findings suggested that the responses could be represented by 3 categories: (1) invertebrates present only at sites of low NO_2^- concentrations *e.g. Ephemerella* (mayfly); (2) invertebrates present only at sites of elevated NO_2^- *e.g. Asellus* (hog-louse), *Polycelis* (flatworm); and (3) invertebrates present at most sites, irrespective of NO_2^- levels *e.g. Gammarus* (freshwater shrimp). *Daphnia* (water-flea) and *Hexagenia* (American mayfly) were included in the study to represent standard test organisms for toxicity testing in the United Kingdom and America, respectively. These invertebrates employ a wide array of gaseous transport mechanisms: (1) because of the large surface area of *Polycelis*, simple diffusion is sufficient to meet gaseous requirements; (2) *Hexagenia* and *Ephemerella* use the unique insect tracheal system; (3) the crustaceans *Asellus* and *Gammarus* use the respiratory pigment haemocyanin while *Daphnia* employs haemoglobin (Barnes, 1987).

Asellus, Polycelis, Gammarus and *Ephemerella* were netted from local streams in Northern Ireland. Wild populations of the American mayfly, *Hexagenia,* were obtained from Aquatic Research Organisms Inc, 1 Lafayette Rd, Hampton, New Hampshire, U.S.A. whilst *Daphnia* were taken from in-house cultures. All invertebrates were acclimatised for one week at temperatures similar to that in the environment in which they were collected employing Lower Bann river water (NO_2^- < 0.03 mg N L^{-1}; NH_4^+ < 0.1 mg N L^{-1}; NO_3^- = 0.2 mg N L^{-1}; Cl^- = 31 mg L^{-1}) which was piped directly into the laboratory at Coleraine. Reagent grade sodium nitrite (Sigma, U.K.) diluted in the same water was used to obtain the desired NO_2^- concentrations. Although a limited amount of food was available in this water, specimens were not fed during either the pre-test acclimatisation period or the 96-hour tests.

Ten individuals of the same invertebrate species were added in duplicate to each test concentration contained in two 250 ml test chambers. To maintain a constant temperature, the test chambers were placed in a 18°C water bath. Mortality, based on a failure to respond to mechanical stimulation, was recorded initially at time 0 and then daily for 4 days. Dead invertebrates were delicately removed daily. Nitrite in the water was monitored spectrophotometrically (Rider and Mallon, 1946). The 24, 48, 72 and 96 hour LC_{50} values were derived using the trimmed Spearman-Karber method (Hamilton *et al.*, 1977) available on SAS for Windows, release 6.10 (SAS Institute Inc., 1994).

2.2.2 Chronic Exposure. To provide information on the sublethal effects of NO_2^- on survival, growth and reproductive success of daphnids, a sublethal chronic life-cycle toxicity test was conducted. A proportional diluter system (Mount and Brungs, 1967), in conjunction with a Hamilton Microlab Model 420 dual syringe dispenser delivered concentrations of 0, 2.5, 10, 20 and 40 mg N L^{-1} sodium nitrite, prepared from a concentrated stock solution, diluted with Lower Bann river water. The solutions were received in 5 receptacles, each of which was split into 4 replicate 1 L glass chambers, with notched overflows. The solution in the test chambers was renewed approximately every 5 hours. These test chambers were maintained in a waterbath at 18°C and the system was allowed to reach an equilibrium for 24 hours prior to test commencement.

The test was initiated with the introduction of 5 neonate daphnids (< 24 hours old, from the same brood stock) to each test chamber. The *Daphnia magna* employed in this experiment were obtained from the same source as those used for determination of the LC_{50} acute study and were distributed on a random basis, in accordance with the values obtained from the SAS random program. Each test chamber was fed 2 mls of a culture of the algal species, *Selenastrum capricornutum* Printz (cell density = 10^8 cells ml^{-1}) and 1 ml trout chow / yeast supplement. Water samples for NO_2^- determination were taken immediately prior to feeding. Mortality was observed daily for 15 days. The date of the first young produced in each chamber was recorded. Subsequently three times a week, adult test daphnids were removed by a smooth glass pipette to an aliquot of test solution. The instars were collected by gently pouring the remaining test solution through a fine-mesh screen and rinsing the young into a petri dish. One ml of Lugol's solution (60 g potassium iodide and 40 g iodine in 1.0 L distilled water) was added to immobilize and stain the young which were subsequently counted. Upon termination of the test, the number of young per adult reproduction day (YAD) was calculated as:

$$YAD = \frac{\text{total number of young}}{(\text{number of adults}) \; X \; (\text{number of days from first brood in control})}$$

The length (distance from apex of helmet to base of the posterior carapace spine) to the nearest mm was measured for each first-generation daphnid that remained using a dissecting microscope equipped with an ocular scale. The dry weight of these individuals (dried at 80°C for 24 hours) was also determined.

Employing GENSTAT 5 release 3.1 (Payne, 1993), analysis of variance (ANOVA) with fitted linear and quadratic effects were determined for survival (at day 15), average adult age to first brood release, growth (weight and length) and reproductive data.

Table 1 Summary of results from BIOENV analysis of the Upper Bann river system (1990 to 1995). Combinations of abiotic variables that give the highest rank correlation for a particular year are shown along with the Spearman rank correlation coeffient (ρ_s)

Year	ρ	k	BOD	&DO	Cond	Temp	pH	SRP	TSP	TP	NH_4^+	Free NH_3	No_2^-	NO_3^-
1990	0.493	2	✓		✓	✓								
1991	0.360	4		✓							✓		✓	
1992	0.570	4		✓		✓	✓						✓	
1993	0.622	3		✓	✓	✓						✓		
1994	0.510	2			✓	✓								
1995	0.426	4			✓		✓						✓	✓

3 Results

3.1 Native Invertebrate Communities

As anticipated from previous studies, the presence and/or absence of macro-invertebrate species in the Upper Bann river system was influenced by % DO, conductivity and temperature. However, it was unexpected to find that NO_2^- too, has an important effect on particular members of the macro-invertebrate community (Table 1). The effects of NO_2^- on these populations were evident during years where NO_2^- concentrations were abnormally high, as illustrated in Figure 1 (1991, 1992 and 1995). The mean concentration of NO_2^- in the Upper Bann of 0.065 mg N L^{-1} clearly exceeds the E.C. guidelines for rivers suitable for sustaining fish (European Economic Community, 1978) and is greater than twice the maximum recommended concentration for drinking water (European Economic Community, 1980).

3.2 Laboratory Toxicity Studies

3.2.1 Acute Studies. The relationship between NO_2^- concentration and its potential to produce invertebrate fatalities was examined in 96h LC_{50} toxicity

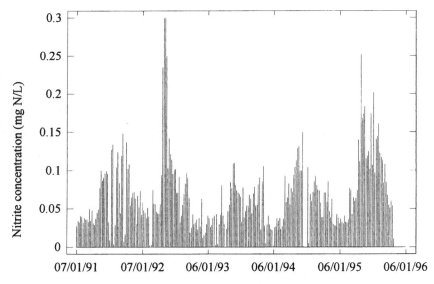

Figure 1 *Concentrations of NO_2^- at the mouth of the Upper Bann river during 1991–1995*

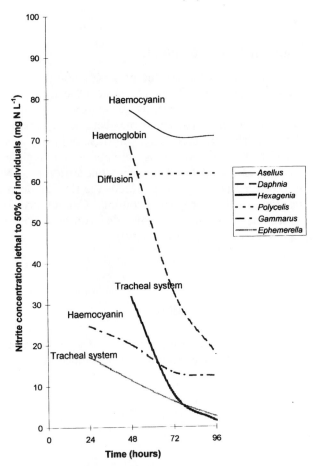

Figure 2   *Toxicity curves for 96-hour tests*

studies. The resulting toxicity curves for each invertebrate species are presented in Figure 2.

The concentration of NO_2^- tolerated by most invertebrate species gradually decreased with exposure time, with mortality in the controls remaining well below the critical limit of 10% (American Public Health Association, 1992). Employing a t-test, the results from the present study were in close agreement ($p = 0.81$) with those of an earlier study by Glass (1994) (Figure 3).

The most sensitive invertebrate species were *Ephemerella* and *Hexagenia*, both Ephemeroptera (may-flies), having 96h LC_{50}'s of 2.5 mg NO_2^--N L^{-1} and 1.4 mg NO_2^--N L^{-1}, respectively. To attempt to draw generalisations concerning the pattern of NO_2^- induced mortality in haemocyanin pigmented invertebrates is not straight forward: *Gammarus* was shown to be the next most sensitive species, with a 96h LC_{50} of 12.3 mg NO_2^--N L^{-1} while *Asellus* was the most tolerant, with mortality only recorded in the 100 mg NO_2^--N L^{-1} treatment. Initially *Daphnia,* a crustacean containing haemoglobin,

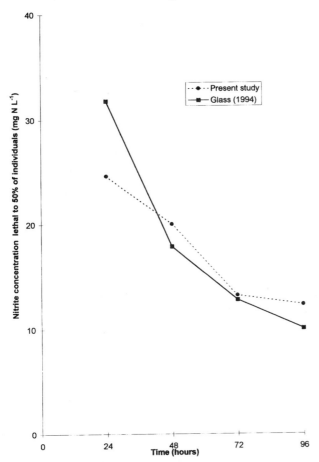

Figure 3 *Comparison of 96h LC$_{50}$ for* Gammarus *employed in the present study with Glass (1994)*

demonstrated resistance to NO_2^-. However, after 48 hours, once the initial fatalitites were detected, a sharp decline in the LC_{50} was apparent. *Polycelis*, a platyhelminth, was insensitive to the NO_2^- concentrations imposed in this study, with the LC_{50} remaining constant at 61.6 mg NO_2^--N L^{-1}.

3.2.2 Chronic Study. By day 3, 100% daphnid mortality was achieved in the 40 mg N L^{-1} treatments but fatalities in other treatments were not significantly different during the 15 day study period. Although NO_2^- induced mortality was not evidenced in the 0–20 mg N L^{-1} treatments, sub-lethal effects were pronounced.

The length and dry body weight of daphnids, both parameters indicative of growth, were significantly reduced in a linear fashion on exposure to increasing NO_2^- concentrations (Equations 1 and 2). The reproductive output also showed a negative linear relationship with NO_2^- concentration (Equation 3).

$$\text{Length of adult daphnid} = 4.253 - (0.0023 X NO_2^-)$$
$$r^2 = 0.769; \quad p < 0.001 \tag{1}$$

$$\text{Dry weight of adult} = 0.804 - (0.0211 X NO_2^-)$$
$$r^2 = 0.971; \quad p < 0.001 \tag{2}$$

$$\text{Number of young per adult per reproductive day} = 7.658 - (0.224 X NO_2^-)$$
$$r^2 = 0.821; \quad p = 0.01 \tag{3}$$

where length is measured in mm, weight in mg and NO_2^- in mg N L^{-1}. The haemoglobin content of daphnids exposed to NO_2^- concentrations of 10 and 20 mg N L^{-1} was greatly increased, as manifested by a dark red body coloration. Unfortunately there is no available test to quantify the haemogobin in such small animals.

4 Discussion

The structure of invertebrate communities in Northern Ireland rivers are shaped by the NO_2^- concentrations of the aquatic environment in which they reside (Table 1). Although the fauna present in rivers are more commonly exposed to relatively low concentrations of NO_2^-, higher concentrations are needed to induce mortality on a short-term basis (Figures 1 and 2). The 96h LC_{50} (an estimate of the short term NO_2^- tolerance of organisms) of *Daphnia*, *Ephemerella*, *Gammarus* and *Hexagenia* were comparable to those reported for *Oncorhynchus mykiss* (rainbow trout) under similar environmental conditions (Lewis and Morris, 1986). *Polycelis* and *Asellus* were more resistant to the effects of NO_2^- pollution (approximately 50 times greater than *Hexagenia*). This is not unexpected given that they thrive in polluted sections of streams (Macan, 1962).

Protection against NO_2^- toxicity in fish is afforded by Cl^- ions present in the ambient water. It has been proposed that the Cl^- competes with NO_2^- for active transport across the gill cells and into the body cavity (Perrone and Meade, 1977; Tomasso *et al.*, 1979). The Ephemeroptera (*Hexagenia* and *Ephemerella*) do not rely on Cl^- for osmoregulation, instead organic molecules *e.g.* free amino acids are utilized (Barnes, 1987). It is tempting to speculate that the high susceptibility of Ephemeroptera to NO_2^- induced mortality may be due to the lack of a Cl^-/NO_2^- competition system to reduce the entrance of NO_2^- into the body.

The Cl^-/NO_2^- competition system is believed to be operational in *Gammarus*. Glass (1994) demonstrated that a significant protective effect was evident in individuals exposed to elevated NO_2^- concentrations in water with an artificially supplemented Cl^- concentration of 70 mg L^{-1}. Although the

Cl^- concentration used in the present study was only 30 mg L^{-1}, a t-test showed the results were not significantly different ($p = 0.81$) from Glass (1994) (Figure 3). This is in general agreement with previous studies where Cl^- concentrations of even 1 mM (37.5 mg L^{-1}) provide protection against all but abnormally high NO_2^- concentrations (Bath and Eddy, 1980; Eddy et al., 1983; Gutzmer and Tomasso, 1985; Glass, 1994). In Northern Ireland, the major rivers associated with the Lough Neagh catchment have a mean Cl^- concentration of 19.5 mg L^{-1}. According to the European Inland Fisheries Advisory Commission report (1984) in order to support salmonid fisheries, NO_2^- concentrations should not exceed 0.12 mg N L^{-1} in rivers with a Cl^- content of 20 mg L^{-1}. Unfortunately in the Upper Bann river this limit is frequently exceeded in the summer months (Figure 1) so that it is likely that the biota are under stress in this river system.

Although Cl^- provides some protection against the transport of NO_2^- into aquatic organisms, it is not 100% successful and some does penetrate. The toxicity of NO_2^- in fish and crustaceans is believed to act through the oxidation of the respiratory pigments (forming methaemoglobin from haemoglobin or methaemocyanin from haemocyanin), which are then incapable of oxygen transport (Lewis and Morris, 1986; Eddy and Williams, 1994). However, there was no evidence in the present study to suggest that invertebrates containing these pigments were more susceptible to NO_2^- than those obtaining oxygen through other mechanisms. Considering methaemocyanin formation is unimportant under physiological conditions (Tahon et al., 1988) one is led to speculate that the production of methaemocyanin does not contribute significantly to mortality. Supporting this argument is the observation that the NO_2^- tolerance of haemocyanin pigmented crustaceans tested in the present study were vastly different (Figure 2). Evidence for methaemoglobin formation was observed in the chronic *Daphnia* study where the body of daphnids exposed to NO_2^- concentrations $\geq$ 10 mg N L^{-1} were a deep red colour, indicating an increase in the volume of haemoglobin as a manifestation of oxgyen related stress. Further study is needed to resolve if methaemoglobinaemia in *Daphnia* is the fatal mode of action of NO_2^- or whether other factors work in assocation with this mechanism.

Sub-lethal exposure of *Daphnia* to NO_2^- has other toxic effects on the organism. Regression analysis showed that there is a significant ($p < 0.001$) linear supression of growth of individuals even in a 15 day period (length and weight measurements, Equations 1 and 2) and the sustainability of daphnid populations is affected, with a significant reduction in brood size ($p = 0.01$) accompanying increasing concentrations (Equation 3). Over a certain time period, these effects in themselves will alter the population structure, but with so much variation in physiology between invertebrates it is difficult to extrapolate these results to other species.

The 96h LC_{50} toxicity tests are valuable in making predictions regarding the toxicity of a chemical, yet insufficient to determine an absolute lethal concentration for invertebrates. The United States Environmental Protection Agency (1973) recommends a conversion factor of 0.05 be applied to convert the 96h

LC_{50} to concentrations that are safe to freshwater organisms. This would translate the lowest 96h LC_{50} observed in the present study to a concentration of 0.071 mg N L^{-1}, a value which is constantly exceeded in Northern Ireland rivers during the summer period (Figure 1). On the other hand the Cl^- concentrations present in these rivers are sufficient to offer most biotic species protection against NO_2^- concentrations up to 0.12 mg N L^{-1} (European Inland Fisheries Advisory Commission, 1984). Aquatic insects which make up an average of 38% of faunal species (compared to fish which constitute 13%) of unpolluted streams (Roback, 1974), do not have this protection, hence regulatory bodies will need to consider their sensitivity as a separate issue. Because the standard 96h toxicity test cannot evaluate sub-lethal or cumulative effects, it must not be forgotten that these effects are also detrimental to the sustainability of the invertebrate population structure. As macro-invertebrates occupy such a key position in the aquatic food chain, it is important that more research on NO_2^- toxicity is performed to aid their conservation.

5   Conclusion

The health hazards ascribed to nitrate which have been the focus of EU water quality legislation occur as a result of the metabolic conversion of nitrate to nitrite. Although the effects of NO_2^- toxicity on freshwater fish is widely documented, no research has been instigated on their food supply, *e.g.* the aquatic invertebrates. Results from a multivariate BIOENV programme which relates invertebrate assemblages to specific environmental parameters suggest that NO_2^- influences invertebrate community structure in Northern Ireland rivers.

The major toxic effect of NO_2^- on fish is reported to be hypoxia, caused by reducing the oxygen binding affinity of the respiratory pigment, haemoglobin. There is some speculation that the same mechanism occurs with the crustacean respiratory pigment, haemocyanin. In the present study aquatic invertebrates which employ a variety of different gaseous exchange mechanisms were subjected to toxicity testing (96hr LC_{50}). There was no evidence to suggest that crustaceans with haemocyanin *e.g. Asellus, Gammarus* or haemoglobin *e.g. Daphnia* were more susceptible to NO_2^- than insects *e.g. Ephemerella, Hexagenia* which obtain oxygen through the tracheal system or platyhelminthes *e.g. Polycelis* whose respiratory requirements are met by gaseous diffusion from the body surface. Sub-lethal effects of NO_2^- *e.g.* reduction in body weight and reproductive potential, accompanied by an increase in haemoglobin content, were observed in *Daphnia* exposed to NO_2^- for 15 days.

The primary effects of NO_2^- on invertebrate populations may not be related to oxidation of the respiratory pigment as in fish, but instead through other mechanisms. As invertebrates occupy a key position in the food chain of freshwater ecosystems, it is vital that more detailed information is assembled to assess the long term effects of NO_2^- on these communities.

Acknowledgements

We are very grateful to the staff of the freshwater laboratory of the Agricultural and Environmental Science Division for assistance with chemical analysis of the water samples. This work was supported in part by a studentship to B.H.L.K. from the Department of Agriculture for Northern Ireland.

References

American Public Health Association. In *'Standard methods for the examination of water and wastewater'*, A. E. Greensbury, L. S. Clesceri and A. D. Eaton (Eds.), American Public Health Association, New York, 1992, 18th edition, p. 8.1–8.82.

Arillo A., E. Gaino, C. Margiocco, P. Mensi and G. Schenone, *Environ. Res.*, 1984, **34**, 135–154.

Barnes R.D. *'Invertebrate Zoology'*, Saunders College Publishing, Philadelphia, 1987, 5th edition.

Bath R. N. and F. B. Eddy, *J. Exp. Zool.*, 1980, **214**, 119–121.

Clarke K. R. and M. Ainsworth, *Mar. Ecol. Prog. Ser.*, 1993, **92**, 205–219.

Clarke K. R. and R. M. Warwick. *'Change in Marine Communities: an Approach to Statistical Analysis and Interpretation'*, Natural Environment Research Council, UK, 1994.

Colt J. E. and D. A. Armstrong. In *'Proceedings of the Bio-Engineering Symposium for Fish Culture'*, L. J. Allen and E. C. Kinney (Eds.), Fish Culture Section of the American Fisheries Society, Washington D.C., 1981, p. 34–47.

Eddy F. B. and E. M. Williams. In *'Water Quality for Freshwater Fish'*, G. Howells (Ed.), Gordon and Beach Science Publishers, Switzerland, 1994, p. 117–143.

Eddy F. B., P. A. Kunzlik and R. N. Bath, *J. Fish Biol.*, 1983, **23**, 105–116.

Environmental Protection Agency, *Ecol. Res. Ser. EPA.R.73.033, 1973*.

European Economic Community, *Off. J. Eur. Comm.*, 1978, **L222**, 34–54.

European Economic Community, *Off. J. Eur. Comm.*, 1980, **L229**, 11–29.

European Inland Fisheries Advisory Commission, *EIFAC Technical Paper*, 1984, **46**, 1–19.

Foy R. H. and M. Kirk, *J.CIWEM*, 1985, **9**, 247–256.

Glass D. M., *Nitrite toxicity and its effects on biological monitoring data in the Upper Bann catchment, Northern Ireland*. M.Sc. Thesis, Kings College, University of London, London, 1994.

Gutzmer M. P. and J. R. Tomasso, *Bull. Environ. Contam. Toxicol.*, 1985, **34**, 369–376.

Hamilton M. A., R. Russo and R. V. Thurston, *Environ. Sci.Tech.*, 1977, **11**, 714–718.

Hilmy A. M., N. A. El-Domiaty and K. Wershana, *Water, Air and Soil Pollut.* 1987, **33**, 57–63.

Jensen F. B., *J. Exp. Biol.*, 1990a, **152**, 149–166.

Jensen F. B., *Aquat. Toxicol.*, 1990b, **18**, 51–60.

Jensen F. B. In *'Toxicology of Aquatic Pollution – physiological, cellular and molecular approaches'*, E. W. Taylor (Ed.), Cambridge University Press, Cambridge, 1996, p. 169–186.

Jensen F. B., N. A. Anderson and N. Heisler, *J. Comp. Phys.*, 1987, **157B**, 533–541.

Lewis W. M. and D. P. Morris, *Trans. Am. Fish. Soc.*, 1986, **115**, 183–195.

Macan T. T., *Schweiz. Z. Hydrol.*, 1962, **24**, 386–407.

Margiocco C., A. Arillo, P. Mensi and G. Schenone, *Aquat. Toxicol.*, 1983, **3**, 261–270.

Mount D. I. and W. A. Brungs, *Wat. Res.*, 1967, **1**, 21–29.

Parrish P. R. In *'Fundamentals of Aquatic Toxicology'*, G. M. Rand and S. R. Petrocelli (Eds.), Taylor and Francis, Bristol, 1985, p. 31–57.

Payne, R. W. *'Genstat 5 Release 3 Reference Manual'*, Oxford Science Publications, Clarendon Press, Oxford, 1993.

Perrone S. J. and T. L. Meade, *J. Fish. Res. Board Can.*, 1977, **34**, 486–492.

Rider B. F. and M. G. Mallon, *Ind. Eng. Chem. (Anal)*, 1946, **18**, 96–99.

Roback S. S. In *'Pollution Ecology of Freshwater Invertebrates'*, C. W. Hart and S. L. H. Fuller (Eds.), Academic Press, New York, 1974, p. 313–376.

Russo R. C. In *'Fundamentals of Aquatic Toxicology'*, G. M. Rand and S. R. Petrocelli (Eds.), Taylor and Francis, Bristol, 1985, p. 455–471.

SAS Institute Inc. *'The SAS system for Windows, Release 6.1'*, Cary, USA, 1994.

Smith C. E. and W. G. Williams, *Trans. Am. Fish. Soc.*, 1974, **2**, 389–390.

Smith R. V., *Wat. Res.*, 1977, **11**, 453–459.

Smith R. V., R. H. Foy, S. D. Lennox, C. Jordan, L. C. Burns, J. C. Cooper and R. J. Stevens, *J. Environ. Qual.*, 1995, **24**, 952–959.

Smith R. V., L. C. Burns, R. M. Doyle, S. D. Lennox, B. H. L. Kelso, R. H. Foy and R. J. Stevens, *J. Environ. Qual.*, 1997, **26**, 1049–1055.

Tahon J. P., D. van Hoof, C. Vinckier, R. Witters, M. de Ley and R. Lantie, *Biochem. J.*, 1988, **249**, 891–896.

Tomasso J. R., B. A. Simco and K. B. Davis, *J. Fish. Res. Board. Can.*, 1979, **36**, 1141–1144.

Weiss J. In *'Handbook of ion chromatography'*, E. L. Johnson (Ed.), Dionex Corp., California, 1986.

Wickins W. F. In *'Recent advances in aquaculture'*, J. F. Muir and R. J. Roberts (Eds.), Croom Helm, London, 1982, p. 87–178.

Wolff I. A. and A. E. Wasserman, *Science*, 1972, **177**, 15–19.

13

Sensitive Areas, Vulnerable Zones and Buffer Strips: a Critical Review of Policy in Agricultural Nitrate Control

Hadrian F. Cook

ENVIRONMENT DEPARTMENT, WYE COLLEGE, UNIVERSITY OF LONDON, ASHFORD, KENT TN25 5AH, UK

Abstract

The issue of nitrate loading in waters affects both aquatic ecosystems and drinking water destined for human consumption. To combat the problem there exists a range of agri-environmental schemes designed to reduce nitrate concentrations in surface water bodies and groundwaters. Success will depend largely on the measures adopted, in terms of protection zone designation and controls over arable land management or changes in land use.

Water quality standards are supra-nationally defined, having their origins in European Union Directives which are translated into legislation by the UK as an individual Member State. The drive for improvements to water quality is co-incident with current reviews of the Common Agricultural Policy, giving an opportunity for the review of current practices.

While there has been great progress in terms of protection zone implementation since the Water Act 1989, it is argued that measures should increasingly incorporate compulsory protection zone designation for groundwaters, targeted buffer strips to protect surface water bodies, take account of contamination from non-agricultural sources, increase grassland or woodland areas and maximise the use of environmental survey data and hydrological modelling procedures.

Areas of uncertainty remain, including discussions regarding the extent of resource protection to be sought while the imposed legal requirements demand that decisive action be taken. Protection measures adopted should therefore be as scientifically credible as possible, enhance 'sustainable development' of water resources, and be defensible in the planning process.

All these considerations make the setting of confidence limits extremely difficult when planning for the protection of waters. Against a background of uncertainty, it is a reappraisal of protection zone designation which is the subject of this paper.

1   Introduction: Some Issues

The basis for the problem of nitrates found in excess concentration in waters is an increase in the farmed area linked with intensification of agricultural inputs. Within these parameters, long-term use of fertilisers increased until the mid-1980s, and then levelled-off (House of Lords, 1989). In a recent review published by the Royal Society of Chemistry, Addiscott (1996) ends not with a scientific statement of the problem, but finds the origins of the problem are 'sufficiently political to require a political solution'. He calls for a switch from supporting production, to supporting environmental goods, the proper use of scientific information and the need for long-term strategic planning. This paper aims to discuss these issues.

A modest switch of UK Government funds away from agricultural support to supporting Agri-Environmental Programmes has already occurred (*e.g.* Cook and Norman, 1996). In future, there is likely to be a 'prominent role for agri-environmental instruments to support sustainable development of rural areas' (European Commission, 1997). The complexity of the problem involves decoupling production goals from environmental benefits, itself a massive social and economic issue incorporating notions of food security, the future of certain rural communities, market intervention and reforms of the Common Agricultural Policy (CAP). The latter are aimed to improve the competitiveness of EU agriculture through improvement of market balances and moving towards world market prices. This paper will, however, concentrate upon issues of scientific and technical protection of waters, and discuss certain longer-term planning issues.

First some complexities require discussion.

'The environment' can be seen as a process between humans and the natural world which is complicated, hence *sustainable development* may be a concept whose time has come, yet which defies accurate description (Redclift, 1994). The paradigm of sustainable development arises from the Rio Summit of 1992, however, a conference in The Hague during November 1991 recognised a need for the management and protection of groundwater on a sustainable basis (NRA, 1992a). Ideally, systems thus managed should embrace notions of inter-generational equity, and be able to withstand environmental shocks without suffering long-term damage.

For water, we are able to set limits supposed not to exceed the 'carrying capacity' of the system (Gardiner, 1994) and this may prove adequate for the present in seeking prescriptions for sustainable development. Carrying capacity is readily appreciated in terms of defining river flows, either in setting Minimum Acceptable Flows, or better, by determining a seasonally defined minimum flow regime which better recognises the natural variability. In this context, limits may be set for concentrations of nitrogen compounds, and Water Quality Objectives may be met, which reflect the optimum (permissable) loading of pollutants in the system and the use to which a water body is put. Bulk uses of water should also be sustainable; that is long-term abstraction should not exceed the ability of a surface water catchment or aquifer system to function in defined ways.

We may look back on more than a century of hard hydrological, soils and hydrogeological information, be it in spatial (*i.e.* map) form, or as time series of surface flows, groundwater levels and hydrochemical data (Cook, *in press*). It defies logic if maximum use is not made of *available environmental information*. Since the 1970s this information is greatly enhanced by the widespread adoption of process-based numerical simulation models of surface and groundwater flows and contaminant behaviour.

No predictions, be they of geographical vulnerability to contamination, or of contaminant behaviour, are without *uncertainty*. Aquifer protection measures are a classic illustration. For England and Wales, Skinner and Foster (1995) propose zones be defined within an 'uncertainty framework'. These are a 'best estimate catchment area' which satisfies groundwater balance considerations and published for public policy considerations, a 'zone of confidence' arising from the overlap of all plausible combinations, and a 'zone of uncertainty' comprising the outer envelope formed by all plausible combinations. In the UK, mathematically derived statements about uncertainty are not yet forthcoming.

Issues of agro-chemical contamination become ensnared in *goal conflicts*. There are calls for land diverted from agricultural use, with the aim of reducing production, to be targeted in order to maximise environmental benefits, specifically landscape, soil, water and ecological conservation. It is argued that this is the most economically beneficial approach in terms of value for (public) money and environmental gain (Potter *et al*, 1993). Targeting set-aside land for conservation of soil and waters have come to little. The area of compulsory enrolment is now set to fall to five percent, with a possible longer-term aim of reduction to zero (European Commission, 1997). It is also conceivable that goal conflicts arise with respect to water use, be it for natural ecosystems, human consumption or other abstraction purpose.

Changing objectives also present problems, although these need not be so serious as goal conflicts. For example, it is clear that effective groundwater *source protection* measures are in place and probably enforceable (Cook and Norman, 1996). These, given time, should protect groundwater supplies from a range of potential contaminants, but a cursory examination of the literature suggests that some may prefer wider-ranging aquifer *resource protection*, probably of entire unconfined aquifer systems (Adams and Foster, 1992). Furthermore, the presence of whole-aquifer vulnerability maps (*e.g.* NRA, 1994a) designed for general-purpose aquifer protection to be incorporated in the planning process, suggests that somewhere there may be a hard-core of hydrogeological zealots set on minimising contaminant loading on the saturated zone at any cost. For surface waters, catchment-wide protection measures may be appropriate for resource protection.

There are *differing approaches* in seeking solutions to any problem. For example, there could be widespread restrictions on the use of nitrogen fertilisers and the impact upon the farming economies of large areas of central and eastern England might be profound (House of Lords, 1989). A figure of four million hectares has been mentioned. A dichotomy emerges, on the one hand there are those who would wish to emphasise a de-intensification of arable (and grassland) production practices, incorporating less agrochemical use. For example, there may be improvements for water quality if an 'Arable Incentive Scheme' were adopted by MAFF (following EU Agri-Environmental Regulation 2078/92) in order to increase biodiversity through changing arable management prescriptions (English Nature *et al,* 1997). On the other hand, others argue that targeting vulnerable land areas as water protection zones and buffer strips is more efficient, yet it must be acknowledged there is an ever-present threat of compensatory intensification elsewhere.

With competing demands on the water environment and EU agri-environmental considerations, a strictly Positivist, one-way approach of problem identification, development of research based solutions and policy implementation is inadequate. Faced with a multi-faceted problem requiring a pluralist approach (Pretty, 1995 ch.1), we must decide what is economically acceptable in terms of agricultural yield loss, social cost to the farming community, how much land to divert out of agriculture and how much water contamination is to be tolerated. The nitrate issue is driven by many goals.

2   Protection from Nitrates: the Historical Dimension

Much environmental loading of nitrates has occurred since the Second World War. It is attributed to the ploughing of grassland, and to the indirect effect of fertiliser application to intensive arable; the proportion leached directly from fertiliser is normally small (Addiscott, 1996). The overall result is noticeable in both overall surface and groundwater nitrate concentrations (NRA, 1992b). However, the Nitrate Coordination Group noted upward trends in concentration in surface waters from the 1920s to the 1970s, with a tendency to stabilise in the 1980s (House of Lords, 1989).

The oft quoted cliche regarding groundwater protection is that 'out of sight' has also meant 'out of mind' (NRA, 1992b). Cook (*in press*) traces the institutional, legal and scientific perceptions of groundwater development in England, noting there has been a serious lag of protection from both contamination and over-abstraction behind the actual problems which emerged. This sad state of affairs contrasts historically with surface water protection measures which were credibly understood before the twentieth century.

A recent case, that of *Cambridge Water Co vs Eastern Counties leather plc (1994)*, concerned groundwater contamination with industrial solvents. Here evidently, out of sight had meant out of mind, and the outcome effectively upheld earlier judgements. Although it was found possible to establish strict liability for the escape of the contaminants, the judgement imposed a foreseeability requirement as a pre-requisite to liability (Leeson, 1995, p.160). Consequently, it is likely to avoid a rush of retrospective claims against those held responsible for groundwater contamination.

The control of nitrate is primarily seen as a problem of diffuse pollution in catchments. In this respect, measures taken, excepting sewage discharges or accidental discharges of agricultural origin, have to be catchment-wide, or at least identify areas of potential contaminant loading. It would furthermore be extremely difficult to identify those responsible in any legal sense.

In the UK, land-use measures to protect surface waters date from the late nineteenth century, when the Birmingham Corporation Water Act 1892 enabled the City to acquire commons in the catchment of its upland supply in Wales (Sheail, 1982). Preventative measures for the protection of groundwaters have long been sought, but often this had to involve land ownership in the catchment to be effective. Essentially, the practice of the protection of ground-waters from agriculture dates from the inter-war years, when the Brighton Corporation purchased land to be managed under grazing in order to protect the town's water supply from speculative urban development (Sheail, 1992). Earlier, the Margate Corporation Act 1902 had permitted measures for the protection of groundwater supplies from sewage.

In the post-War era, land-use measures employed on local authority owned land around the 'Cornish' source operated by the Eastbourne Waterworks provided encouragement. From a concentration of 71 mg.l^{-1} nitrate (expressed as NO_3^-), concentrations were above the limit of 50 mg.l^{-1} NO_3^- set for the EU Drinking Water Directive. Fertiliser restrictions around the source, from 1952, caused a halving of groundwater nitrate concentrations during the 1970s. The EU limit is also used as a benchmark for surface water abstraction (DoE, 1986; Headworth, 1994).

Nitrates in surface waters are a contributor to eutrophication alongside P, and are especially important in lowland river and lake systems, a notable example being the Norfolk Broads. Here nitrate contamination arises from sources such as artificial fertilisers, silage liquor, and sewage. Combined with P from sewage effluent (including that derived from detergents) there has been a long-term deterioration in water quality and aquatic ecosystems, notable the demise of many species of emergent macrophytes (BA, 1993). The solution to nitrate pollution sought is to define Broadland an Environmentally Sensitive Area. The Waveney catchment furthermore contains a Nitrate Vulnerable Zone. Fertiliser use on agricultural land is thereby restricted, and hence regulation by Statutory Code for the storage and disposal of agricultural organic waste.

3   Operation of Buffer Strips

Fertiliser-free buffer strips set parallel to water courses might be considered 'good practice' for a range of reasons.

Haycock and Burt (1993) found strips were able to attenuate nitrate in subsoil groundwater beneath grassland, over distances as small as 8 m. Research at Wye College, University of London, has shown the efficacy of, narrower (5-m) grass strips acting as unfertilised, grass buffers between fertilised arable land, and water courses. The sites were on reclaimed alluvial marshland on the Walland Marsh Site of Special Scientific Interest (SSSI). During the period November to April, drainage in these soils is predominantly from the field to the dykes. This includes the winter 'drainage period', and defines the period of maximum transport of agrochemical pollutants to watercourses.

Under a strip, peak nitrate concentrations in shallow groundwaters beneath arable land were reduced to between 3 and 6 percent beneath the edge of a 5-m wide strip bordering a watercourse; here maximum concentrations seldom exceed 3 mg NO_3^-/l (Moorby and Cook, 1992). Concentrations entering the marshland dykes were therefore small. Where a strip was absent the concentrations entering the dyke waters were appreciable greater. Soils were of the Newchurch-Walland complex, typically textures are silty clays and silty clay loams between 0 and 1.0 m, but with irregular sandy layers and heavily gleyed 'unripe', fluid subsoils below about 1.5 m depth.

White *et al* (*in press*) report a similar investigation for soils of the Agney and Guildford series, predominantly loamy with well developed intrusions of sandy subsoil horizons which would be expected to permit greater horizontal transport of shallow groundwaters. Groundwater concentrations in the saturated zone were broadly similar; however detailed hydrogeological measurements also permitted loadings of nitrate to be calculated. For a site with no 5-m strip, only a narrow grass bank bordering the dyke, around 1.32 Kg N/100m of dyke was lost in 128 days of the 'drainage period' 1994/5. This represented a small proportion of the nitrogen applied annually to the winter wheat crop, but was considered sufficient to create water quality problems in the receiving waters. There, peak concentrations all exceeded the EU limit, although water drained from a far wider catchment area than that monitored at the field site.

Where a 5-m strip was present, loading during the drainage season was less than seven percent that at the unbuffered site. Mechanisms of nitrate reduction beneath a strip are considered to be uptake by the root system during warmer spells in the winter, autumn and early spring; anaerobic, microbial denitrification in the saturated zone, immobilization by the soil biomass coupled with regular cutting and removal of grass to minimise the build-up of soil biomass.

It is probable that, where de-nitrification and uptake operate on shallow groundwaters where horizontal flow to dykes predominates, strip emplacement is beneficial in reducing nitrate loading to surface waters. In other hydrogeological circumstances, such as where deep percolation occurs over aquifer systems which do not permit anaerobic denitrification, or in the presence of underdrainage, the results may not be so impressive because the strip may be effectively 'underpassed'.

However, in strip emplacement, other benefits can accrue. These include the attenuation of different contaminants in subsoil saturated flow, a reduction of pesticide spray drift, avoidance of soil erosion through the trapping of (coarser) sediment in runoff and the cessation of ploughing at a distance from the edge, preventing soil-borne pollutant loading through bank failure (*e.g.*

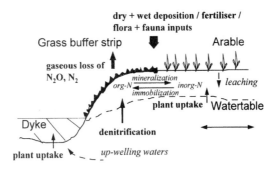

Figure 1 *Possible nitrogen pathways beneath a fertiliser-free strip at the edge of an arable field (Source: White et al, 1998)*

Muscutt *et al*, 1993). There is improved access to the field around the margin and benefits to wildlife will result from the provision of a vegetated barrier between water and crop; this may furthermore act as a corridor for species migration.

Strips achieve a reduction in acreage under intensive management proportional to the area under the buffer. Ideally, outside areas designated SSSI, this might be incorporated in acreage specified as Set-Aside. Provided there is no corresponding intensification in areas not under buffer, agro-chemical loading on the local environment would in any case be reduced together with a fall in production.

4 Protection Zone Definition: Technical Considerations

Since the adoption of widespread numerical modelling in the 1970s to surface and groundwater and contaminant flow, and Geographical Information System (GIS) technology in the 1980s, there has arisen the possibility of designating protection zones on the basis of computer modelling procedures. These may be either process based models (which model rates of water and contaminant flow) or the collation of overlay – based 'spatial models' using data derived from soil, geological, land-use, climate and other spatially derived datasets (*e.g.* Cook and Norman, 1996). Spatial models aim to predict vulnerable locations where, for example, combinations of shallow, well drained soil, shallow groundwater depth, low rainfall (causing low dilution in groundwaters) and aquifer fissuring make the waters contained within aquifers especially vulnerable to groundwater contamination from nitrates. This allows for the 'targeting' of areas where certain land uses might be avoided (Cook, 1991).

As modelling procedures improved, fixed-radii designations around sources were abandoned in favour of modelled transport times. The hydrogeological criteria of aquifer characteristics (storativity and transmissivity), pressure head distributions, recharge rates and abstraction rates used to calculate travel times have come to dominate zone designation to the exclusion of a range of other factors, notably those connected with the soil and unsaturated zone transport of contaminants.

Designation of groundwater Nitrate Vulnerable zones (NVZs), Nitrate Sensitive areas (NSAs) and general purpose groundwater Source Protection Zones (SPZs), following the groundwater Directive, are established upon travel times of water towards a source (NRA, 1992a). These are adopted from continental practice (Adams and Foster, 1992) and the NRA (now Environment Agency) adopted travel times of 50 days (Zone I), 50 to 400 days (Zone II) and that required to support abstraction at the source in the long term.

Ideally, in constructing a spatial model from overlays of environmental models all likely factors should be taken into consideration for the contaminant in question. Figure 2 shows such a scheme. Although the groundwater

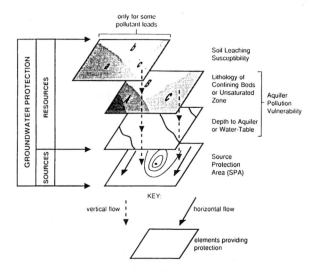

Figure 2   *Overlay techniques for identifying areas vulnerable to groundwater contamination (Source: Adams and Foster, 1992)*

vulnerability maps produced for whole-aquifer protection do incorporate soil and aquifer vulnerability on a matrix basis and ascribe a weighting (Skinner and Foster, 1995), the approach is not used in UK *source protection*. Areas designated for the latter are limited to zonation based upon saturated zone travel times. The main objection is the unknown weighting to be applied to such spatial variables as soil leaching vulnerability, aquifer fissuring and pollutant attenuation in the unsaturated zone. It is argued that this policy needs re-consideration.

The appropriateness of particular approaches to given situations (based upon local conditions and data availability) do require flexibility. For example, the designation of Pesticide Management Zones in the United States can take either an indexed overlay approach, or a dynamic modelling approach depending upon local conditions and data availability (NRC, 1993). Problems of ascribing weighting (*i.e.* attributing relative importance) of overlays is an acknowledged problem, yet decades of research in soil science has shown the importance of variable such as depth, texture and organic carbon content in defining topsoil liability to nitrate leaching.

An approach which combines saturated zone travel times with overlay procedures combining soil leaching liability, degree of aquifer confining, fissuring, depth to water table and so on has been published by Osborn and Cook (1997). The results are shown in Figure 3.

It can be seen that many areas deemed vulnerable to nitrate contamination (including considerable agricultural land) using a comprehensive spatial model, fall outside the designated area, and hence are not subject to enforcement of 'Good Agricultural Practice' (GAP).

Here, on the Isle of Thanet, arable and horticultural land use and development linked to contamination by sewage have led to problems of source contamination identified for over a century (Headworth, 1994). Indeed, the coastal areas of Thanet are heavily urbanised. However policy should evolve towards tighter targeting and land-use prescriptions incorporating all that is known from technical and historic information, and should consider the vulnerability to pollution from non-agricultural sources.

5 Protection Zone Designation: Policy Considerations

Schemes employed for the protection of waters from nitrate contamination are the product of differing goals (*e.g.* protection of natural waters, their habitats and water for abstraction) and are operated by several public bodies. Watson *et al.* (1996) find some progress regarding inter-agency co-ordination, and propose that further improvement could be achieved by developing a mandatory policy referral system involving the DoE, MAFF, EA, local government, water undertakings and farming interests. Catchment management planning in the UK would seem a vehicle of integration through the Environment Agency (EA), yet there no little effective control over rural land use, in stark contrast with urban areas (*e.g.* Cook, 1993).

Emplacement of fertiliser-free cut grass buffer strips, a practical way of reducing nitrate loading from shallow watertables in surface waters has been a policy instrument of English Nature since the mid-1980s under SSSI agreements, and looks set to expand (Cook *et al*, 1997). It is also desirable that 'Water Fringe Areas' of the Habitat Scheme (MAFF, 1994a) allows for the withdrawal from agricultural production, strips of land with average width 10 to 30 m 'adjacent to a designated watercourse or lake'; both arable and permanent pasture are eligible and enrolment is for 20 years. Countryside Stewardship, instigated by the Countryside Commission (CC, 1994), has since passed to MAFF, and has similar provisions for riparian land.

In Scotland, 'Flexible Set-Aside', under 1995 rules has been promoted by the Scottish Agricultural College, Farming and Wildlife Advisory Group and the Scottish Environmental Protection Agency (SEPA). These state

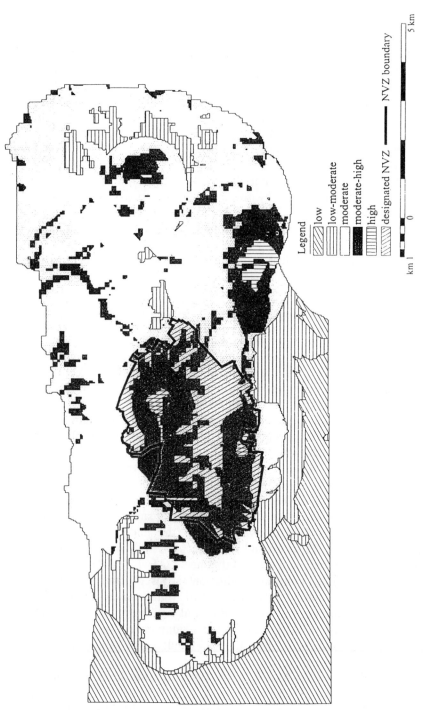

Figure 3   *Combined vulnerability model for groundwater contamination from nitrates and the designated NVZ, Isle of Thanet Kent, England (Source: Osborn and Cook (1997)*

that margins should have a minimum width of 20m, and permanent vegetation must be established between farmland and watercourse (SAC, *undated*).

Defining protection zones is a hydrological/hydrogeological matter, quite properly the responsibilities of EA and SEPA. They also deal with pollution incidents. For groundwater protection zones, consulting engineers running proprietary software were generally used to model travel times for the saturated aquifer. The agricultural side, including advice and registration and financial aspects was organised by MAFF in England and Wales.

EU Directives such as 'The Groundwater Directive' (80/68/EEC), 'The Drinking Water Directive' (80/778/EEC) and 'The Nitrate Directive' (91/676/EEC) form the supra-national legal framework in place for Member States. These are translated into their respective legislation and regulatory arrangements. Two are directly concerned with nitrates, however the intention is that large abstractions for potable supply be covered by the Surface Water Directive Abstraction (75/440/EEC) and subsequent domestic legislation.

There is provision under the Water Resources Act 1991 to define protection zones. NSAs have their origins in the 1980 Drinking Water Directive which concerns the quality of water for human consumption (MAFF, 1994b). Participation is voluntary, and farmers are compensated for loss of income by reducing nitrogen fertiliser input to arable land, or by conserving or restoring low input grassland. The initial scheme operated between 1990 and 1995 in areas where the EU limit was approached, or exceeded. Initial public consultation resulted in ten being defined, since then a further 22 have been designated. Nitrate Advisory Areas were designated in 1990 and farmers received detailed advice, but no compensation. The NSA options include schemes for the conversion of arable land to zero or low input nitrogen land use, for example the Premium Arable Scheme for the restoration of grassland (MAFF/WOAD, 1995). Otherwise management prescription varying from unfertilised and ungrazed to 150 kg/ha of total N fertiliser with grazing optional. There is gathering evidence that NSA prescriptions may be proving effective, especially for arable land converted to extensive ('low-input') grassland (MAFF, 1996).

NVZs are to encourage agricultural practices which are environmentally beneficial (Grey *et al*, 1995) and are compulsory in that nitrogen management prescriptions are laid down by Statutory Code. The objective is to avoid nitrate contamination of waters (mostly based upon surface catchments) from a range of agricultural sources including fertilisers (both natural and artificial), silage, farmyard washings, and so on. NVZs designation is compulsory (MAFF/WOAD, 1994), and follows the 1991 Nitrate Directive (91/676/EEC) requiring member states to identify vulnerable areas and set in place action plans to reduce nitrate concentrations below the EU limit. Certain other Member States, notably The Netherlands and Denmark, by contrast, merely defined their entire territories effectively NVZs.

The Directive requires Member States to:

1. **Designate as NVZs** all known areas of land which drain to waters where nitrate concentrations exceed, or are expected to exceed, the EC limit of 50 mg/l NO_3.
2. **Establish action programmes** which will become compulsory in these zones at a date to be agreed between 1995 and 1999.
3. **Review the designation** of NVZs at least every four years.

Certain other Member States, notably the Netherlands and Denmark, designated their entire territories.

Designation concerns itself with both vulnerable surface and groundwaters, and with eutrophic freshwater, estuarine, coastal and marine waters. It takes account of physical and environmental characteristics of the environment (such as a reasonable proportion of the theoretical catchment required for recharge), and especially considers the behaviour of nitrogen compounds, and current understanding of the impact of remedial action. Regular monitoring of waters is required. For convenience, the proposed boundaries are 'hard features' such as field boundaries or roads. Within the zones thus defined, farmers are, for example, required to:

1. Limit their application of organic manure, initially 210 kg/ha N over four years, reducing to 170 kg/ha after four years. These limits, more stringent than for NSAs, can be revised upward in certain circumstances.
2. Manure must be adequately well stored, and limited in application to levels which are consistent with the net nitrogen requirement of the crop.
3. Farmers are required to maintain field by field records of fertiliser use, and keep these records available for inspection.

The measures applied for enrolled farmers are complicated, and voluntary in NSAs while NVZs, by contrast, generally cover larger areas. Following consultation, a total of some 600 000 hectares in England and Wales are designated in around 70 locations. These also protect certain zones between bores, and extensive designation (rather than individual designation) is justified on the basis of uniformly high nitrate concentrations and because of the importance of the sources for public supply. Following consultation, small boundary changes were made to 31 of the proposed zones, and three deferred for further discussion (MAFF/WOAD, 1995). Both NSA prescriptions, and the Code prevent organic manure application within 10 m of a watercourse, and 50 m of a well or spring.

Areas which come under both schemes have to comply with the most stringent limits for nitrogen applications. In NVZs, farmers are obliged to operate in accordance with GAP (MAFF/WOAD, 1991), with no compensation for supposed loss of productivity resulting from limits upon fertiliser use;

as such NVZs, differs from the NSA scheme. The reason for not going beyond GAP is the stated reason for not compensating within NVZs, yet some livestock farmers are concerned that the impacts will create loss in productivity and hence compensation may be required. NVZ prescriptions have been criticised on account of the absence of an option to convert arable land to zero or low-input nitrate land use under grass or woodland, no provision for the establishment of green cover crops or prevention of ploughing of permanent pasture. It has been argued that there is a requirement to be more stringent than GAP to be effective in nitrate control (Osborn and Cook, 1997).

6 Conclusions

UK nitrate policy has certainly moved on apace since the Water Act 1989. Use is made of modelling procedures, hydrogeological information and hydrochemical data. There remains room for improvement.

On the technical level there is a need to widen the kinds of environmental information employed in protection zone emplacement. Only saturated zone travel times are considered in catchment definition and zonation around groundwater sources destined for human consumption. There is a lack of consideration relating to unsaturated zone residence times, mixing of waters and (where plausible) solute attenuation in source protection zone designation for nitrate. Furthermore, there is an appreciable literature relating soil properties to nitrate behaviour (Carter *et al.*, 1987; Cook, 1991). Farmers readily identify with soil type, making farm scale- or 'micro-' targeting easier and full use of information should be more readily defensible in the wider planning process.

Buffer strip emplacement looks to be a promising option where hydrogeological conditions permit; however, care must be taken to avoid situations where a strip might be inappropriately located. Such situations include the deep-percolation of waters through aquifers which do not de-nitrify and yet feed vulnerable surface watercourses, and the presence of artificial underdrainage which underpasses the strip to feed a watercourse.

There is arguably a problem of 'culture' between the agronomically minded soil scientists and agricultural policy makers, and deeper-probing water engineers and hydrogeologists. The latter come from a background where considerations of aquifer storage, transmission and yield calculations dominate, somewhat detached from activities on the surface! The will is there to protect the resource, but essential environmental linkages, such as soil and unsaturated zone properties tend to be overlooked.

Moving from source to resource protection, for both surface and groundwaters, removes elements of uncertainty, especially where ensuring that catchment definition is sufficient hydrographically to contain the problem. There remains a problem with atmospheric deposition of nitrogen compounds whose control remains outside land use considerations. However, blanket designations of aquifers and surface catchments may greatly assist in reducing uncertainty, and some EU member states have effectively declared their territories to be NVZs.

We may identify a progression from free advice in NAA to compensation in NSA and finally compulsion in NVZs. The degree of 'carrot' and 'stick' proffered to the farmer will probably always be debated, and compulsory measures are certainly required. Yet, because the geographical origins of diffuse pollution are difficult to pin down, one is tempted towards the carrot side of the equation, even if it entails cost to the public purse. In any case, the likely cost of ill-will (involving policing, prosecutions and environmental damage) in situations of compulsion is unknown. Ongoing reform of the CAP should begin to address such issues, especially regarding farmer compensation.

The lack of arrangements for compensation under NVZ will have to be reviewed, especially because of the need to restore very vulnerable areas to unfertilised, zero or low density grazing, or even establish trees in place of farmland. There are schemes for farm woodland planting which should be targeted into protection zones.

In surface water protection, there needs to be a consistent and coherent policy for the emplacement of buffer strips. At the present state of knowledge there remains room for further research into modelling their operation and optimisation of effectiveness through their location. Emplacement on certain SSSIs has been the responsibility of English Nature. MAFF administers both the Habitat Scheme and Countryside Stewardship in England and Wales. There is more than a glimmer of hope if EU policy (European Commission, 1997) looks to abolish compulsory set-aside. However, it is desirable that appropriate provision is made under some alternative Directive.

Arrangements in NVZs and NSAs allow designation by one Competent Authority and management by another, and it seems to work. The EA, with its wide-ranging brief to prevent pollution, its Catchment Management Plans (currently evolving into Local Environment Agency Plans), is well placed to identify vulnerable watercourses; especially those where high input farming is adjacent to floodplains. Encouragement for the emplacement of buffer strips is given (e.g. NRA, 1994b), but incentives are lacking. Alternatively, it might even be desirable to make the emplacement of narrow, unfertilised, grass strips compulsory for certain watercourses.

General purpose groundwater SPZs provide a framework for zoning of land, even if the land-use prescriptions are absent (Cook, 1993). These might be used to target agricultural land uses deemed less prone to leach agrochemicals than more intensive uses. Indeed, present arrangements for the pollutant specific NVZs and NSAs only really aim to control nitrates *of agricultural origin*. History teaches us that there are other sources, notably leaky sewers and cesspools which cause nitrate contamination.

This paper is not a polemic against UK nitrate control policy. There are presently grounds for optimism, but as we move into the twenty-first century agri-environmental policy needs streamlining, and there remains an implicit need to maximise our use of scientific data and modelling in the planning process.

7 Acknowledgements

The author wishes to acknowledge his co-workers and students who were involved in much of the research quoted in this paper. They are Dr J. Lyn Garraway, Dr Hilary Moorby, Dr Charlotte Norman, Ms Susie Osborn, Dr Clive Potter and Dr Stephanie White. Technical advice and assistance was provided by Ms Sarah Brocklehurst, Mr Keith Hart, Ms Trudy Krol and Mr Kris Roger. Financial and other forms of support has been provided by the Economic and Social Research Council, English Nature, the former National Rivers Authority, the Ministry of Agriculture, Fisheries and Food and Wye College, University of London. The co-operation of Walland Marsh farmers, particularly Mr Larry Cooke, is also gratefully acknowledged, as is that of the Environment Agency.

Opinions expressed are those of the author, and do not necessarily reflect those of other individuals or organisations.

References

Adams, B. and S.S.D. Foster, 1992, Land surface zoning for groundwater protection, *J. Inst. Water and Env. Man.* **6**(3), 312–320.

Addiscott, T.M., 1996, 'Fertilisers and nitrate leaching.' In: R.E Hester and R.M. Harrison (eds). *Issues in environmental science and technology*, no. 5, The Royal Society of Chemistry, Cambridge, UK, 1–26.

Broads Authority, 1993, *No easy answers, the draft Broads plan.* The Authority, Norwich 158pp.

Carter,A.D., R.C. Palmer and R.A. and Monkhouse, 1987, 'Mapping the vulnerability of groundwater to pollution from agricultural practice, particularly with respect to nitrate.' In: W. Van Duijvenbooden and H.G. van Waegervingh (eds) *Vulnerability of soil and groundwater pollutants.* Proceedings of an international conference at Noordwijk ann Zee, The Netherlands, 30 March-3 April 1987.

Cook, H.F., 1991, Nitrate protection zones: Targeting and land use over an aquifer. *Land Use Policy* **8**(1), 16–28.

Cook, H.F., 1993, Progress in water management in the lowlands *Prog. in Rural Policy and Plan.* **3**, 91–103.

Cook, H.F. 1998, *The protection and conservation of water resources.* Wiley, Chichester.

Cook, H.F. 1998, Groundwater Development in England, *Env. and Hist*, in press.

Cook, H. and C. Norman, 1996, Targeting Agricultural Policy: and analysis relating to the use of Geographical Information Systems. *Land Use Policy* **13**(3), 217–228.

Cook, H.F., J.L. Garraway, H. Moorby and S.K.and White, 1997, 'Operation and establishment of fertiliser-free grass strips in the protection of watercourses fed by shallow watertables.' Report to English Nature, Environment Department, Wye College, University of London.

Countryside Commission, 1994, 'Countryside Stewardship: Handbook and Application form.' Countryside Commission Pamphlet no. 453, Cheltenham.

Department of the Environment, 1986, *Nitrate in water*, Pollution paper no. 26, HMSO, London.

English Nature, The Game Conservancy Trust and the Royal Society for the Protection of Birds, 1997, 'Crops and Biodiversity: A proposal for an arable incentive scheme'.

European Commission, 1997, *Agenda 2000: Final Volume 1, for a stronger and wider union.* COM(97)2000, 23–34, Brussels.

Gardiner, J.L, 1994, Sustainable Development for River Catchments. *J. Inst. Water and Env. Man.* **8**(3).

Grey, D.R.C., D.G. Kinniburgh, J.A. Barker and J.P. Bloomfield, 1995, *Groundwater in the UK,* Groundwater Forum Report FR/GF 1.

Haycock, N.E. and T.P. Burt, 1993, The sensitivity of rivers to nitrate leaching: The effectiveness of near- stream land as a nutrient retention zone. In: Thomas, D.S.G. and Allison R.J. (eds), *Landscape sensitivity,* Wiley, Chichester, UK, 261–272.

Headworth, H.G., 1994, 'The groundwater schemes of Southern Water, 1970–1990. Recollections of a golden age'. Southern Science, Crawley, Sussex, August, 1994.

House of Lords, 1989, *Nitrate in water.* House of Lords Select Committee on the European the European Communities, Rpt. no 16, paper 73–1, London, HMSO.

Leeson, J.D., 1995, *Environmental Law.* Pitman, London 479pp.

Ministry of Agriculture, Fisheries and Food, 1994a, The Habitat Scheme: Water Fringe Areas. MAFF, London, PB 1732/1 HS/FSA/1.

Ministry of Agriculture, Fisheries and Food, 1994b, 'The Nitrate Sensitive Areas Scheme'. Information pack PB1729A, The Ministry, London.

Ministry of Agriculture, Fisheries and Food, 1996, 'Nitrate Sensitive Areas Scheme'. Results: Winter 1995/96, The Ministry, London.

Ministry of Agriculture, Fisheries and Food/Welsh Office Agriculture Department, 1991, *Code of Good Agricultural Practice for the protection of water.* The Ministry, London.

Ministry of Agriculture, Fisheries and Food/Welsh Office Agriculture Department, 1995, 'Government Response to the Consultation on the Designation of Nitrate Vulnerable Zones in England and Wales'. The Ministry, London.

Moorby, H. and H.F. Cook, 1992, Use of fertiliser-free grass strips to protect dyke water from nitrate pollution. *Aspects of Applied Biology* **30**, 231–234.

Muscutt, A.D., G.L. Harris, S.W. Bailey and D.B. Davies, 1993, Bufferzones to improve water quality: a review of their potential use in UK agriculture. *Agric., Ecosystems and Env.,* **45**, 59–77.

National Research Centre,1993, *Ground Water Vulnerability Assessment: Predicting relative contamination potential under conditions of uncertainty.* National Academy Press, Washington DC, p. 28.

National Rivers Authority, 1992a, Policy and Practice for the protection of groundwater, the Authority, Bristol.

National Rivers Authority, 1992b, *The Influence of Agriculture on the Quality of Natural Waters in England and Wales,* Rpt. of the NRA, Water Qual. Ser. no. 6, The Authority, Bristol.

National Rivers Authority, 1994a, *Groundwater Vulnerability 1:100 000 map Series sheet 47.* East Kent. The Authority, London, HMSO.

National Rivers Authority, 1994b, 'Kentish Stour Catchment Management Plan Consultation Report', NRA Southern Region, Worthing, October 1994.

Osborn, S. and H.F. Cook, 1997, Nitrate Vulnerable Zones and Nitrate Sensitive Areas: A policy and technical analysis of groundwater source protection in England and Wales. *J. Env. Plan. and Man.* **40**(2), 217–233.

Potter, C., H.F. Cook and C. Norman, 1993, The targeting of rural environmental Policies: an assessment of agri-environmental schemes in the UK. *J. of Env. Plan. and Man.,* **36**(2), 199–215.

Pretty, J.N., 1995, *Regenerating Agriculture.* Earthscan.

Redclift, M. (1994). Reflections on the 'sustainable development' debate. *Int. J. Sustainable Development and World Ecol.* **1**, 3–21.

Scottish Agricultural College, Farming and Wildlife Advisory Group, Forth River Purification Board and Rhone Poulenc (undated) 'Buffer Strips'.

Sheail, J., 1982, Underground water abstraction: indirect effects of urbanisation on the countryside. *J. Hist. Geog.*, **8**, 395–408.

Sheail, J., 1992, The South Downs and Brighton's water supplies – an inter-war study of resource management. *Southern Hist.* **14**, 93–111.

Skinner, A. and S. Foster, 1995, 'Managing Land to protect water; the British experience in groundwater protection' Paper XXVI *Congress of the International Association of Hydrogeologists*, Edmonton, Canada, June, 1995.

Watson, N., B. Mitchell, and G. Mulamoottil, 1996, Integrated Resource management: Institutional Arrangements Regarding Nitrate Pollution in England. *J. of Env. Plan. and Man.* **39**(1), 40–64.

White, S.K., H.F. Cook, J.L. and Garraway, in press, Fertiliser-free grass buffer strips to attenuate nitrate input to marshland dykes. *Water and Env. Man.*, **12**, 54–59.

14

Considerations of the Amino Nitrogen in Humic Substances

Thomas M. Hayes[1], Michael H. B. Hayes[2] and André J. Simpson[1]

[1] THE UNIVERSITY OF BIRMINGHAM, SCHOOL OF CHEMISTRY, EDGBASTON, BIRMINGHAM B15 2TT, UK
[2] DEPARTMENT OF CHEMICAL AND ENVIRONMENTAL SCIENCES, FOUNDATION BUILDING, UNIVERSITY OF LIMERICK, LIMERICK, IRELAND

Abstract

Humic substances (HS) in a grassland soil, and in the drainage waters from the soil were fractionated and aspects of the compositions of the different fractions were investigated. Our data indicate that the compositions of the less transformed humic fractions can be related to the plants of origin, and are the most rich in sugars and amino acids (AAs). As humification progresses the AA and sugar residues are metabolised. Eventually the HS released in the drainage waters have the least contents of AA and sugar residues. Products of the metabolism of the AAs (ammonia and nitrates) are utilised by growing plants, but when rainfall exceeds evaporation, and especially when plant growth is slow or not occurring, these metabolites are lost in the drainage water. Our data cannot be interpreted in terms of the fate of the AAs associated with the HS in the drainage waters, or that of the other forms of organic N (such as heterocyclic N) in these substances.

1 Introduction

Amino nitrogen is the most abundant form of organic nitrogen (N) in soil organic matter (SOM) that is readily available for soil microorganisms. This N form can persist during the transformations in the soil environment of the organic residues of plant, animal, and microbial tissues. In the course of the transformations to humic substances (HS) much of the N is converted to ammonia and to nitrate. There is, however, a degree of protection against microbial (enzymatic) attack when amino N is associated with the soil HS.

HS are operationally defined as the humic acids (HAs), or the fractions of SOM soluble in aqueous base but precipitated at pH 1, the fulvic acids (FAs), which are soluble in aqueous solvents at all pH values, and the humin materials which are, in the classical definitions, humified substances that are insoluble in all aqueous solvents. In mineral soils humins are usually considered to be associated with clays. When removed from the clays these are found to have components similar to HAs and FAs (*e.g.* Clapp and Hayes, 1996).

Amino sugars in soil polysaccharide structures are lesser contributors to the organic N of SOM. So too are the bases of the nucleic acid and the porphyrin structures. Heterocyclic N components, that may or may not be related to those in nucleic acids and porphyrins, can also be associated with HS. In order to understand how N is contained in SOM, and its eventual conversion to nitrate, it is important to be aware of the kinds of structures which contain the N and of the possibilities for the release of that N. Attention will focus here mainly on the amino N associated with the HS in a selected soil and in the drainage waters from that soil.

Table 1 *Location and description of sampling sites*

Sample	Sampling date	Weather	Location	Soil description	Plot descriptions
Devon 1 (D1)	05-03-1991	Sample taken at the end of a rainfall event, the drains were beginning to slow.	AFRC, Institute for Grassland and Environmental Research Station, North Wyke, Devon	Hallsworth Series, clayey non-calcareous pelostagnogley (Armstrong and Garwood, 1991). Grassland,	Permanent pipe drains at 85cm depth. Addition of 400 kg N ha^{-1} ann^{-1} (Tyson *et al*, 1992)
Devon 2 (D2)	end-02-1992	Sample taken during a dry period, very slow drain flow.	(SX 650 995)	experimental site.	
Devon 3 (D3)	end-11-1992	Sample was taken during a downpour. The drains were fast flowing.			Surface interceptor drain at 30 cm depth. Addition of 400 kg N ha^{-1} ann^{-1}

2   The Site, Soil, and Samplings

Information relevant to the samplings is given in Table 1. Three soil drainage water samples were collected at the AFRC Institute of Grassland and Environmental Research (IGER) farm at North Wyke, Okehampton, Devon. Two samples (D1 and D2) were collected from a specially drained one hectare field. That field is drained in such a way that surface or runoff water is obtained independently of the drainage waters, and a schematic cross-section of the drainage system is given by Armstrong and Garwood (1991). The third sample was collected as surface runoff from an undrained plot. Further site details are provided by Hayes *et al.* (1997).

Water samples from the drainage interceptors were collected in 50 L aluminum barrels.

2.1   Preparation of Water Samples

The water was pressure filtered through Whatman WCN cellulose nitrate (0.45 μm pore size) or Sartorious cellulose acetate (0.2 μm pore size) membranes using 142 mm Millipore membrane filter holders. The pH of the filtrate was adjusted to 2 and the waters were then passed through XAD- and XAD-4 resins in tandem, and HAs, FAs and XAD-4 acids were isolated as described by Hayes *et al.* (1996, 1997), and by Hayes (1996). The preparation of the XAD-8, XAD-4, and IR-120 resins (to prepare the H^+-exchanged humic preparations from the back eluates in base from the XAD columns) is described by Thurman and Malcolm (1981) and by Hayes (1996).

2.2   Isolation of Soil Humic Fractions Using XAD-8 and XAD-4 Resins in Tandem

The soil was exhaustively extracted in series using sodium pyrophosphate (Pyro, 0.1M) neutralised to pH 7 (with phosphoric acid), Pyro, pH 10.6, and Pyro + 0.1M NaOH at pH 12.6. The procedure which involved the uses of XAD-8 and XAD-4 resins in tandem was similar to that for the water samples. Filtrates, diluted with distilled water to an organic matter concentration < 15 mg L^{-1}, were adjusted to pH 2, using 6M HCl. Then the solutions were pumped though the XAD-8 resin at a flow rate of ca 40 mL min^{-1}, and the column effluent was pumped at the same flow rate through the XAD-4 resin column (see Hayes, 1996). The procedure is summarised in Figure 1.

Distilled water at pH 2 was pumped through the columns to wash out residual materials not sorbed by the resins. The XAD-8 column was then back eluted with 0.1m NaOH, the centre cut eluate was adjusted to pH 1 (6m HCl), and HAs were let precipitate overnight. The supernatant (FA at pH 1) was filtered (0.2 μm membrane), and desalted by pumping onto the XAD-8 at a flow rate of 4 mL min^{-1}. After one column volume of water at pH 1 had passed through, distilled water was pumped on until the conductivity of the effluent was < 250 mS cm^{-1}. Back elution was carried out using 0.1m NaOH. The effluent was passed through IR-120 (H^+-form) and the eluate (FA) was freeze dried.

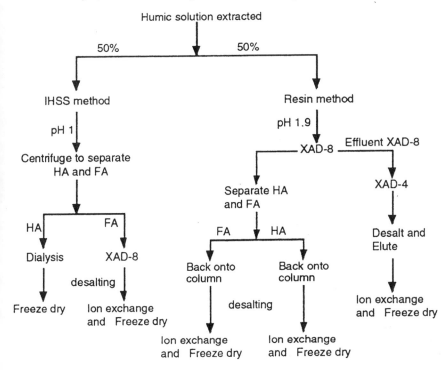

Figure 1 *An outline of the procedures used to isolate HS using resins in tandem and the procedure used to isolate the IHSS HA and FA standards from soils*

The HA precipitate was redissolved in 0.1m NaOH, diluted to give a HA concentration of ca 15 mg L^{-1}, and the pH was lowered (stepwise) to 2.0. A precipitate which formed between pH 2.5 and 2.0 was recovered by centrifugation. It was not possible to recover the precipitate for every HA fraction. The HAs remaining in solution at pH 2 were pumped back onto the XAD-8 column (at a flow rate of 4 mL min^{-1}) for desalting. Water (pH 2) was pumped down the column prior to desalting, and the sorbed HA was recovered as described for the FA fraction. The HAs which were precipitated (at pH 2.5) were dialysed [Visking dialysis tubing with a molecular weight cutoff (MWCO) of 12 000 daltons] against distilled water till chloride free, and the retentates were then freeze dried.

Material sorbed on the XAD-4 resin was desalted as described for the FA, eluted with 0.1m NaOH, and passed through IR-120 (H^+-form) prior to freeze drying. Finally, the XAD-8 and XAD-4 resins were washed with distilled water prior to the recovery of the neutrals fractions by soxhlet extraction with ethanol as the solvent.

HAs and FAs were isolated from a second batch by the procedure used to isolate Standard HAs and FAs of the International Humic Substances Society (Swift, 1997).

2.3 Elemental Compositions, Moisture and Ash Contents, and $\delta^{13}C$ and $\delta^{15}N$ Values

Carbon, hydrogen and nitrogen were determined by Mr. K.R. Scott of the School of Chemistry using a Perkin Elmer 240 Elemental Analyser. Values for $\delta^{13}C$ and $\delta^{15}N$ were obtained in the laboratories of Professor C.E. Clapp of the Department of Soil, Water, and Climate of the University of Minnesota, USA. Analyses were carried out, at least in triplicate, on solid samples of about 100 mg using an elemental analyser (Carlo Erba, model NA1500) and a stable isotope ratio mass spectrometer (Fisons, Optima model) continuous flow system. Results of the isotope analyses are expressed in terms of δ values (‰) where

$$\delta^{13}C \text{ or } \delta^{15}N = 100(R_{sample}/R_{standard}-1)$$

and R = ratio of $^{13}C/^{12}C$ or of $^{15}N/^{14}N$. The $\delta^{13}C$ values were calculated relative to Pee Dee Beleminite (PDB) as an original standard. Urea, with a $\delta^{13}C$ value of -18.2%, served as a working standard. The standard for $\delta^{15}N$ is atmospheric N_2 ($\delta^{15}N = 0$). Acetanilide (from Carlo Erba Instruments), $\delta^{15}N = -1.1\%$, was used as the working standard.

2.4 Potentiometric Titrations and Cation Exchange Capacity (CEC) Data

The procedure used is described by Hayes (1996), and is outlined in Hayes *et al.* (1996).

2.5 Nuclear Magnetic Resonance (NMR) Spectroscopy

The solid state CPMAS ^{13}C NMR spectra were acquired compliments of Drs R.S. Swift and J.O. Skjemstad of CSIRO Division of Soils, Glen Osmond, S. Australia. Aspects of the procedure are outlined by Hayes *et al.* (1996).

Details of the one- and two-dimensional NMR experiments, using Bruker AMX-600 and Bruker DRX-500 instrumentation, are given by Simpson *et al.* (1997).

2.6 Determinations of Amino Acids (AAs)

Amino acids (AAs) were determined using a procedure based on that of Turnell and Cooper (1982). Samples were hydrolysed for 24 h in 6M HCl at 115 °C under dinitrogen gas. Free amino acids were extracted into 0.1M perchloric acid, and separated and quantified (as *o*-phthalaldehyde derivatives) by HPLC using a 15 cm × 4.6 mm reverse phase Spherisorb ODS II 3 μm column (Phase Separations Ltd.), and fluorescence detection. Norvaline was used as the internal standard.

3   Results and Discussion

3.1   Yields of HS from Soil Extracts

The data in Table 2 indicate that the bulk of the HS isolated from the soil were in the 0.1m pyrophosphate, pH 7, isolate, and the lowest amounts were in the pH 10.6 isolate. Table 3 gives the yields from elutions in ethanol (after the humic fractions had been isolated).

Table 2   *Yields (mg) of HAs, HA precipitates at pH 2.5 (HApt), FAs, and XAD-4 acids from the Devon soil using the XAD-8 and XAD-4 resins in tandem (corrected for moisture and ash)*

Extraction solution	HA (mg)	HApt (mg)	Total HA (mg)	FA (mg)	XAD4 (mg)	Total (mg)
0.1M Na$_4$P$_2$O$_7$, pH 7	114.1	193.6	307.7	567.0	219.7	1094.4
0.1M Na$_4$P$_2$O$_7$, pH 10.6	8.6	–	8.6	21.3	31.7	61.6
0.1M Na$_4$P$_2$O$_7$, + 0.1M NaOH, pH 12.6	36.8	33.1	69.9	50.1	46.4	166.4
Grand Total	159.5	33.1	386.2	638.4	297.8	1322.4

Table 3   *Yield (mg) of neutrals eluted (EtOH) from the XAD resins used to process the Devon soil extracts*

Resin	Yield (mg)
XAD-8	300[*]
XAD-4	268
Grand Total	568

[*]Figure is based on a recovery of neutrals from *ca.* 25% of the ethanol used in the resin extraction process.

The overall yield by the resin procedure (1893 mg) was more than double that from the IHSS method (905 mg). The difference may be partly explained by the contribution to the resin-isolated total (568 mg) from the XAD-4 acids and the XAD-4 neutrals. These would have been lost in the IHSS isolation procedure. Furthermore, considerable losses were experienced during the dialysis of the HA fraction.

Clearly, there are differences in the operationally defined HA and FA fractions obtained by the two procedures.

3.2   Yields of HS from Drainage Waters

DOC values were not measured for the original samples. Thus it is not possible to give a full picture of the distribution in the different DOC fractions. The D1 and D2 samples were from a drained one hectare plot, while the D3 sample was from a site where field drains were not installed. On this (undrained plot) > 80% of the hydrologically effective rainfall (H.E.R.) could be accounted for with 'V' notch weirs sampling surface runoff, whereas on the plots drained by mole (55 cm depth) and pipe drains (85 cm depth), about 85% of H.E.R. was monitored through the drainage system. Both the water content and height of the water table are lower in the drained soil over the whole year (Armstrong and Garwood, 1991). The OM content of the undrained plot is higher than for the drained soil (Schofield, 1995). Stevenson (1994) indicated that undrained soils can contain up to 10% or more OM than drained soils. Results in Table 4 show that the humic OM content of the waters from the undrained soil was significantly greater than for the drained soil. The ratio of FA:HA:XAD-4 acids is seen to be very different for the undrained (D3) sample. Typically the ratio of FA:HA:XAD-4 acids is approximately 9:1:1 for surface waters (Malcolm, 1991).

Table 4 *Volume of sample processed (L) and weights of humic substances recovered (g), on a dry, ash free basis*

Sample	Vol. water	FA (g)	HA (g)	XAD-4 acids	Total (g)
Devon1 (D1)	1500 (L)	2.125	0.359	0.339	2.822
Devon2 (D2)	2000 (L)	4.6	0.2542	0.4053	5.2595
Devon3 (D3)	2000 (L)	15.0	7.24	1.0	23.24

In the case of the D3 sample, that ratio (15:7:1) shows a significant deviation from the normal distribution. The D3 sample contained 4 times more humic material than samples D1 or D2. That would imply that much more DOC (and humic material) was lost from the undrained soil, and yet the SOM content was higher in that soil. The high HA content is significant because the HA fraction had a greater influence (than the FA and XAD-4 fractions) on the behaviour of anthropogenic organic chemicals (Hayes and Hayes, 1997). The HS isolated from the D3 sample were similar to the HS isolated from the parent soil whereas those in D1 and D2 were characteristic of aquatic HS (vide infra).

3.3 Elemental Analyses

Results for the moisture, ash, elemental, $\delta^{13}C$ and $\delta^{15}N$ analyses are given in Table 5. In general, the carbon content followed the order: HA > FA > XAD-4 acids and the oxygen content followed the order XAD-4 acids > FA > HA, but the hydrogen contents did not follow any noticeable trend, and all of the hydrogen values were very similar.

The nitrogen contents for the water samples followed the order XAD-4 > HA > FA, whereas those for the soil fractions decreased in the order HA > XAD-4 acids > FA, and the hydrogen contents decreased in the order XAD-4 acids > HA >FA. The soil HAs had a larger N content compared to the drainage water HAs, and this increased N was probably due to the higher AA contents of the soil HAs. The soil FAs had a lower C content, similar N and H contents, and a higher O content than the drainage water FAs. The higher O content was due to the much higher concentration of neutral sugars (NS) and non-carbohydrate material which resonates in the 65–110 ppm region of the NMR spectrum (vide infra). The concentration of AAs in the soil FAs was also higher, but the overall N contents were similar. The distribution of the AAs in the soil FAs varied. The AAs accounted for 20–30% of the N in the soil FAs, but only ca 5.0% of that in the drainage water FAs. The XAD-4 acids from the soil were significantly different from those from the drainage waters. The soil acids had lower C, and higher H, N, and O contents. These differences were reflected in the NMR spectra.

The soil XAD-4 acids contained a higher proportion of NS and AAs than the XAD-4 acids from the drainage waters. The contribution of AAs to the N in the soil XAD-4 acids was higher than the AA N contribution in the XAD-4 acids from the drainage water.

The ash contents of the HA, HApt, XAD-4 acids, and neutrals fractions were high. Fe and Al were the main contributors. Piccolo *et al.* (1990) isolated HS from an organic soil using NaOH (under N_2) and $Na_4P_2O_7$ at pH 7. They used centrifugation, and filtration through glass wool as the cleanup, a treatment that was similar to the procedure used in this study. The ash contents of their HAs prior to HCl/HF treatment were also high, and they concluded that the HCl/HF treatment decreased the resonances in the aliphatic region of the NMR spectrum, suggesting that the purification eliminated the weakly bound long alkyl chains from the raw humic materials. To avoid artifact formation.the HCl/HF treatment was not used in this study.

The elemental analyses data for the soil FAs indicate that the C content was greatest in the isolates at pH 12.6, and lowest in those at pH 10.6. The O content decreased as the pH of the extracting solution increased, varying by ca 5%. The N and H contents were highest in the isolates at pH 10.6, but similar in the FAs extracted at pH 7 and at pH 12.6.

The elemental compositions of the XAD-4 acids were similar for each fraction. The O and N contents decreased, while the H contents increased as the pH of the extractant increased, but the changes were small. The compositions of the neutral fractions were similar to those of the HAs, but NMR data indicated that these were structurally different.

The high oxygen content of the XAD-4 acids is to be expected since these samples had the highest carboxyl contents, as determined by potentiometric titration (Table 6), and also had the greatest areas in the 190–160 ppm (carboxyl) resonances in the NMR spectra (Table 7). Titration and NMR data indicate that the FA oxygen contents were intermediate.

The carbon contents for the HAs isolated from each of the water samples were similar (all in the range 53–57%). The oxygen contents lay in the range 33.3 to 44.9%, hydrogen in the range 1.3–5.7%, and nitrogen was in the range 2.0–5.4%.

Table 5 *Elemental data for the HS from the waters (WATER) from the drained plot (D1 and D2), and from the undrained plot (D3), and for the HA, FA, XAD-4 acids isolated by the resin technique from the soil (SOIL) at pH 7 (R7), pH 10.6 (R10), and pH 12.6 (R12), and for the XAD-8 neutrals (XAD-8N) and XAD-4 neutrals (XAD-4N). I7, I10, and I12 are samples isolated at pH 7, pH 10.6 and pH 12.6 using the IHSS procedure. R7pt refers to the HA fraction precipitated between pH 2.5 and 2.0. Values for C, H, N, and O are expressed as percentages of the total weight, and are calculated on a dry, ash-free basis. Ash contents are given on a moisture free basis*

Sample	Moisture	Ash	C%	N%	H%	O%	$\delta^{13}C(\%)$	$\delta^{15}N(\%)$
Water								
D1 HA	8.5	43.67	54.95	3.08	3.06	38.91	−30.8	2.79
FA	8.5	1.92	54.83	2.71	1.31	41.15	−29.83	1.18
XAD4	8.1	6.44	48.15	4.16	2.85	44.84	−28.64	2.79
D2 HA	10	6.81	56.6	3.84	3.89	35.67	−30.99	3.11
FA	8.3	0.82	56.2	2.83	4.26	36.71	−30.15	1.3
XAD4	10.5	3.2	51.34	5.63	3.93	39.1	−28.88	3.22
D3 HA	10.4	1.11	56.04	3.36	3	37.6	−31.76	1.57
FA	10.5	0.23	55.81	3.16	2.89	38.14	−30.63	0.83
XAD4	10.6	4.24	55.7	4.25	3.65	36.4	−28.96	3.63
Soil								
HA								
R7	6.39	24.38	56.08	3.84	3.52	36.56	−29.3	2.48
R7pt	6.7	45.15	55.51	5.77	5.17	33.55	−28.15	3.06
I7	9.7	11.01	55.75	4.53	3.59	36.12	−28.98	2.06
R10	8	30.38	50.12	5.64	4.14	40.11	−28.3	2.08
I10	6.8	29.1	51.23	6.10	5.32	37.35	−27.72	1.98
R12	5.6	15.54	52.46	4.94	5.39	37.20	−28.7	2.87
R12pt	6	13.04	52.50	5.64	5.89	35.98	−28.59	3.2
I12	6.5	10.84	54.52	5.31	5.35	34.82	−28.93	3.02
FA								
R7	8.1	1.85	51.25	2.47	3.63	42.65	−29.1	1.61
I7	4.2	1.61	47.73	2.42	3.04	46.81	−28.8	1.53
R10	8	18.77	44.57	3.29	3.43	48.71	−27.7	3.12
I10	7.65	10.31	46.09	3.32	3.18	47.41	−27.8	3.32
R12	6	11.35	52.21	2.66	2.98	42.14	−29.4	2.78
I12	5.9	6.05	51.13	2.94	4.97	40.96	−29.1	2.91
XAD4								
R7	7.86	16.32	43.95	4.50	4.00	47.55	−28.4	3.6
R10	8	47.74	43.93	4.16	5.64	46.27	−27.6	3.15
R12	6	32.12	44.46	3.70	6.92	44.93	−27.3	4.22
XAD-8 N	8	9.56	51.78	3.13	5.72	39.37	−27.73	7.57
XAD-4 N	6.4	40	50.53	5.93	6.19	37.34	−25.17	6.06

3.4 $\delta^{13}C$ Data

The $\delta^{13}C$ data (Table 5) indicate that the inputs to the humic materials were predominantly from C3 plants. C3 plants tend to have $\delta^{13}C$ values around $-27‰$ (Deines, 1980; Clapp *et al.*, 1997). Sugars, cellulose and hemicellulose show $\delta^{13}C$ values close to the mean plant isotopic composition, Pectin appears to be enriched in ^{13}C, while lignin and lipids are depleted in this isotope relative to the total plant (Deines, 1980).

In all cases (for soil and water HS) the HAs were the most negative, with the XAD-4 acids the least negative (or most enriched in ^{13}C). Clapp *et al.* (1997) have shown that the $\delta^{13}C$ values for HAs and FAs were similar for samples from the same sources. They also noted that the XAD-4 acids from the same sources had values which were less negative, and this might indicate the microbial involvements in the transformation processes. There were no significant variations in the $\delta^{13}C$ values for each humic fraction from the same source isolated over the period of the study. The values for the same fractions of the D1, D2, and D3 samples were similar. The $\delta^{13}C$ values for the XAD-4 acids for the samples were in the range -28.64 to -28.96, and these might be considered to reflect inputs from microbial metabolism to the C of the samples. These fractions also had the highest carbohydrate contents which would have values close to the plant mean.

The data also show that the $\delta^{13}C$ values for the soil HAs and FAs were very similar to the same fractions isolated from the drainage waters. XAD-4 acids isolated from soil were slightly less negative than those isolated from the drainage waters. That might indicate that the drainage water XAD-4 acids had undergone more extensive microbial transformations, whereas those from soil retained more of the parent plant signature.

The $\delta^{13}C$ values for the HAs and FAs indicate that carbohydrate type materials increased as the pH of the extracting solutions increased. This is based on the fact that there was a slight decrease in the $\delta^{13}C$ value (less negative) in the isolates at pH 12.6. This was confirmed by the NS analyses. Although the values for the XAD-4 acids remained constant as the pH of the extracting solution increased, the NS contents did change. This suggests that the XAD-4 acids isolated at pH 12.6 were 'newer', and had not progressed along the path of microbial transformations.

The $\delta^{13}C$ values for the XAD-4N fraction indicated that this material was most enriched in ^{13}C, but we do not have sufficient supplementary data to rationalise that observation. The XAD-8N fraction had a $\delta^{13}C$ value close to that of the XAD-4 acids.

3.5 $\delta^{15}N$ Data

A discrimination between the lighter ^{14}N and heavier ^{15}N isotopes occurs during biological and chemical processes (Delwiche and Steyn, 1970), and this leads to an increase in the $\delta^{15}N$ of the unreacted fraction of the substrate. Studies have shown that nitrogen isotopes are useful for discriminating between nitrate sources in ground water (Komor and Anderson, 1993). Ratio values for animal wastes range from 10–22‰, those for organic material in soil range from 4–9‰, and those for commercial fertilisers from −4 to +4‰.

There are limitations in interpretations of the data. $\delta^{15}N$ ratios from nitrogen sources cover a range of values which do not have distinct boundaries. It is possible, for example, to have a sample with a $\delta^{15}N$ value of +6‰ made up of a sample which contains 100% nitrogen with a $\delta^{15}N$ value of +6‰, or as an average of a mixture composed of 50% N with a $\delta^{15}N$ of 0 and 50% with a $\delta^{15}N$ value of +12‰. This is important when the N inputs are mixed (for example, mixing commercial fertilisers and animal wastes).

The $\delta^{15}N$ data (Table 5) did not follow the trends seen for the $\delta^{13}C$ values for the various humic fractions, and there was no apparent trend. All the values lay in the range +0.8 to +4.7‰ for the water humics, and the values for those from the soil were in the range +2.8 to +5.8‰. The $\delta^{15}N$ values for the soil and water HS decreased in the order XAD-4 acids > HA > FA. Values for the HA and XAD-4 (2.8 to 3.2‰) fractions from the waters were higher than those for the FAs, indicating that commercial fertilisers applied to the land may have had a significant input to the N content of the FAs. The higher values for the HAs and XAD-4 acids indicate greater inputs from animal wastes.

The fractions isolated from the surface water of the undrained plot (D3), although following this trend, had values which were significantly different. The $\delta^{15}N$ values for the HAs (1.6‰) and FAs (0.8‰) would indicate significant contributions from commercial fertilizer applications. The XAD-4 (3.6‰) acids had inputs which diluted the effect of the artificial fertilizer. Both of the sampling plots received high inputs of N, in the order of 400 kg N ha^{-1} per annum (Tyson *et al.*, 1992). This indicates that more of the fertilizer N was incorporated into the HA and FA fractions than in the XAD-4 fractions.

In general, the $\delta^{15}N$ values increased as the pH of the extracting solutions increased. The data also show similarities between HS isolated by the 'resin' and IHSS methods.

3.6 Potentiometric Titration Data

Results for the cation exchange capacity (CEC) calculations for the drainage water samples are given in Table 6. All of the CEC values were calculated on a dry-ash-free basis.

CEC values for the HAs from the three samples were similar. This is in agreement with the NMR data (Table 7). Also, values for the FAs and XAD-4 acids were similar.

The CEC values for the FAs were significantly greater than for the HAs. Carboxylic acids and strong acids contributed most to these, which is consistent with the NMR data.

It is clear that there was a significant contribution to the acidity of the HA fraction from weak acid groups (phenols). Again, this is consistent with the NMR data.

Table 6 *Cation exchange capacity (meq g^{-1}) values calculated at various pH values of the humic (HA), fulvic (FA), and XAD-4 (XAD4) acids from the Devon water drained (D1 and D2) and undrained (D3) plots*

					pH					
Sample	3	4	5	6	7	8	9	10	11	12
D1 HA	–	1.35	2.05	2.61	2.84	3.10	3.21	4.20	4.84	4.95
D1 FA	1.91	3.45	4.50	5.20	5.50	5.85	5.90	6.20	6.24	6.34
D1 XAD4	2.50	4.30	5.50	6.25	6.98	7.04	7.25	7.50	7.75	7.93
D2 HA	0.9	1.30	2.06	2.56	2.90	3.05	3.31	4.15	4.85	5.04
D2 FA	1.85	3.51	4.26	5.01	5.50	5.86	5.98	6.10	6.15	6.30
D2 XAD4	2.60	3.98	5.50	6.30	6.84	7.01	7.50	7.64	7.98	8.05
D3 HA	–	1.22	1.98	2.42	2.74	2.93	3.13	4.04	4.61	4.64
D3 FA	1.85	3.15	4.38	5.14	5.51	5.52	5.75	6.04	6.11	6.24
D3 XAD4	2.78	4.18	5.45	6.26	6.75	7.03	7.17	7.73	7.95	8.137

The CEC and (acidity) patterns for the XAD-4 acids would indicate that the acidity was largely contributed by the carboxylic acids, but there was a contribution also from weaker acids (possibly enols). The NMR data indicated that aromaticity was low, and the absence of resonances for phenols indicate that phenols contributed little to the acidities.

3.7 NMR Data

The integrated areas from the CPMAS ^{13}C NMR spectra for seven resonance regions for the fractions isolated from the soil and waters are given in Table 7. Detailed discussion of these data, and of the relevant spectra is being prepared for publication, and only aspects related to the N contents of the samples are emphasized here.

The divisions made were arbitrary. Resonances due to carbohydrate material (at ~60–65 ppm) were not included with the data for the 65–110 ppm region. The overlap of resonances between the different regions would limit applications of accurate quantitative analyses of the data, and the integrated areas might best be used for comparisons.

The ^{13}C NMR spectra for the HApt formed at pH 2–2.5 from the isolates from the soil at pH 7 and at pH 12.6 indicate that these fractions had characteristics different from those of the other HAs and FAs isolated from the soil. The HApt isolated at pH 7 showed very little aromatic character, and was aliphatic in nature. Also, it did not resemble the drainage water HAs. The HApt isolated at pH 12.6 was very similar to the HA isolated at that pH. However, chemical analyses data indicate that the sample had a relatively low NS (6.6%) content, but a high AA (14%) content. (The lack of an anomeric peak, but the strong resonance in the 50–65 ppm region indicates this.) That fraction had a high concentration of non-carbohydrate material resonating in the 60–100 ppm region.

However, the presence of a doublet at 18 and 22 ppm, with a smaller

Table 7 *Integrated areas of seven regions in the CPMAS ^{13}C NMR spectra of the humic acids (HA), fulvic acids (FA) and XAD-4 acids isolated from the Devon water from the drained (D1, D2) and the undrained (D3) plots, for these fractions fractionated by the resin-in-tandem procedure, and isolated from the soil at pH 7 (R 7), pH 10.6 (R 10), and pH 12.6 (R 12) and by the IHSS method (I 7, I 10, I 12), and for the HA precipitates (R 7pt HA and R 12pt HA) formed at pH 2.5 to 2 in the cases of the extracts isolated at pH 7 and pH 12.6*

Region Sample	220–190 Ketonic/ aldehyde	190–160 Carboxyl	160–140 O-Aryl	140–110 Aromatic	110–65 O-Alkyl	65–45	45–1 Alkyl C	fa
D1 HA	4	14	5	23	10	7	37	28
D2 HA	1	13	5	23	15	12	31	28
D3 HA	3	15	9	27	13	13	21	36
D1 FA	2	16	4	11	15	–	51	15
D2 FA	4	17	1	16	13	–	49	17
D3 FA	4	20	5	21	13	7	30	26
D1 XAD4	1	24	2	6	29	–	39	8
D2 XAD4	2	18	1	6	28	14	32	7
D3 XAD4	3	19	2	9	36	11	21	11
Devon Soil HS isolated using the XAD resins								
R 7 HA	4	16	7	22	18	12	21	29
R 7pt HA	2	17	3	8	36	–	35	11
R 10 HA	4	14	5	10	34	–	34	15
R 12 HA	2	12	4	13	17	13	39	17
R 12pt HA	0	13	3	10	18	13	43	13
Devon Soil HS isolated following the IHSS method								
I 7 HA	3	14	5	22	16	13	28	27
I 10 HA	1	15	1	7	42	–	34	8
I 12 HA	2	11	3	12	18	13	41	15
Devon Soil HS isolated using the XAD resins								
R 7 FA	5	21	4	21	17	11	22	25
R 10 FA	3	17	5	14	26	8	28	19
R 12 FA	3	15	4	15	25	12	27	19
Devon Soil HS isolated following the IHSS method								
I 7 FA	4	22	3	19	20	10	23	22
I 10 FA	4	15	5	14	42	–	21	19
I 12 FA	2	14	4	10	29	12	30	14
Devon Soil HS isolated using the XAD resins								
R 7 XAD4	1	19	1	6	53	–	21	7
R10XAD4	3	12	3	6	58	–	18	9
R12XAD4	2	9	1	4	72	–	13	5

*f*a Total aromaticity

contribution at 32 ppm, indicates that a large proportion of the aliphatic material was of microbial origin.

All of the NMR spectra for the XAD-4 acids were similar, and all of the ^{13}C NMR spectra all exhibited resonances in the carboxyl, aromatic-C, O-C-O, carbohydrate, and aliphatic regions. The concentration in each region changed, depending on the pH of the extracting solution. There was little evidence for tannin-type structures in the XAD-4 acids. There resonance at 140–160 ppm was small, and there was little evidence for methoxyl. Quinone structures were unlikely to contribute in the carboxyl resonance. Thus, the AAs may have contributed to the resonance at 160–190 ppm.

The aromatic signals accounted for only about 5–7% of the humic material, and these were at 115–145 ppm, characteristic of alkyl substituted aryl carbon, and to non-substituted aromatic carbon (Hammond *et al.*, 1985). The spectra in this region were simple, and so it is difficult to discuss the types of aromatic materials present.

There was a strong resonance in the 100–110 ppm region, characteristic of anomeric C (C-1 of sugars). Inputs from tannin substances were unlikely because of the lack of resonance at 140–160 ppm, and around ~58 ppm (methoxyl C).

NMR spectroscopy revealed compositional differences between the humic fractions isolated at the different pH values. These spectra showed that the HAs and FAs isolated at pH 7 had more aromaticity and acidic character than those isolated at the higher pH values. On the other hand, the carbohydrate, and (possible) carbohydrate-related components, enriched the fractions isolated at the higher pH values. The FAs extracted at pH 10.6 and at 12.6, and the HAs isolated at pH 12.6, were relatively rich in O-aromatic substituents. That would indicate that the HS isolated at pH 7 were more highly humified, and those from the more alkaline solutions were plant-derived and in the process of transforming to more highly oxidised HS. The spectra also indicated similarities in the HAs and FAs isolated at pH 7 and at pH 12.6, using the IHSS and the XAD-8 and XAD-4 resin in tandem methods. The components of the XAD-4 acids were mainly aliphatic, highly carboxylic, and were low in aromaticity. Again, there were differences between the XAD-4 acids isolated at the different pH values. In general, the contents of carbohydrate, or of carbohydrate-related materials increased with the pH of the extracting solution.

Traditionally, resonances at 65–110 ppm have been assigned to carbohydrates. It is clear that structures other than carbohydrates also contributed to the intensities in this region, such as aliphatic C bonded to OH groups, and ether C occurring in saturated five and six membered rings, as well as α-carbons or β-carbons of some amino acids (Breitmaier and Voelter, 1987). As stated, the soil FAs contained 3–6% AAs, and the HAs contained 9–16% AAs. The soil FAs contained 4–11% NS, and the HAs contained 4–7% of these. Integrations of the NMR spectra indicated that 15–42% of the humic materials resonated in the 65–110 ppm region of the spectrum. Clearly all of the HAs and FAs in the present study contained significant amounts of modified or non-carbohydrate materials.

The D3 samples were different from the others. These had similar carboxyl (190–160 ppm), larger aromatic (110–140 ppm), and lower aliphatic hydrocarbon (10–65 ppm) contents compared to the other fractions. These also had significantly larger resonances in the 65–110 ppm region, which is consistent with the NS analyses because both of these samples had significantly higher NS contents. The D3 XAD-4 acids sample had the highest NS (5.3%) content, and had the largest integrated area in the 65–110 ppm resonance, although all of the area could not be accounted for as NS. AAs and higher alcohols could also contribute to resonances in this region, but it is unlikely that these could account for the differences observed.

The 160–190 ppm resonance is indicative also of carboxyl, or of secondary amide linkages of peptides and proteins. Because the HAs and FAs in the present study contained significant amounts of N (up to 4.25%), and since AAs constituted 30–50% of this nitrogen, it is probable that amide linkages contributed to the 160–190 ppm resonance.

The FAs and HAs contained 3–6% and 9–16% AAs, respectively. Both sets of samples contained significant amounts of aromatic functionalities (15–29%), and NMR spectra suggested lignin and tannin-type structures containing quinone groups were significant contributors to functionality.

Distinct peaks were visible in some ^{13}C NMR spectra of the HAs and FAs (and were overlapped in others) around 21–22, 26–28, and 31–32 ppm in the 0–48 ppm aliphatic resonances. At 40 ppm alkyl C and AA C were significant contributors to the resonance, and AA C also contributed to the broad resonance at 53 ppm, and C in aliphatic methyl ethers contributed to the peak at 58 ppm (Breitmaier and Voelter, 1987). Schnitzer and Preston (1983) assigned the peaks at 53 and 56 ppm to OCH_3 in aromatic esters and ethers, and the peak at 58 ppm can be attributed to AA C (Piccolo *et al.*, 1990). The resonance at 63 ppm had contributions from carbon in amino acids and carbohydrates (Schnitzer and Preston, 1983).

A five fold decrease was found in the AA N in the drainage water HS compared to the contents in the soil HS. Hydrolysis of AAs linking the humic 'core' structures in the soil would release the 'core' materials which were then soluble in the waters. The N in the AAs would convert to soluble nitrates. Although the direct detection of nitrogen through NMR experiments is difficult, due to the insensitivity of the ^{14}N nucleus to the NMR experiment and the very low natural abundance of the sensitive ^{15}N isotope, it is possible to detect N indirectly through couplings with the sensitive ^{1}H nucleus. One-dimensional (1–D) ^{1}H NMR chemical shifts can help identify areas where N of various types should appear, but with the overlap of H-C chemical shifts across the spectra it is difficult from 1–D ^{1}H NMR alone to know which responses are from N-H protons and which are from C-H protons. Two dimensional NMR experiments are useful for understanding how N is incorporated in AA residues which in turn will determine their release to the environment.

In separate studies we have also carried out 1–D NMR studies on HS isolated from a moss climax vegetation using 600 MHz instrumentation and

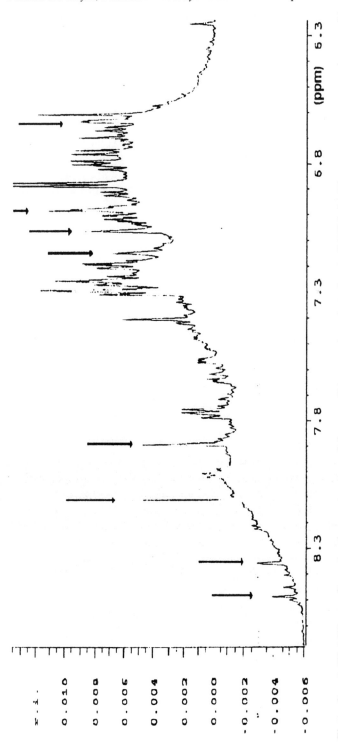

Figure 2 *A 600 MHz one dimensional proton spectra of the moss FA aromatic region taken in DMSO d₆. The arrows represent peaks caused by amide protons these peaks are seen to disappear with the addition of D₂O*

DMSO d_6 solvent. There was evidence (Figure 2) for peptide-type N in the aromatic region of the spectrum. That evidence was removed when D_2O was added.

In addition to the information obtained on nitrogen, these NMR experiments and others including (1–D) DEPT ^{13}C, 2-D ^1H-^1H COrrelation SpectroscopY (COSY), 2-D ^{14}H-^{13}C Heteronuclear Multiple Quantum Coherence (HMQC), and 2-D ^1H-^{13}C Heteronuclear Multiple Bond Coherence (HMBC) are extremely powerful tools for the determination of 'backbone' humic structures. Experiments utilizing these techniques and well fractionated samples will allow different isomers, aromatic substitution patterns, fatty acids, and sugars to be identified, and in some cases whole areas of structures to be mapped. The results will help provide an understanding of the full role of nitrogen in the soil environment.

3.8   Results of the Amino Acid (AA) Analyses

The soil HAs, FAs, and XAD-4 acids had, on average, 5, 7, and 5 times more AAs, respectively, than did the same fractions in the drainage waters. The total AA contents, in nmoles mg^{-1}, followed the order HA > XAD-4 > FA. These findings agree with those for other aquatic HS (Thurman, 1985, Watt *et al.*, 1996a; 1996b). AA contents were similar to the NS contents, and were ca 2.3 to 3.3% of the masses of the HAs isolated, and accounted for ca 12% of the N in the HA fractions, and only 5% of that in the FAs and XAD-4 acids. Although the AA contents of the FAs were similar for all samples, the total AA contents of the XAD-4 fractions varied, and those of the XAD-4 acids of the D3 fraction were exceptional. AAs accounted for 11% of the humic N in that fraction.

The abundances of the AAs in the HA fractions tended to decrease in the order glycine (GLY) > aspartic acid (ASP) > alanine (ALA) > glutamic acid (GLU) > threonine (THR) > valine (VAL) > leucine (LEU) > serine (SER) > isoleucine (ILE) > lysine (LYS) > phenylalanine (PHE) > tyrosine (TYR) > arganine (ARG) > histidine (HIS) > methionine (MET). That order has two of the hydrophobic acids (GLY, ALA, containing the amino and carboxyl groups) in high abundances, and with lesser, but more or less equal amounts of the polar dicarboxylic AAs (ASP and GLU). The abundances of the three other hydrophobic acids (LEU, ILE, VAL), and of the hydroxyamino acids (THR and SER) were intermediate. The basic AAs, LYS and ARG, were in relatively low abundance, as were the aromatic acids, PHE and TYR, and the sulfur-containing compounds HIS and MET. In general, the abundances of the first four AAs decreased in the order GLY > ASP > ALA > GLU for all of the fractions studied. The distributions of the last five generally decreased in the order, PHE > TYR > ARG > HIS > MET, and those of the remaining varied, depending on the humic fraction and source. The VAL and THR contents were reversed for the FAs (compared to the HAs), and the other AAs generally decreased in the order, LEU > ILE > PHE > LYS.

The concentration of AAs in the isolates at any particular pH tended to decrease in the same order as for samples from the drainage waters: HA > XAD-4 acids > FA. The HApts generally had the highest AA contents. For the most part the total acidic (TA), total basic (TB), total neutral hydrophobic (TH$_o$), and total neutral hydrophilic (TH$_i$) AAs followed the order: TH$_i$ >> TA > TH$_o$ >> TB. Only the XAD-4 acids from D2 deviated from that trend (TH$_i$ >> TH$_o$ > TA >> TB). Some distinct trends were observed in relation to the overall compositions of the AA subgroups. The TH$_i$% always followed the order XAD-4 acids > FA > HA. The TB% and TH$_o$% generally followed the order HA > FA > XAD-4. The TA% distribution varied, with the XAD-4 acids having the highest content, while the HA and FA contents interchanged.

Ishiwatari (1985) has outlined differences in AA compositions between aquatic and soil HS. In the cases of soil HS, he showed that the relative abundances of the TB AAs and of the TH$_i$ AAs increased in the order: FA < HA < humin. We found similar trends for the HAs and FAs of drainage waters.

Our data for the soil HS followed the general trends reported by Schnitzer (1982), and also those found by Alberts *et al*. (1992) for AAs in HA and FA isolates from the OM in a salt marsh estuarine sediment. They (Alberts *et al*) also found that the AAs associated with the sediment HS were a factor of 10 higher than those in FAs and HAs from streams, and they have shown that the concentrations of AAs in the living plants (precursors of the HAs and FAs) had lower AA contents, while the contents of the dead plants were higher. Watanabe *et al*. (1994) extracted HS from soils using the traditional IHSS method and the XAD-8 resin. Their total AA contents in the HA and FA fractions were similar to those found in the present study, and the total AA contents were similar in the humic fractions isolated by both methods. Our data indicate that the AA contents of the HAs isolated by the IHSS (I) and resin (R) methods were broadly comparable. However, we found that as the pH of the extractant was increased, the AA contents of the humics isolated increased. The HA isolated from soil at pH 10.6 using the IHSS procedure had a greater AA content than did the resin isolated sample. The HA isolated at pH 7, and processed following the I procedure included the R pH7 HApt (R7HApt) fraction. This fraction contained significantly more AAs (14%) than the R7HA fraction (7%). The I7HA (which includes the R7HApt fraction) contained 9% AAs, and the additional content (compared with the R7HA) included the contribution from the R7HApt.

The HAs isolated at pH 7 had the lowest AA contents, 7%, but at pH 12.6 the HA contained 16.3% AAs. This is reflected in the NMR spectra at about 55 ppm resonance. The HA at pH 7 contained significant methoxy carbon (from tannins), and had a sharp peak at around 58 ppm, and there was still a sharp peak ~57 ppm in the NMR of the HA isolated at pH 12.6. The concentration of tannin-type components in the extracts at pH 12.6 was not as great; thus AAs probably contributed significantly to the resonance at ~57 ppm in the NMR spectrum of the HAs isolated at pH 12.6.

The soil FAs also contained significant quantities of AAs, 3–6%. Generally the AA content increased as the pH of the extracting solution increased. Normally, the isolates at pH 12.6 had the highest AA contents, but in the cases of the XAD-4 acids the fraction isolated at pH 10.6 had the highest AA content (7.4%) followed the same trends. The NS content for this fraction was also higher than for the extractants at pH 7 and at pH 12.6.

The TH_i and TA AAs in the XAD-4 acids generally constituted about 80% of the total AAs. The concentration of the TH_i and TA AAs was generally greatest in the isolates at pH 12.6, and lowest in those at pH 7. The reverse trend was followed for TH_o samples, while the concentration of the TB AAs was always highest in the extracts at pH 10.6.

The TH_i and TA AAs constitute about 70% of the HA isolates. The concentration generally decreased as the pH of the extracting solution increased. The TH_o and TB fractions followed the reverse trend. The FAs exhibited similar trends, but the TH_i and TA fractions accounted for > 85% of the isolates at pH 7, and > 70% of those at pH 12.6.

The distributions of the TH_i and TA AAs indicate that six AAs (ASP, GLU, THR, SER, GLY, and ALA) were the major components of the AAs found in the HS

The TH_o AAs in the soil HS decreased in the order: HA > FA > XAD-4 acids. The abundance of the TH_i AAs generally followed the order: XAD-4 acids > FA > HA. However, the isolates from the soil at pH 7 and pH 12.6 followed the order: HA > XAD-4 acids > FA. The distribution of the TA between the isolates at the different pH values did not follow any set pattern, but the general order HA > XAD-4 acids > FA held for most fractions. Although these general patterns were followed, the concentrations of the AAs within each subgroup varied, depending on the procedure which was used to isolate the humic material. The RHAs generally contained more of the acidic and neutral H_i AAs compared to the IHAs, while the reverse was true of the basic and NH_o AAs. The FAs isolated following both procedures also exhibited differences.

The AAs accounted for significantly more of the N content of the soil HS, compared with the drainage water HS. AA N accounted for about 40% of the N in the HApts isolated from the soil. The AAs account for ~25% of the N in the HA fraction, but in the case of the R12HA sample, these accounted for 52% of the N. The N in the IHA fractions was composed of ~34% AAs, although the HApt material resulted in this increase (the HApt fraction is an integral part of the HAs isolated using the IHSS procedure). AAs contributed ~20–30% of the N in the FA fractions. In general, the %N due to AA increased as the pH of the extracting solution increased.

Analysis of the individual AAs gives some information about their distribution in the different humic fractions. Seven or eight AAs were the major contributors to the AA contents; however, no definite trends in order of abundances were established. ASP and GLU were the predominant AAs in the HAs isolated. The six least abundant AAs tended to decrease in the order, ILE > ARG > PHE > HIS > TYR > MET. No definite trend was evident for the other AAs in the HAs isolated from the soil at the different pH values, but

GLY, ALA, THR, SER, and LEU gave significant contributions. The abundances of the last six AAs were similar for the FAs; ASP was generally the most abundant, with major contributions from GLY, GLU, ALA, THR, and VAL. The abundances of the AAs in the XAD-4 acids varied but ASP and GLY were dominant species, followed by GLU, ALA, THR, SER, VAL, and LEU. These distributions were different from those described for the HAs, FAs, and the XAD-4 acids from the drainage waters.

4   Conclusions

This study has investigated the compositions of humic substances (HS) in a grassland soil, and in the drainage waters from the soil. Two procedures were used to isolate humic fractions from the soil. One used a sequential and exhaustive extraction at different pH values, and two resins in tandem to recover the humic fractions. The other followed that used to isolate the soil Standards of the International Humic Substances Society. The resin procedure can be recommended where more extensive fractionations are required.

The compositions of the drainage waters and of the soil HS were different, especially with regard to the sugar and amino acids (AAs) contents.

The fundamental theme of the study focused on the extents to which HS containing organic precursors (especially AAs) of nitrate and ammonia can enter aquatic systems.

The novel fractionation procedure used has shown that the least transformed (most plant like) and less soluble soil HS were the most enriched in AAs and neutral sugars (NS) residues. This suggests that:

1.   as humification proceeds the plant AAs become mineralized, and the AAs associated with the HS are likely to have microbial origins;
2.   the HS which enter the drainage waters are linked in the soil through peptide and carbohydrate components, and the soluble humic fractions are released to the soil solution on hydrolysis of these components; or
3.   the humic components are held in the soil in non-covalently linked associations with saccharides and peptide-type structures, and with themselves. In time the associations are disrupted and the humic fractions are solubilised.

The study indicates that microbial processes are important for the release of HS to drainage waters. The release is greatest in the drainage flush when rainfall exceeds evaporation and the SOM has wetted (become swollen) after the summer period, and the ambient temperature is conducive to microbial activity. Release is least at the end of the winter period following the period of low microbial activity.

The relatively low content of AAs in the drainage waters would suggest that significant metabolism of these took place before the HS were released. During the growing season the nitrates and ammonia released will be utilized by the growing vegetation. However, such nutrients, when released throughout the non-growing season, would enter the drainage waters. Our experience does not encompass the fate of the AAs associated with the HS in the waters, or of the other forms of organic N in the organic matter in soils and waters.

References

Alberts, J.J., Filip, Z., Price, M.T., Hedges, J.I. and Jacobsen, T.R. (1992). CuO-oxidation products, acid hydrolysable monosaccharides and amino acids of humic substances occurring in a salt marsh estuary. *Org. Geochem.* **18**, 171–180.

Armstrong, A.C. and Garwood, E.A. (1991). Hydrological consequences of artificial drainage of grassland. *Hydrol. Process.* **5**, 157–174.

Breitmaier, E. and Voelter, W. (1987). *^{13}C NMR Spectroscopy.* 3rd ed. Verlag Chemie, Weinheim, West Germany.

Clapp, C.E. and Hayes, M.H.B. (1996). Isolation of humic substances from a mollisol using a sequential and exhaustive extraction process. pp. 3–11. *In* C.E. Clapp *et al.* (eds), *Humic Substances and Organic Matter in Soil and Water Environments: Characterization, Transformations, and Interactions.* IHSS, St. Paul.

Clapp, C.E., Layese, M.F., Hayes, M.H.B., Huggins, D.R. and Allmaras, R.R. (1997). Natural abundance of ^{13}C in soils and water. pp. 158–175. *In* M.H.B. Hayes and W.S. Wilson (eds) *Humic Substances, Peats, and Sludges: Health and Environmental Aspects.* Royal Society of Chemistry, Cambridge.

Deines, P. (1980). The isotopic composition of reduced organic carbon. pp. 329–406. *In P. Fritz and J.Ch. Fontes (eds), Handbook of Environmental Isotope Geochemistry, Volume 1.*. Elsevier Scientific Publishing Company, Amsterdam,

Delwiche, C.C. and Steyn, P.L. (1970). Nitrogen fractionation in soils and microbial reactions. *Environ. Sci. Technol.* **4**, 929–935.

Hammond, T.E., Cory, D.G., Ritchey, W.M. and Morita, H. (1985). High resolution solid state ^{13}C NMR of Canadian peats. *Fuel* **64**, 1687–1695.

Hayes, T.M. (1996). Study of the humic substances from soils and waters and their interactions with anthropogenic organic chemicals. PhD Thesis, Univ. of Birmingham.

Hayes, T.M. and and Hayes, M.H.B. (1997). Considerations of the influences of water soluble humic substances on the solubilities of anthropogenic organic chemicals. pp. 655–669. In J. Drodz *et al.* (eds), *The Role of Humic Substances in the Ecosystems and Environmental Protection.* Polish Society of Humic Substances, Wroclaw.

Hayes, T.M., Hayes, M.H.B., Skjemstad, J.O., Swift, R.S., and Malcolm, R.L. (1996). Isolation of humic substances from soil using aqueous extractants of different pH and XAD resins, and their characterisation by ^{13}C NMR. pp. 13–24. *In* C.E. Clapp, *et al.*, (eds), *Humic Substances and Organic Matter in Soil and Water Environments: Characterization, Transformations, and Interactions.* IHSS, St. Paul.

Hayes, T.M., Watt, B.E., Hayes, M.H.B., Clapp, C.E., Scholefield, D., Swift, R.S., and Skjemstad, J.O. (1997). Dissolved humic substances in waters from drained and undrained grazed grassland in SW England. *In* M.H.B. Hayes and W.S. Wilson (eds), *Humic Substances, Peats, and Sludges: Health and Environmental Aspects.* Royal Society of Chemistry, Cambridge.

Hopkins, D.W., Chudek, J.A. and Shiels, R.S. (1993). Chemical characterisation and decomposition of organic matter from two contrasting grassland soil profiles. *J. Soil Sci.* **44**, 147–157.

Ishiwatari, R. (1985). Geochemistry of humic substances in lake sediments. pp. 147–180. *In* G.R. Aiken *et al.* (eds), *Humic substances in Soil, Sediment and Water, Geochemistry, Isolation and Characterisation.* Wiley, New York.

Komor, S.C. and Anderson, H.W. Jr. (1993). Nitrogen isotopes as indicators of nitrate sources in Minnesota Sand-Plain aquifers. *Ground Water* **31**, 260–270.

Malcolm, R.L. (1991). Factors to be considered in the isolation and characterisation of aquatic humic substances. pp. 369–391. *In* H. Boren and B. Allard (eds), *Humic Substances in the Aquatic and Terrestrial Environment.* Wiley, Chichester.

Piccolo, A., Campanella, L. and Petronio, B.M. (1990). Carbon-13 Nuclear Magnetic Resonance spectra of soil humic substances extracted by different mechanisms. *Soil Sci. Soc. Am. J.* **54**, 750–756.

Schnitzer, M. (1982). Organic matter characterisation. pp. 581–594. In A.L. Page *et al.* (eds), *Methods of Soil Analysis. Part 2.* (2nd ed). ASA-FSSA, Madison, Wis.

Schnitzer, M. and Preston, C.M. (1984). Effects of acid hydrolysis on the ^{13}C NMR spectra of humic substances. *Plant and Soil* **75**, 201–211.

Schofield, D. (1995). Personal Communication. AFRC Institute for Grassland and Environmental Research, North Wyke, Devon, UK.

Simpson, A.J., Burdon, J., Graham, C.L., and Hayes, M.H.B. (1997). Humic substances from podzols under oak forest and a cleared forest site II. Spectroscopic studies. pp. 83–106. *In* M.H.B. Hayes and W.S. Wilson (eds), *Humic Substances, Peats, and Sludges: Health and Environmental Aspects.* RSC, Cambridge

Stevenson, F.J. (1994). *Humus Chemistry, Genesis, Composition, Reactions.* Wiley, NY.

Swift, R.S. (1997). Soil organic matter. Chapter 35. *In* D.A. Sparks *et al.* (eds), *Methods of Soil Analysis: Chemical Methods.* ASA-FSSA, Madison, Wis.

Thurman, E.M. (1985). Humic substances in groundwater. pp. 87–104. *In* G.R. Aiken, D.M. McKnight, R.L. Wershaw and P. MacCarthy (eds), *Humic Substances in Soil, Sediment and Water, Geochemistry, Isolation and Characterisation.* Wiley, NY.

Thurman, E.M. and Malcolm, R.L. (1981). Preparative isolation of aquatic humic substances. *Environ. Sci. Technol.* **15**, 463–466.

Turnell, D.C. and Cooper, J.D.H. (1982). Rapid assay for amino-acids in serum or urine by precolumn derivatization and reversed-phase liquid chromatography. *Clinical Chem.* **28**, 527–531.

Tyson, K.C., Garwood, E.A., Armstrong, A.C. and Schofield, D. (1992). Effects of Field Drainage on the Growth of Herbage and the Liveweight Gain of Grazing Beef Cattle. *Grass and Forage Sci.* **47**, 290–301.

Watanabe, A. Itoh, K., Arai, S. and Kuwatsuka, S. (1994). Comparison of the Composition of Humic and Fulvic Acids Prepared by the IHSS Method and NAGOYA Method. *Soil Sci. Plant Nutr.* **40**, 23–30.

Watt, B.E., Malcolm, R.L., Hayes, M.H.B., Clark, N.W.E., and Chipman, J.K. (1996a). The chemistry and mutagenic potentials of humic substances from different watersheds in Britain and Ireland. *Water Res.* **32**, 1502–1516.

Watt, B., Hayes, T.M., Hayes, M.H.B., Price, R.T., Malcolm, R.L. and Jakeman, P. (1996b). Sugars and Amino Acids in Humic Substances Isolated from British and Irish Waters. pp. 81–91. *In* C.E. Clapp *et al.* (eds), *Humic Substances and Organic Matter in Soil and Water Environments: Characterization, Transformations, and Interactions.* IHSS, St. Paul, Minn.

15

Fertilizer and Nitrate Pollution in India

Himanshu Pathak

DIVISION OF ENVIRONMENTAL SCIENCES, INDIAN
AGRICULTURAL RESEARCH INSTITUTE, NEW DELHI 110 012,
INDIA

Abstract

Fertilizers are major contributors to increased crop production in India. At present, the country consumes about 14 million tonnes of fertilizer per annum of which more than 75 per cent is nitrogenous fertilizer. As India is trying to increase its food grain production, fertilizer consumption is expected to increase. But concern has been expressed that over-reliance on mineral fertilizers may cause unsustainable environmental penalties. The main problem associated with excess nitrogenous fertilizer use is its movement to water bodies causing eutrophication of surface water and nitrate (NO_3^-) pollution of groundwater. However, there are very few cases of nitrate contamination of groundwater by fertilizer in India. In Punjab, where the nitrogenous fertilizer consumption is about 200 kg ha^{-1}, the average concentration of nitrate-N in groundwater was about 4–5 mg l^{-1}. However, a very high level of nitrate-N (>350 mg l^{-1}) in groundwater is reported from a few urban pockets of India. This could be attributed to the dumping of animal manures, organic wastes from industries and sewage on to soil. Indian soils are very poor in N content and malnourished except for a few tracts of Punjab and Haryana. Thus, there is little scope for fertilizer-related nitrate pollution in India, However, care should be taken to anticipate the long-term pollution problems.

1   Introduction

Indian agriculture, until the middle of the twentieth century, relied mostly on organic manure. When India gained independence in 1947, ammonium sulfate was the only N fertilizer known in the country. Introduction of urea in 1956 and diammonium phosphate in 1967 has revolutionized the fertilizer application in the country (1). The green revolution in India owes much of its success to fertilizer application. During the post green revolution phase there has been a phenomenal rise in fertilizer consumption in India (Fig. 1). At present the country consumes about 14 million tonnes of fertilizer per annum and more than 75 per cent of it is nitrogenous fertilizer (2). Average consumption of fertilizer in the country is about 70 kg ha^{-1} yr^{-1}, which is much less in comparison to many other countries like China, South Korea, North Korea, Japan, United Kingdom, USA and Netherlands, where the consumption of N, P and K is often more than 250 kg ha^{-1}yr^{-1}. However, there is regional disparity in India in terms of fertilizer consumption (Table 1). Farmers of Punjab use nearly 250 kg fertilizer ha^{-1} yr^{-1}, whereas in the North East Regions it is only 5–10 kg ha^{-1} yr^{-1}. Most importantly, irrespective of the agro-climatic zones there is excess mining of nutrients *i.e.*, removal of N, P and K by crop uptake is more than their replenishment with fertilizers (Table 1). In 2000 AD, removal of N, P and K is expected to be about 32 million tonnes and the consumption is expected to be 22 million tonnes.

Considering 50% efficiency, the contribution of fertilizers towards the total removal will be 11 million tonnes creating a gap of 21 million tonnes. Though organic manures, biological nitrogen fixation, precipitation and irrigation water will contribute to some extent, a major portion of this nutrient will be mined from soil. Therefore, as India is trying to achieve the second 'Super Green Revolution', the fertilizer consumption needs to increase to keep the soil fertile and make its agriculture sustainable. But concerns have been expressed about the environmental impact of fertilizer use. There are some claims of dire

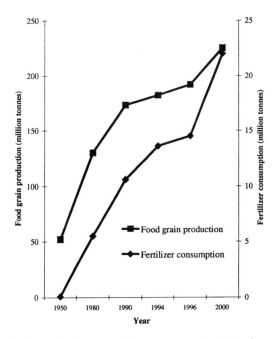

Figure 1 *Food grain production and fertilizer consumption in India*

Table 1 *Consumption, removal and range of N, P and K use ('000 tonnes) in different agro-climatic regions of India during 1988–89*

Region	Removal	Consumption	Gap
Western Himalayan	309	160	149
Eastern Himalayan	445	110	335
Lower Gangetic Plain	911	570	341
Middle Gangetic Plain	1815	1300	515
Upper Gangetic Plain	2553	1400	1153
Trans Gangetic Plain	2540	1670	870
Eastern Plateau & Hills	1326	360	966
Central Plateau & Hills	2449	620	1829
Western Plateau & Hills	2149	860	1289
Southern Plateau & Hills	2636	1570	1066
East Coast Plains & Hills	1861	1310	551
West Coast Plains & Ghats	664	400	264
Gujarat Plains & Hills	860	650	210
Western Dry Region	265	50	215
The Islands Region	–	–	–
All India	20,783	11,030	9,753

Source: (29)

consequences with fertilizer use and more particularly against N fertilizers. This makes sense as fertilizer use efficiency by crops seldom exceeds 50 per cent. The remainder is lost from the soil system by leaching, run-off, denitrification and volatilization and pollutes the soil, water and air, the vital resources of nature. Evidence exists that crop yields are stagnating, physical and chemical properties of soil are deteriorating, water is being polluted with nitrate and the atmosphere is being enriched with nitrous oxide (N_2O) posing a great threat of global warming and ozone layer depletion due to over reliance on mineral fertilizer. The health problems associated with nitrate rich water have been discussed elsewhere in this publication. The article attempts to review the work done in India on the nitrate pollution in the water bodies and the role of fertilizer in augmenting this pollution.

2   Problems of Nitrate Pollution in India

The welfare of any civilization depends heavily upon the availability of fresh water. Unfortunately, though India receives on an average 1000 mm annual rainfall, fresh drinking water is still not available to a considerable portion of its citizens. In this country surface water, sometimes even untreated, as well as groundwater, mostly untreated, are used for human and animal consumption. Under those circumstances, it will be highly critical to let this free gift of nature be polluted in the name of development. But recent studies indicated that pollutants such as nitrate contributed by various anthropogenic sources is posing a menace to this valuable resource. In India the studies on pollution of water bodies with nitrate remained dormant until the mid eighties. Some studies thereafter, have reported high nitrate content, more than 100 mg l^{-1}, in groundwater in a significant number of samples (3, 4, 5, and 6). Handa (3) analyzed 2000 samples from dug wells and 500 samples from deep tube wells and observed that a considerable number of samples had high (>50 mg l^{-1}) nitrate content. The nitrate in deep water was only about 1–2 mg l^{-1} whereas the comparable level in shallow water was up to 100 mg l^{-1} in humid areas and up to 1000 mg l^{-1} in arid and semiarid regions. He concluded that high nitrate was due to pollution either from human or animal sources or from run-off from agricultural fields dressed with chemical fertilizers. A rapid reconnaissance of nitrate in shallow groundwater (Table 2) was carried out by the

Table 2 *State wise consumption of nitrogenous fertilizers and nitrate in groundwater*

State	N fertilizer consumption (kg ha^{-1} yr^{-1})	Mean nitrate in groundwater (mg l^{-1})	Maximum nitrate reported (mg l^{-1})
Andhra Pradesh	11.60	13	208
Bihar	23.60	21	350
Gujarat	22.16	50	410
Haryana	91.05	99	1800
Himachal Pradesh	4.20	8	120
Jammu and Kashmir	3.10	7	275
Karnataka	20.43	15	200
Madhya Pradesh	8.40	30	473
Maharashtra	10.59	45	–
Northern States	11.2	8	275
Orissa	8.53	15	800
Punjab	162.33	55	562
Tamil Nadu	30.70	26	1030
Uttar Pradesh	52.55	23	634
West Bengal	43.00	14	480

Source: (6)

Central Ground Water Board (6). It showed that the samples from Haryana and Punjab, the states where the fertilizer consumption is high, contained high amount of nitrate in groundwater. Kumar and Singh (7) also reported that the groundwater samples from Mahendragarh district of Haryana, contained high proportion of nitrate (>45 mg l^{-1}) in 75% of the samples. Subsequently, National Environmental Engineering Research Institute, Nagpur, Maharashtra, analyzed groundwater from selected districts of 17 states of India and reported nitrate content was above the World Health Organisation (WHO) limit in 1290 out of 4696 samples. Other studies including that of Mathur and Ranganathan (8) from Jodhpur, Rajasthan, Singh *et al.* (9) from Lucknow city, Uttar Pradesh, Tamta *et al.* (10) from Bangalore, Karnataka also reported high nitrate content (>50 mg l^{-1}). A study from Guhahati, Assam, showed that only 6% of samples contained more than 45 mg NO_3^- l^{-1} (11). Handa

(12) reviewed the work done on nitrate pollution in India and reported that in the tube well (confined aquifer) water samples, about 60% of the samples had less than 1 mg nitrate-N l^{-1} and less than 5% had over 5 mg nitrate-N l^{-1}. In dug wells, however, many places contained anomalously high concentrations of nitrate. He also observed that the nitrate content of surface waters in India was quite low. Nearly 50% of the stream waters analyzed had less than 0.7 mg nitrate-N l^{-1}. Only in 10% of the stream waters, was the nitrate-N content more than 10 mg l^{-1}.

3   Fertilizer and Nitrate Pollution

Pollution of groundwater from fertilizer is caused by leaching, whereas for surface water it is by run-off. The magnitude of loss of N through leaching depends upon soil conditions, agricultural practices, agro-climatic conditions, type of fertilizers and methods of application. The time taken by nitrate to move from the root zone to the water-table, therefore, varies considerably. In sandy soils with high water table and high rate of fertilizer application, it may reach the water table in matter of days whereas in heavy soils, low rainfall and low rate of application with deep water table, it may take years. There are limited studies, which correlate the nitrate pollution of groundwater and the use of N fertilizers. Bajwa *et al.* (13) studied the influence of fertilizer application on nitrate content of groundwater in some districts of Punjab, where the fertilizer application rate is the maximum in the country. They observed that 78.4% and 21.6% of 21 to 38 m deep tube well water samples contained <5 and 5–10 mg nitrate-N l^{-1}, respectively (Table 3) and none of the samples contained nitrate-N concentration of more than 10 mg l^{-1}. In the groundwater samples collected from 9 to 18 m deep hand pumps located at village inhabitants, 64% and 2% samples contained 5–10 and >10 mg nitrate-N l^{-1}, respectively (Table 4). They concluded that animal wastes dumped in the inhabited areas could be the possible cause of high nitrate in hand pumps.

Table 3   *Nitrate   concentration in ground water samples from tube wells (21 to 38 m deep) located in cultivated areas of Punjab*

Block	No. of samples	N-Fertilizer application (kg ha^{-1} yr^{-1})	Mean nitrate-N (mg l^{-1})
Dehlon	84	249	3.84 +/− 1.61
Ludhiana	33	258	3.12 +/− 1.40
Sudhar	43	242	3.93 +/− 1.66
Kartarpur	34	193	2.73 +/− 1.22
Jandialaguru	24	172*	4.13 +/− 1.18
Malerkotla	18	151*	3.93 +/− 1.18

*Vegetable growing area, along with fertilizers 15–20 t ha^{-1} of farm yard manure containing 75–100 kg N ha^{-1} yr^{-1}.
Source: (13)

Table 4 *Nitrate concentration in ground water samples from hand pumps (9 to 18 m deep) located in village inhabited areas of Punjab*

Block	No. of samples	Percentage of samples containing		
		0–5 mg N l^{-1}	5–10 mg N l^{-1}	>10 mg N l^{-1}
Dehlon	177	17	80	3
Ludhiana	46	43	57	–
Sudhar	52	54	46	–
Kartarpur	56	52	46	2
Jandialaguru	14	50	50	–
Malerkotla	22	27	72	–

Source: (13)

They also found that higher nitrate concentrations, though less than the WHO limit, in groundwater were observed where rice, maize, orchards and vegetables are grown. They suggested that inclusion of deep rooted crop in a rotation during the rainy season could effectively decrease nitrate leaching beyond the root zone. Singh *et al.* (14) have reported that in extensively irrigated coarse textured highly percolating soils of Central Punjab, where 40–50 percent of applied nitrogen was lost due to leaching, only 10% of the samples contained nitrate-N concentration of more than 10 mg/l. Though several studies *e.g.*, Bulusu and Pande (15) from Gulbarga district, Karnataka, Gupta (16) from Udaipur district, Rajasthan, Mehta *et al.* (17) for dug well waters of Ganjam district of Orissa, have attributed high nitrate content to N fertilizer use but no conclusive evidence was put forward. Lunkad (6) made an attempt to correlate the nitrate level in groundwater with the consumption of N fertilizers (Table 2). He observed that the maximum nitrate content reported in some ground-water samples from Haryana, Punjab, and Uttar Pradesh in the north, Tamil Nadu in the south, Orissa (Ganjam district) and Bihar in the east and Gujarat on the west also parallels high average nitrate and high N fertilizer consumption. He, however, also observed several exceptions. For example, in West Bengal, where the average nitrate level was low in spite of high dosage of N fertilizers. He concluded that amongst the three basic physiographic-geologic divisions of India, *i.e.*, Indo-Gangetic plain, Peninsular plateau and north and north east India, the Indo-Gangetic plain has the greatest nitrate pollution risk for groundwater as it is reliefless, and comprises of a thick pile of unconsolidated and permeable alluvial sediments. He suggested that in this region fertilizer application must be accompanied by good drainage facilities, which are, unfortunately, lacking in Punjab and Haryana.

Understanding of the fate of applied N, particularly its leaching pattern will help us to estimate the contribution of nitrate in groundwater. Krishnappa and Shinde (18) studied the fate of ^{15}N labeled urea applied at the rate of 100 kg N ha^{-1} under tropical flooded rice culture and observed that leaching losses amounted to be 7.5% during a sequence of three crops and two inter crop fallows. The maximum contribution of the ^{15}N pulse to the nitrate-N content of the groundwater was about 2% and that occurred in the first crop season and it declined below 0.2% by the third crop season. Therefore, this study indicated that fertilizer is not a major contributor of nitrate in groundwater. Arora *et al.* (19) monitored the nitrate content of tube well waters in the farm of Indian Agricultural Research Institute for three years (1975 to 1978) and observed that it varied from 5 mg l^{-1} to 35 mg l^{-1}, the higher concentration in the rainy season. They noted that the nitrate content started rising during the rainy season up to October November and the fell during January–February, which they attributed to the percolation of accumulated rainwater. They observed that as much as 50% of the applied fertilizer was leached below 50 cm after the first crop, about 15% leached below 150 cm soil depth during a cropping season in a coarse loamy Ustocrept. However, they did not analyze how much of the applied N actually joined the stream of groundwater.

The movement of N to surface water takes place mainly through run-off. Magnitude of loss of N by run-off depends upon soil conditions, management practices, rate of fertilizer application and most importantly weather conditions. As during rainy season most parts of India experiences heavy torrential rain, and this season ironically the main growing season, there is every chance of lot of run-off loss of applied fertilizer. Bijoy-Singh (20) reported that about 90 per cent of nutrients to surface waters come from non-point sources of which 80 per cent is from agricultural lands. Padmaja and Koshy (21) observed that 70 per cent of applied urea N was lost when the field was drained on the day of urea application. Draining of the land after 48 hours of urea application reduced the loss to 44 per cent and after 5 days the loss was negligible. As urea was hydrolyzed and NH_4^+ was adsorbed or fixed by the clay colloids, the loss was less with passage of time.

4 Causes of Nitrate Pollution

Several reasons have been put forward for the high concentration of nitrate in certain pockets. Some of the most cited causes are given below.

4.1 Geological Sources

Weathering of rocks and minerals may release small amounts of ammonium present in them. However, there are no known nitrate deposits in the Indian subcontinent, which could explain the occurrence of high nitrate in ground waters (12). Study of nitrate level in relatively pollution-free zones of India such as high altitude lakes and rivers and snow clad mountains might help in understanding the anthropogenic nature of this pollutant in groundwater (Table 5). In Central Himalayan snow and ice nitrate is about 0.5 mg l^{-1} which is close to its content in Himalayan rivers such as Bhagirathi and Alakananda and in Ganges at Rishikesh before it enters the plains. No nitrate is detected in the rain water around the Dal Lake of Kashmir but the lake water contains an average nitrate concentration of 1.07 mg l^{-1}, which reflects its anthropogenic contribution.

Table 5 *Nitrate content in Indian terrestrial waters in relatively pollution free areas*

Locality	Nitrate $(mg\ l^{-1})$
Central Himalayan snow	0.496
Central Himalayan ice	0.436
Himalayan rivers:	
Bhagirathi	0.310 to 0.992
Alakananda	0.992
Ganga (at Rishikesh)	0.806
Dal lake	1.07
World average river water	1.00
World average ocean water	0.67

Source: (6)

4.2 Atmospheric Sources

Nitrate content in the rain waters in several parts of India revealed that it is normally less than 1.0 mg l^{-1} (12). Therefore, the contribution of this (atmospheric) source could not be very high. Fixation of atmospheric N by legumes and nonlegumes, is generally utilized by the crops. Hence this possibility is also ruled out.

4.3 Soils

Indian soils, being located in the tropical region, are poor in N content, except some soils in the hilly areas. Therefore, their contribution towards the groundwater nitrate is expected to be small (22).

4.4 Irrigation with Treated or Untreated Sewage

In India sewage, being a good source of plant nutrient, is widely applied to land for irrigation. Persistent leaching of the dissolved nitrate possibly enriches the aquifer. Also topographic depressions, when remain filled up with the sewage effluents, act as most potential repositories of uninterrupted percolation of nitrate.

4.5 Use of Industrial Effluent

Several industrial effluents such as distillery effluent, sugar mill effluent and paper mill effluent are used by farmers to supply the crop with water and nutrients. However, excessive use of these effluents, particularly in sandy, shallow depth soil might cause leaching of its N. Several places in India, in and around the factory areas, have been affected by a nitrate pollution problem (23).

4.6   Poor Sanitation and Drainage Facilities

The localities which show very high concentration of nitrate, poor drainage facilities, leakage from septic tanks are likely to be the other sources of this problem.

4.7   Use of Organic Manure

The higher levels of nitrate in some water samples near agricultural fields might also be attributed to extensive use of farmyard manure. In the backdrop of cattle population of nearly 50 million, which is the highest in the world, huge amounts of farmyard wastes are generated and this bears considerable significance in relation to the total nutrient consumption in the country's agriculture (24). United Nation Environment Programme (25) reported that (Fig. 2) in similar situations the leaching loss is more from soil N (25%) and organic N (25%) as compared to that of fertilizer N (15%). Therefore, where soil is a major contributor of nitrate in water, fertilizer is unlikely to contribute the major portion. However, fertilizer N, after immobilization, becomes a part of soil N and undergoes more leaching, but the same is true for organic N as well.

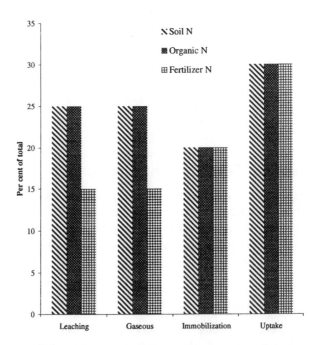

Figure 2   *Removal of N through various pathways*

4.8 Application of Fertilizers

Application of fertilizers may also be an additional factor for nitrate pollution of groundwater. Although the above discussion suggests that there is no specific evidence of nitrate pollution due to fertilizer application, it is apparent that it can contribute substantially in the places of high rate of application, high rainfall, light textured soil and shallow water table. Therefore, as the fertilizer consumption in the country will increase day by day, there is need to develop strategies to control the pollution.

5 Pollution Abatement Strategies

There could be two major approaches to minimize possible environmental pollution due to fertilizer application. Firstly, to improve the ability of plants to compete with the processes which led to losses of nutrients from the soil-plant system to the environment and secondly, reduction of losses themselves. While the first one can be achieved by genetical and biotechnological means, the second one calls for improved fertilizer-use efficiency by improving soil and crop management practices. In the context of emerging biotechnologies it can be tried to develop new plant types to use nutrients more efficiently so that there would not be much loss of applied nutrient. New strains of nitrogen fixers (*Rhizobium* and blue green algae) with higher N fixation capacity can be developed so that dependence on fertilizer will reduce and thereby reducing the threat of fertilizer pollution. Integrated plant nutrition supply systems, the basic principle of which is the maintenance of soil fertility, sustainable agricultural productivity and improving farmers' profitability through judicious and efficient use of mineral fertilizer, organic matter, green manuring, biological N fixation and other inoculants, have to be practised. At the same time measures have to be taken to control erosion and other losses of fertilizer nutrients from soil systems. Use of nitrification inhibitors can possibly play some role in protecting the groundwater from nitrate pollution (26). Though the studies on the effect of nitrification inhibitors on nitrate leaching are rather limited, some studies, particularly by Rudert and Locascio (27) and Timmons (28) reported that nitrapyrin, a nitrification inhibitor, can reduce the leaching of nitrate to a great extent. Lunkad (6) suggested some preventive measures, which amongst many others, included mandatory preparation of aquifer maps for all Indian towns, district headquarters and cities. Such maps, he emphasized, should depict the depth and aerial extent of the aquifers, distribution of shallow and deep wells through them, annual fluctuation of water table, and major pollution criteria such as nitrate content in water. Sewage network and industrial effluent disposal points should be super-imposed on these maps and such maps should be updated regularly. In view of the fluctuating monsoon over India, farmers in different regions should be guided by media-diffusion about manuring period in a routine manner along with weather forecasting. The Government could also introduce legislation restricting fertilizer application to a recommended rate.

6   Conclusion

The main reasons for high nitrate concentration in water seemed to be the dumping of organic wastes, urban sewage and industrial effluent in soil. Fertilizer also accelerates the problem in places of its non-judicious use. But for India the situation is critical. Although excessive fertilizer use does create environmental problems, the concern is also of the degradation of soil due to excess mining of soil fertility caused by the inadequate replenishment of nutrients removed by successive harvests. Therefore, a balance has to be made between these two antagonistic processes. The maximum permissible limit of fertilizer application, *i.e.* beyond which it will be harmful to the environment, should be estimated. The emphasis should be given to increase the fertilizer-use efficiency to prevent the long-term pollution problems and help the resource-poor farmers in saving the money they invested in fertilizer application. There is a lack of sufficient scientific data on nitrate content in water sources in different zones of India and their temporal variation necessary for evaluation of health risks. Hence, more studies on these aspects need to be carried out.

References

1. Goswami, N.N. and R.K. Rattan, *Fertilizer News*, 1990, **35** (9), 15.
2. Fertilizer Association of India, New Delhi, India, *Annual Report*, 1997.
3. Handa, B.K. *Proc. International symposium on the geochemistry of natural waters*, Burlington, Ontario, Canada, 1975, 34.
4. Kakar, Y.P. *Proc. Seminar on Water Quality and Its Management, Central Board of Irrigation and Power*, 1985, 59.
5. Lunkad, S.K *Bhu-jal News.*, 1988, **3**(3), 18.
6. Lunkad, S.K. *Bhu-jal News.*, 1994, **9**(1), 4.
7. Kumar, S. and J.Singh, *International seminar on hydrology*, Vishakhapatnam, India, 1988, 11.
8. Mathur, A.K. and S.Ranganathan, *Bhu-jal News*, 1990, **5**(2), 16.
9. Singh, B.K., O.P. Pal and D.S. Pandey, *Bhu-jal News.*, 1991, **6**(2), 46.
10. Tamta, S.R., S.L.Kapoor and T.Goverdhanan, *Bhu-jal News.*, 1992, **7**(2), 5.
11. Prakash, R., P.K.Kanwar and A.Kumar, *Bhu-jal News.*, 1992, **7**(4), 24.
12. Handa, B.K. *Fertilizer News*, 1987, **32**(6), 11.
13. Bajwa, M.S., Bijoy Singh, and P. Singh, *Proc. First Agricultural Science Congress*, National Academy of Agricultural Sciences, Indian Agricultural Research Institute, New Delhi.1993, p. 223.
14. Singh, I.P., Bijoy Singh and H.S. Bal, *Indian. J. Agric. Econ.*, 1987, **42**(9), 404.
15. Bulusu, K.R. and S.P. Pande, *Bhu-jal News.*, 1990, **5**(2), 39.
16. Gupta, S.C. *Bhu-jal News.*, 1992, **7**(2), 17.
17. Mehta, B.C., R.V.Singh, S.K.Srivastava and S.Das, *Bhu-jal News.*, 1990, **5**(2), 44.
18. Krishnappa, A.M. and J.E. Shinde, 'Soil as fertilizer or pollutant', IAEA, Vienna, 1980, p.127.
19. Arora, R.P., M.S.Sachdev, Y.K.Sud, V.K.Luthra and B.V.Subbiah, 'Soil as fertilizer or pollutant', IAEA, Vienna, 1980, p. 3.
20. Bijoy Singh, 'Nitrogen Research and Crop Production' (H.L.S.Tandon Ed), FDCO, New Delhi, 1996, p. 159.

21. Padmaja, P. and D.R. Koshy, *J. Indian Soc. Soil Sci.* 1978, **26**, 74.
22. Pathak, H., H.C. Joshi, A. Chaudhary, N. Kalra and M.K. Dwivedi, *Proc. National Symposium on recent development in soil science* held at PAU, Ludhiana, India, 1995, p. 63.
23. Pathak, H. and N. Ahmed, *Asia Pacific J. Environment Development*, 1995, **2**(2): 73.
24. Joshi, H.C. and D.L. Deb, *Proc. 81st Indian Science Congress Association*, Jaipur, Rajasthan, India, 1994.
25. United Nations Environment Programme, *Proc. Regional FAO/NAP seminar on fertilization and environment*, Bangkok, Thailand, 1992.
26. Prasad, R. and J.F.Power, *Advances in Agronomy*, **54**, 233.
27. Rudert, B.D. and S.J. Locascio, *Agronomy J.* **71**, 487.
28. Timmons, D.R. *J. Environ. Qual.* 1984, **13**, 305.
29. Biswas, B.C. and R.K. Tewatia, *Fert. News,* 1991, **36** (6), 13.

16

Status of Nitrate Contamination in Ground Water near Delhi: a Case Study

Navindu Gupta and H. Chandrasekharan

DIVISION OF ENVIRONMENTAL SCIENCES WATER TECHNOLOGY CENTRE, INDIAN AGRICULTURAL RESEARCH INSTITUTE, NEW DELHI – 110 012, INDIA

Abstract

Intensive use of fertilizers for the ever increasing farming and agro-industry have revolutionized crop production. At the same time, contribution of the nutrients such as nitrates from applied fertilizers to surface and ground water is a major concern. ^{15}N isotope studies have shown that for most crops less than 10 per cent of applied fertilizer is lost by leaching and the major source of leached nitrates can originate from the soil organic reserves and mineralized nitrogen. The present study has been undertaken to assess the level of nitrate and to explore the possible source of the same in some of the villages around Delhi since most of the inhabitants of these villages use this water for drinking and domestic purposes. The ground water samples collected in May-July (Summer) 1996 and October-November (Autumn) 1996 and analyzed for major halides and oxy-ions using Ion Chromatography (DX-300, Ion Chromatography, USA).

Results revealed that levels of nitrate concentration in ground water are above the maximum permissible limit (45 mg/L nitrate). One of the sources of nitrate is arable agriculture where the intensification of agricultural practices in the last two to three decades has increased causing nitrate leaching in soil-water systems. The distribution of nitrate in ground water is controlled by some of the hydro-meteorological factors like precipitation, irrigation, vegetation up-take, redox reaction, denitrification.

The excessive concentration of nitrate in drinking water can cause methaemoglobinaemia (blue baby syndrome) in infants and stomach cancer in adults. It is now established that nitrate itself is not that toxic but its reduced species; nitrites were found to be more active biologically and are sources of concern for human health. The study provides information to the planners for management of water resources as well as for judicious application of fertilizers with special reference to nitrogenous, in order to prevent further deterioration of soil and ground water.

1 Introduction

The problem of leaching of nutrients (especially nitrate) from applied fertilizers and/or organic/inorganic nitrogen reserves from soil to ground water is a matter of concern. The levels of NO_3^--N are associated with source availability and regional environmental factors. The increase in levels of NO_3^--N can also arise in intensively cultivated areas under horticultural crops and under animal feed lots. Some studies have been conducted in this regard (CAST, 1985); but comprehensive information on the concentration of nitrates in the soil profiles and ground water under different systems of land management and heavily fertilized cropping systems is lacking in farming areas of Delhi. The geological source of nitrate is ruled out and large quantities of industrial nitrate are seldom discharged into the environment unless spillage from such industries having activity concerning nitrate occurs.

The mechanism of acute and chronic toxicity of nitrates and nitrites in mammals is not clearly understood. Extensive studies are necessary to establish the direct nexus between ingestion of NO_3^- through drinking water and hypertension, stomach cancer, increased infant mortality, non Hodgkin's lymphoma and other related diseases (Malberg *et al.*, 1978; Hill *et al.*, 1973; Super, *et al.*, 1981 and Spalding, *et al.*, 1993). In addition to the adverse effect of nitrate on a biological system, high nitrate is undesirable in fermenting and dyeing industries as it reacts readily with organic moiety. The nitrate present in the irrigated water is considered to be desirable but its high concentration reduces soil permeability and may thus have adverse effects on fertility (Young,1975). Diminished soil permeability may also impede infiltration rate thereby affecting natural replenishment of underlying aquifers.

Ground water is the major component of base-flow in *creeks* and rivers and hence, the understanding of its contribution to the NO_3^- load in surface water is important. As a result, an effort to protect streams from shallow ground water inputs of NO_3^-, suspended sediments, phosphorus, pesticides, heavy metal. The distribution of NO_3^- in ground water is controlled by some of the hydrological factors like precipitation, irrigation, vegetation uptake, denitrification. Therefore, the present study has been undertaken to study the variation in the levels of nitrate in ground water from a location-specific potential belt of agricultural fields near Delhi.

2 The Study Area

2.1 Physiography and Drainage:

The study area, covering 1485 sq. km., occupies a part of the Indo-Gangetic alluvial plains, transected by a quartzite ridge in the southeastern part (Fig. 1). The drainage of the east of the ridge enters the Yamuna River and to the west

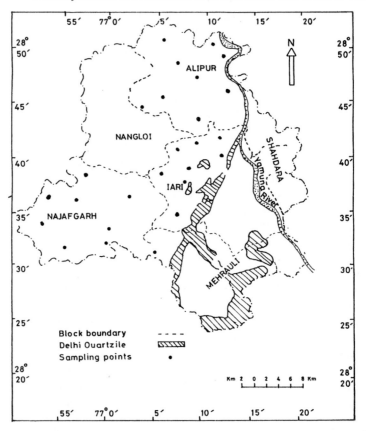

Figure 1   *Study area: part of the Indo-Gangetic alluvial plain*

of it, the south western part receives the surface run-off. The alluvium consists mainly of beds of clay (92–98 per cent), sand and gravel with a thickness not less than 122 m (Sett,1964). A large part of the area has alkaline and saline soils. The climate of the area is semi-arid with average annual rainfall (1931–1995) of 71 cm and mean annual potential evaporation of about 254cm. Soils are mainly saline and alkaline and are predominantly sandy loam and loamy sand in texture.

2.2   Methods and Instrumentation

More than 60 ground water samples (depth varying from 16 to 38m.) from dug wells and tube wells in cultivated and semi-urban areas were collected in May-July, 1996 and October-November, 1966. These samples were grouped into three; namely, Alipur (North), Najafgarh (Northwest) and research farm of IARI (Central) of Delhi state (Fig. 1). Heavy dressings of N fertilizers and farmyard Manure (FYM) are widely applied in these areas. Water samples so collected in 60ml capacity plastic bottles after adding few drops of CHCl₃ to minimize loss of NO₃⁻ due to biological activity.

Concentrations of nitrate in these water samples were measured by Ion Chromatography with TCD (using anion column AS-4A-SC, 4 mm, Dionex-300 system, USA). The carbonate-bicarbonate eluent was used by dissolving 0.1911 g of Na₂CO₃ (for 1.3mM) and 0.1426g. of NaHCO₃ (for 1.7mM) per litre in a deionized water. The flow rate of eluent was maintained at 2.0 ml/min. Typical chromatogram of two water samples representing the two extremes of values of nitrates are shown (Fig. 2).

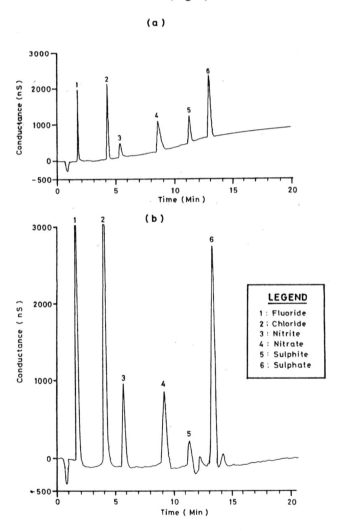

Figure 2   *Ion chromatogram of (a) Standard and (b) Unknown water sample containing halides and oxy - ions*

3 Results and Discussion

The nitrate concentration in the ground water of Delhi area varies from location to location as well as within the small part of the region, showing 18 mg l^{-1} to 70 mg l^{-1} in Najafgarh (Northwest, Fig.3), 20 mg l^{-1} to 78 mg l^{-1} in Alipur (North, Fig. 4) and in research farms of IARI, New Delhi (Central,

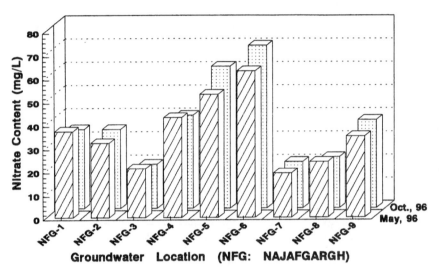

Figure 3   *Here groundwater samples except from NFG-5 & NFG-6 are within acceptable limits for drinking. Monsoon rains (July–September) have marginal effect on nitrate contents in ground water except for NFG-5 & NFG-6*

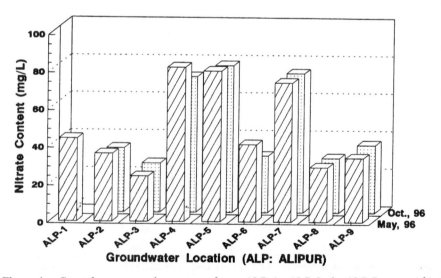

Figure 4   *Groundwater samples except from ALP-4, ALP-5 & ALP-7 are within acceptable limits for drinking. Monsoon rains (July–September) have marginal effect on nitrate contents in all ground water samples*

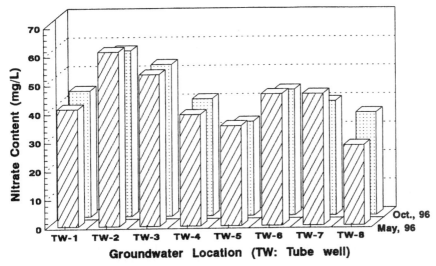

Figure 5 *Figure shows that all ground water samples except from TW-1 and TW-2 are within acceptable limits for drinking. It is obvious that the monsoon rains (July–September) have marginal effect on nitrate contents in ground water except for TW-8*

Fig. 5) having nitrate from 28 mg l^{-1} to 58 mg l^{-1}. These variations of nitrate content of ground water may be due to difference in water recharge process from evaporated irrigation water and surface run off water as well as intermixing of groundwater. Out of eight representing sites at IARI, the two sample sites are showing 50 mg l^{-1} and 58 mg l^{-1} NO_3^- concentrations which are exceeding the general acceptability limit of 45 mg l^{-1} nitrate or 10 mg l^{-1} NO_3^--N (WHO, 1984). The samples from the same sites collected during October, 1996 after monsoon rains have marginal effect on nitrate content in ground water except for TW-8 (Fig.4) which could be explained as this sample site is located in a low lying area and monsoon rain water was standing for a long time and due to surface evaporation, the concentration of NO_3^- is increased. Therefore, it could be said that depending on the degree of evaporation to recharge and the concentration of nitrate in the soil mainly determines the nitrate status in soil-water systems.

The nitrate concentration of ground water in any location is controlled by water recharging through the unsaturated zone and by lateral ground water flow from the surrounding area. The lateral inflow from these areas may also contribute to some extent to the nitrate concentration in the ground water of Delhi area. The direct contribution of rainfall to nitrate contamination is small because the average nitrate content in rain water is quite negligible. The results of these three major regions of Delhi clearly indicates that the Delhi Area is not affected by high concentration of nitrate and generally does not exceed the limit for irrigation purposes. However, some pockets have shown high concentration of nitrate exceeding 45 mg l^{-1}. The following suggestions emerge out of the investigations presented above.

1. Periodical estimation of nitrate in water samples from different sources and analyse them through iso-contours.
2. Judicious fertilizer application in nitrate rich ground water (if used for irrigation).
3. Manure and liquid sludge be sparingly used wherever aquifers are replenished by rains.
4. Blending of low nitrate water be practised wherever possible.

Acknowledgements

Authors wish to thank their colleague, Mr. S.K. Tyagi, for providing water samples.

References

1. CAST, 1985, Council of Agricultural Sciences and Technology, USA. Rep. No. 103.
2. Hill, M.J., G. Hawksworth, and G. Tattersall, 1973,
3. Malberg J.W., L.P. Savage and J. Osteryoung, 1978. Nitrates in drinking water and the early onset of hypertension, Environ. Pollut. 15; 155–160.
4. Sett, D.N., 1964, Ground water geology of Delhi region. Bull. Geol. Survey. India, Ser. B, 16 : 1–35.
5. Spalding R.F and M.E Exner, 1993, Occurrence of nitrate in ground water – a review, J.Environ Qual. 22: 392–402.
6. Super, M., H. Heese, D. Mackenzie, W.S. Dempster, J. duPless and J.J. Ferreira, 1981, An epidemiologic studies of Well-water nitrates in a group of south West African Namibian infants. Water Res., 15: 1265–1270.
7. WHO (World Health Organization), 1984, Guidelines for drinking water quality. WHO, Geneva.
8. WHO/UNEP, 1989, Meybeck, M, D. Chapmans and R. Helmer (Eds), Global Freshwater Quality: A first Assessment, Basil Blackwell, Oxford.
9. Young, K 1975, Geology: The paradox of Earth and Man. Houghton Mifflin, Boston.

Section 3

Nitrate and Health
Introductory Comments

by T. M. Addiscott

IACR-ROTHAMSTED, EXPERIMENTAL STATION, HARPENDEN, HERTS. AL5 2JQ, UK

Nitrate has been in the dock for the last 10–20 years, charged with offences against human health and the environment. This group of papers is concerned with the charges relating to human health, which fall into two main sections. 1) That, by causing methaemoglobinaemia, it endangered the lives of very young infants and killed some of them. 2) That it has caused stomach cancer in adults, and led to many painful and unpleasant deaths.

This very useful conference provided the first real opportunity for those of us working on the agricultural side of the nitrate problem to interact with researchers interested in the medical aspects of 'the infamous anion'. Public perceptions about the agricultural aspects of the problem proved remarkably inaccurate when confronted by scientific investigation, and we learned that much the same was true of the medical side. Most evidence suggests that, far from being harmful to human health, nitrate has some very important functions in our bodies. Indeed, if our bodies do not take in enough nitrate, they manufacture it *in situ.* So how is it that the subject of so many 'shock horror' stories is a good thing after all? And is it good for all of us?

A key element in the hypothesis relating nitrate to stomach cancer was the idea that nitrate was reduced to nitrite in the mouth. (For the non-chemists, the nitrite ion, NO_2^-, differs from nitrate, NO_3^-, only in that it has one less oxygen atom, but the two ions differ considerably in their behaviour.) Several of the papers (Walker, Li *et al,* Golden & Leifert, and Benjamin & McKnight) show that this part of the hypothesis is essentially correct. The reason one's tongue feels rather rough near the back is that it has lots of little pouches in it; as shown by Duncan *et al*, these provide homes for nitrate-reducing bacteria. Nobody knew, prior to the research described in these papers, that our bodies enrich our saliva with nitrate so that these bacteria could reduce it to nitrite and so supply the stomach with nitrite. This is a considerable surprise, because the other element in the nitrate/stomach cancer hypothesis was that the nitrite reacted in the stomach with secondary amines, produced in the digestion of proteins, to form a carcinogenic N-nitrosamines. So does the nitrite produce N-nitrosamines? The answer, according to Walker, is that it can do so. N-nitrosoproline is found in urine, but it may actually be produced in the saliva. So, how great is the threat of stomach cancer from N-nitrosamines formed following the reduction of nitrate? According to Boink's paper, these compounds have proved to be carcinogenic in animals, but epidemiological studies show no relation between nitrate and stomach cancer.

Why supply the stomach with nitrite? The rather surprising answer is, so that it can react with the stomach acid to produce nitric oxide and other killer chemicals. These do not kill us, they kill unwanted visitors in our stomachs – a wide range of gastrointestinal pathogens – as described in the papers by Walker, Golden & Leifert, Benjamin & McKnight and Dykhuizen *et al.* In short, the nitrite produced in the mouth from nitrate is the 'fuel' for our defence system against infectious diseases. There are 3–5 million deaths from gastrointestinal disease each year, according to Golden & Leifert, and they suggest that we need to rethink our negative attitude towards nitrate.

Benjamin & McKnight suggests some other reasons why nitrate is good for you. It seems to reduce platelet function and may thereby protect against heart disease. Indeed, this could be the reason why eating plenty of vegetables helps to prevent heart attacks. Even the nitrate you lose in sweat is useful; it is reduced to nitrite and nitric oxide on the surface of the skin and so helps to prevent skin infections.

But what about the children? Methaemoglobinaemia, whose mechanism is explained in the paper of Boink, is very rare, but it can kill. It is also associated with water from wells rather than from the mains, and therefore with water that may be contaminated by bacteria. Walker noted that healthy babies will tolerate nitrate concentrations that exceed the European Union limit of 50 mg 1^{-1}, while sick ones are affected at less than the limit, and Benjamin & McKnight seemed to see gastroenteritis as a more important factor than nitrate in the occurrence of methaemoglobinaemia. Thus the role of nitrate in the problem of methaemoglobinaemia may not be certain yet.

Could nitrate be responsible for other problems? McKinney *et al.* showed that childhood diabetes seemed to be strongly associated with the concentration of nitrate in drinking water. This is a serious matter, because 20,000 people in the under-twenty age bracket have diabetes, and it is also a puzzle from an evolutionary standpoint. Nitrate seems central to the defence system against gastrointestinal pathogens which is found in other mammals as well as man, and our bodies have mechanisms for producing it and retaining it against excretion that are described by Benjamin & McKnight. These functions presumably evolved over a long period, so how could another bodily system have evolved at the same time such that nitrate was a threat to it? It was recently reported that 99% of American children are malnourished because they eat so much junk food. Could eating habits have been involved as well as nitrate?

The medical side of the nitrate issue has been fascinating to those of us working on the soil and environmental aspects. There are clearly several aspects in the physiological roles of nitrate and nitrite, notably some discussed in the paper of Boink, about which more needs to be learned, but it seems that some changes in our perceptions of nitrate are needed. The European Union limit of 50 mg l^{-1} of nitrate for potable water, which has caused the expenditure of very large sums of money, does not now appear to have, in Walker's words, a very secure basis. Perhaps, as one participant remarked, the toxicology should have preceded the legislation. Also, has the nitrate scare really been good for our health? Has not the incidence of 'food poisoning' from *Salmonella* and similar organisms increased recently? Could this have anything to do with a sign I saw on a stall? It said, 'Hot dogs – no nitrates'.

17

The Metabolism of Dietary Nitrites and Nitrates

R. Walker

SCHOOL OF BIOLOGICAL SCIENCES, UNIVERSITY OF SURREY, GUILDFORD GU2 5XH, UK

Abstract

Nitrites and nitrates occur as dietary constituents being particularly high in certain vegetables. Endogenous oxidation of nitric acid also produces nitrate. Nitrate is readily absorbed and rapidly excreted in urine (60–70%) with an elimination half life of about five hours. Nitrate is secreted by an active transport mechanism (25%) into saliva; and by passive diffusion into breast milk. About 3% nitrate appears in urine as urea and ammonia in humans; in rats this is about 11%. Nitrate is reduced to nitrite by both mammalian enzymes and nitrate reductases present in microorganisms residing in the gastrointestinal tract. In rats approximately 50% of the nitrate reduced to nitrite is produced by mammalian enzymes, the remainder by microorganisms; this ratio is species dependent.

Nitrite is absorbed by the gastrointestinal tract and is rapidly oxidised to nitrate. The plasma half life is less than an hour in most species and consequently nitrite is not normally detected in body tissues and fluids after oral administration. Nitrite is oxidised *via* a coupled reaction with oxyhaemoglobin producing methaemoglobin (ferrihaemoglobin). It appears in certain cases that endogenous nitrite may be a major factor in this process rather than exogenous material. Nitrite or a chemical species derived from it may be involved in the generation of nitrosamines and nitrosamides of toxicological importance, although neither nitrite or nitrate are carcinogenic.

1 Dietary Exposure

Diet constitutes an important source of exposure to nitrite and particularly nitrate. The major dietary source of nitrate is vegetables. Lettuce, spinach, celery and beetroot commonly contain greater than 1000 mg NO_3^-/kg fresh weight and, depending on growing conditions (most notably temperature and light intensity, and to a lesser degree fertilizer use) and may reach concentrations of 3–4000 mg/kg fresh weight [1]. Nitrate concentrations in other vegetables, such as potatoes and cabbage normally fall in the range of 100–1000 mg NO_3^-/kg but the amounts consumed means that these make a substantial contribution to dietary intakes. Nitrite occurs in plants at low concentrations, usually between 1–2 mg NO_2^-/kg fresh weight and rarely in excess of 10 mg/kg. However, potatoes have been reported to contain 2–60 mg/kg nitrite with a mean nitrite concentration of 19 mg/kg [2].

Estimates of mean nitrate intakes ranged from 31 to 185 mg/day in various European countries with vegetables supplying 80–85% [3]. In the U.K. the population mean intake of nitrate is 54 mg/day but in vegetarians this was as high as 185–194 mg/day, emphasising the importance of vegetables as the major source of nitrate [2]. The intake of nitrite is very much lower and averaged 0.7–8.7 mg/day in various European countries with both vegetables and cured meats being major sources.

2   Endogenous Synthesis

In addition to dietary exposure there is considerable endogenous synthesis of nitrate in mammals and during rigorous exclusion of dietary nitrate, human volunteers excrete in urine about 1 mmole NO_3^- per day [4] *i.e.* approximately the same amount as that provided by food. This endogenous nitrate arises from oxidation of nitric oxide, NO, which is produced by a family of nitric oxide synthetases (NOS), some constitutive and some inducible [5]. The constitutive NOS produce NO for short periods (seconds) in response to intracellular messengers like bradykinin while the inducible forms produce much higher levels over periods of hours in response to immunostimulants. This synthesis of NO occurs in activated macrophages but has also been demonstrated in other cell types including endothelial cells, neurons, neutrophils and hepatocytes [6, 7, 8] and is highly variable and much increased during infections.

The endogenous formation of nitrate independently of dietary sources has complicated the studies of the metabolism and pharmacokinetics of nitrate and nitrite, many of which can provide only qualitative or semi-quantitative data on their interconversion *in vivo*. The various pathways involved in the formation and metabolism of nitrate and nitrite are summarised in Figure 1. In order to clarify the fate of orally administered nitrate/nitrite relative to endogenous synthesis may require the use of ^{15}N-labelled sources but, while this may be feasible in mimicking exposure from drinking water, it is less so in relation to questions of relative bioavailability from vegetables.

3   Pharmacokinetics and Metabolism

The pharmacokinetics and metabolism of nitrate and nitrite, and the potential formation of *N*-nitroso compounds *in vivo* are obviously closely linked.

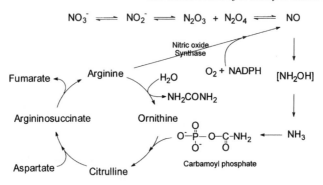

Figure 1   *Metabolism of nitrogen oxides and oxoanions*

Ingested nitrate is readily absorbed from the proximal small intestine [9,10] and rapidly equilibrates with body fluids. In rats, about 50% of an oral dose was detected in the carcass within 1 hour while in humans, peak levels in serum, saliva and urine were achieved within 1–3 hours [11]. There is little absorption from the stomach in most species although this has been reported from the rumen of cattle [12].

In humans and most laboratory animal species except the rat, nitrate is actively secreted in saliva in a dose-dependent manner [13, 14] but Spiegel-halder *et al.* [15] were unable to detect an increase in salivary nitrate concentrations of human volunteers following ingestion of up to 54 mg NO_3^-. The active transport mechanism is common to iodide, thiocyanate and nitrate, in that order of affinity, and smokers who secrete elevated levels of thiocyanate have lower salivary concentrations of nitrate [16, 17]. It has been estimated that, in humans, about 25% of an orally ingested dose of nitrate is secreted in saliva [15, 18] but these estimates are confounded by the great interindividual and diurnal variability in endogenous synthesis and secretion of nitrate.

Although the rat is reported not to possess the mechanism for active salivary secretion of NO_3^- (which has hindered extrapolation of experimental toxicological results to man), it does secrete circulating nitrate into other gastric and intestinal secretions by an active transport process [19] so that entero-systemic cycling of nitrate may occur in this species also. In the dog, after intravenous administration of NO_2^-, in addition to strong salivary secretion large amounts of NO_3^- were excreted in bile, confirming this pathway of excretion as well as oxidation of NO_2^-.

Nitrate appears in milk by a passive diffusion mechanism, and concentrations in human and canine milk did not exceed plasma levels after ingestion of a nitrate-containing meal [20].

Following absorption and equilibration in body fluids, nitrate is rapidly excreted in urine. In humans, independently of dose, about 65–70% of orally administered nitrate is excreted in urine. Excretion is maximal at about 5 hours after dosage and essentially complete within 18 hours [11]. The excretion follows first order kinetics and the elimination half-life has been estimated to be about 5 hours [21]. Some metabolic conversion of nitrate clearly occurs (see Fig.1) since in humans about 3% of a dose of $^{15}NO_3^-$ appeared in urine as urea and ammonia [22]; in rats 11% of the dose appeared as urea and ammonia in urine and faeces [23].

Reduction of nitrate to nitrite *in vivo* may be effected by both enteric bacteria and mammalian nitrate reductase activity. Many species of microorganisms resident in the oro-gastrointestinal tract possess nitrate-reductase activity [24] and this enzyme has been detected in rat liver and intestinal mucosa, although at much lower activity [23]. From comparative studies in germ-free and conventional rats in our laboratory, we concluded that of the 40–50% of a dose of nitrate reduced to nitrite in conventional animals approximately half was effected by mammalian nitrate reductase [25]. However, Fritsch *et al.* [13] were unable to detect such a pathway in dogs and reduction by the oro-gastrointestinal microflora appears to be the most important mechanism in mammals. It appears that the major site of conversion of nitrate to nitrite varies with species and is dependent on the sites of microbial colonisation and absorption of nitrate.

Interestingly, the presence of nitrite in human oral saliva was first reported more than 55 years ago [26] but saliva taken directly from the salivary ducts of man or dog contains only nitrate indicating that a significant amount of reduction occurs in the oral cavity [27]. This is attributed to a stable population of nitrate-reducing bacteria established at the base of the tongue. Stephany & Schuller [28] suggested that the salivary concentration of nitrite was directly related to the orally-ingested dose of nitrate and other workers have reached similar conclusions [15, 29] but Tannenbaum *et al.* [18] produced data which suggest that the reduction process may be saturable at high intakes. On the basis of the (highly variable) salivary levels of nitrate and nitrite following oral ingestion of nitrate by humans, it has been estimated that, of the 25% of ingested nitrate secreted in saliva, 20% is reduced to nitrite (*i.e.* about 5% of the oral dose) and it appears that oral reduction of nitrate is the most important source of nitrite for man and most species which possess an active salivary secretory mechanism.

While the interest in oral reduction of nitrate to nitrite has largely centred around the possible involvement in formation of carcinogenic *N*-nitroso compounds, more recently a physiological role has been postulated in which the generation of nitrite and resultant antimicrobial activity protects against ingress of pathogens by this portal [30].

Gastric pH and hence bacterial populations are low in the stomach of rabbits, ferrets and healthy humans and hence little further reduction of nitrate occurs at this site and nitrite levels in gastric contents are usually low. Conversely, rats and dogs have a higher gastric pH and bacterial colonisation can occur with consequent further reduction of nitrate at this site [31]. In ruminants, the dense population of rumen microflora and relatively high pH make this a major site of reduction of orally ingested nitrate and this leads to the well-documented intoxication (methaemoglobinaemia) by the nitrite produced [12].

In humans subject to achlorhydria, bacterial colonisation of the stomach can occur and the situation then more closely resembles the rat and dog. A strong correlation has been reported between gastric pH, bacterial colonisation and gastric nitrite concentrations in humans over a pH range of 1–7 [32] and elevated levels as high as 6 mg l^{-1} have been reported in achlorhydria associated with pernicious anaemia or hypogammaglobulinaemia [33,34]. The situation in human neonates is less clear. It is commonly asserted that infants under three months of age may be highly susceptible to gastric nitrite production because they have little gastric acid production [35] but Agunod *et al.* [36] examined 12 infants aged 12 hours to three months and found only one with achlorhydria.

With regard to the pharmacokinetics and metabolism of nitrite *per se*, studies of absorption of orally administered nitrite have been made difficult because of the reactivity of $NO_2{}^-$ with dietary constituents/stomach contents and instability at gastric pH levels. Under simulated gastric conditions *in vitro*, nitrite disappeared rapidly at pH<5 and the loss was accelerated by food components [31].

Absorption of nitrite in the rat appears slower than that of nitrate but some gastric absorption has been reported [31]. Intestinal absoption of nitrite is more rapid in the mouse than the rat [19]. Although there is a dearth of information on the absorption of nitrite in man, it can be inferred that it occurs from reports of methaemoglobinaemia following exposure [37].

Nitrite is not normally detectable in tissues and body fluids of animals following oral administration due to its rapid oxidation to nitrate; after intravenous administration in mice or rabbits, rapid equilibration occurs in tissues within 5 minutes and nitrite concentrations in body fluids fall rapidly to low levels within 30 minutes [38]. The plasma half-life in the distribution phase was reported to be 48, 12 and 5 minutes in dogs, sheep and ponies respectively [39]. Nitrite is oxidised in blood by a coupled oxidation reaction with oxyhaemoglobin [40] in which methaemoglobin is produced (Figure 2) leading to the well recognised acute toxicity of nitrite. The reaction rate between nitrite and haemoglobin is species-dependent; in man it is slower than in ruminants but faster than in pigs.

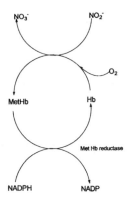

Figure 2

With regard to the methaemoglobinaemia produced by nitrite, drinking water standards for *nitrate* have been drawn up on the basis of levels in drinking water (mainly from wells) associated with infantile methaemoglobinaemia and an assumed threshold below which the risk is minimal. This was apparently based on the assumption that infants are more likely than adults to have a resident gastric microflora capable of reducing nitrate to nitrite and that the phenomenon of infantile methaemoglobinaemia was, in part, a consequence of this. However, this must be questioned as the endogenous synthesis of nitric oxide and subsequently of nitrite can rise dramatically during infantile gastroenteritis. In one study, hospitalised infants with a low nitrate intake (2–7 mg/day) had elevated blood nitrate and methaemoglobin levels associated with acute diarrhoea [24] and in another case, a dyspeptic child had 72% methaemoglobinaemia associated with a nitrate level in drinking water below 50 mg l^{-1} [41] while healthy babies tolerate intakes of up to 21 mg kg^{-1} body weight [42]. In these cases at least, it appears that nitrite of endogenous rather than dietary origin was involved in causing the methaemoglobinaemia together with the increased sensitivity of fetal type haemoglobin to oxidation and immature, low levels of methaemoglobin reductase in neonates.

A further concern relating to the metabolism of dietary nitrate and nitrite is the potential formation *in vivo* of carcinogenic N-nitroso compounds from nitrite, or the nitrosating species derived from this, N_2O_3 and N_2O_4, and dietary amines. This was first postulated more than 30 years ago and since then has been extensively studied using a number of approaches:

(i)   *in vitro* incubation of precursors under simulated oral and gastric conditions;
(ii) analysis of saliva/gastric contents after administration of precursors;
(iii) determination of specific or total N-nitroso compounds in body fluids or excreta following treatment with precursors;
(iv) carcinogenicity studies following co-administration of nitrate/nitrite and amines or amides.

These different approaches have been reviewed previously [1].

With regard to the first approach, many of the studies performed have used unrealistically high concentrations of nitrite and at best provide no more than an indication of a potential for nitrosation to occur. Urinary N-nitroso-proline (N-Pro) has been used as an indicator of total nitrosation *in vivo* since it is non-carcinogenic and is excreted unchanged. N-Pro is excreted at low levels (<7 g) in subjects receiving a low nitrate diet. However, the validity of this approach is somewhat dubious as, in the rat, much of urinary N-Pro (40–90%) was not derived from orally administered ^{15}N-nitrate [43] and in humans there was no correlation between nitrate intake and urinary excretion of N-Pro [44]. In addition, the basal level of excretion of N-Pro was not affected by the inhibitors of nitrosation, ascorbic acid and α-tocopherol, although these compounds did inhibit formation of N-Pro when orally administered along with proline and ^{15}N-nitrate [43]. This indicates that there are at least two sites at which N-Pro may be formed *i.e.* in the stomach under acidic conditions and at some other site(s) by a non-acid catalysed nitrosation, probably by N_2O_3 and N_2O_4 derived from NO in the course of oxidation to nitrate (see Figure 1). These oxides of nitrogen are good nitrosating agents at near neutral pH, as is peroxy nitrite, another postulated intermediate of endogenous origin.

It appears that the role of dietary nitrate in nitrosation *in vivo* may have been greatly overestimated and it is significant that neither nitrate nor nitrite *per se* are carcinogenic in rats or mice [45].

References

1. Walker, R. (1990) *Food Additives and Contaminants*, **7**, 717–768.
2. Ministry of Agriculture, Fisheries & Food (1992) Nitrate, nitrite and *N*-nitroso-compounds in food. Food Surveillance Paper No.32, Her Majesty's Stationery Office, London.
3. Gangolli, S.D., van den Brandt, P.A., Feron, V.J., Janzowsky, C., Koeman, J.H., Speijers, G.J.A., Spiegelhalder. B., Walker, R. & Wishnok, J.S. (1994) Eur. J. Pharmacol. Environ. Toxicol. Pharmacol. Section, **292**, 1–38.
4. Wishnok, J.S., Tannenbaum, S.R., Tamir, S. & de Rojas-Walker, T. (1995) Health aspects of nitrates and its metabolites (particularly nitrite): Proc. Int. Workshop, Bilthoven, 8–10 Nov.1994. Council of Europe Press, Strasbourg.
5. Marletta, M.A. (1988) Chem. Res. Toxicol. **1**, 249–257.
6. Ignarro, L.J. (1990) Annu.Rev.Pharmacol.Toxicol. **30**, 535–560.
7. Snyder, S.H. & Bredt, D.S., (1990) Scientif.Amer. **266**, 68–77.
8. Lancaster, J.R. (1992) Am.Sci. **80**, 248–259.
9. Fritsch, P., de Saint Blanquat, G. & Derache, R. (1979) Toxicol.Eur.Res., **3**, 141
10. Balish, E., Witter, J.P. & Gatley, S.J. (1981) Distribution and metabolism of nitrate and nitrite in rats. Gastrointestinal cancer, Endogenous factors. Banbury Report No.7, 305–319; 337–341.
11. Bartholomew, B.A. & Hill, M.J. (1984) The pharmacology of dietary nitrate and the origin of urinary nitrate. Fd.Chem.Toxicol., **22**, 789–795.
12. Wright, M.J. & Davison, K.L. (1964) Nitrate accumulation in crops and nitrate poisoning in animals. Adv.Agron., **16**, 197–247.
13. Fritsch, P., de Saint Blanquat, G. & Klein, D. (1985) Excretion of nitrates and nitrites in saliva and bile in the dog. Fd Chem.Toxicol., **23**, 655–659.
14. Cohen, B. & Myant, M.B. (1959) Concentration of salivary iodite: a comparative study. J.Physiol., **145**, 595–610.
15. Spiegelhalder, B. Eisenbrand, G. & Preussman, R. (1976) Influence of dietary nitrate on nitrite content of human saliva: possible relevance to *in vivo* formation of N-nitrosocompounds. Fd Cosmet.Toxic.**14**, 545–548.
16. Forman, D., Al Dabbagh, S. & Doll, R. (1985a) Nitrates, nitrites and gastric cancer in Great Britain. Nature, **313**, 620–625.
17. Forman, D., Al Dabbagh, S. & Doll, R. (1985b) Nitrate and gastric cancer risks. Nature, **317**, 676.
18. Tannenbaum, S.R., Weisman, M. & Fett, D. (1976) The effect of nitrate intake on nitrite formation in human saliva. Fd Cosmet.Toxic. **14**, 549–552.
19. Witter, J.P & Balish, E. (1979) Distribution and metabolism of ingested NO_3^- and NO_2^- in germ free and conventional flora rats. App. Env.Microbiol. **38**, 861–869.

20. Green, L.C., Tannenbaum, S.R. & Fox, J.G. (1982) Nitrate in human and canine milk. New Engl.J.Med. **306**, 1367–1368.

21. Green, L.C., Ruiz de Luzuriaga, K., Wagner, D.A., Rand, W., Istfan, N., Young, V.R. & Tannenbaum, S.R. (1981) Nitrate biosynthesis in man. Proc.Natl.Acad.Sci. USA **78**, 7764–7768.

22. Wagner, D.A., Schultz, D.S., Deen, W.M., Young V.R. & Tannenbaum, S.R. (1983b) Metabolic fate of an oral dose of (^{15}N)-labelled nitrate in humans; effect of diet supplementation with ascorbic acid. Cancer Res. **43**, 1921–1925.

23. Schultz, D.S., Deen, W.M., Karel, S.F., Wagner, D.A., & Tannenbaum, S.R. (1985) Pharmacokinetics of nitrate in humans: Role of gastrointestinal absorption and metabolism. Carcinogenesis **6**, 847–852.

24. Hegesh, E. & Shiloah, J. (1982) Blood nitrates and infantile methaemoglobinaemia. Clin.Chim.Acta. **125**, 107–115.

25. Ward, F.W., Coates, M.E. & Walker, R. (1986) Nitrate reduction, gastrointestinal pH and N-nitrosation in gnoto-biotic and conventional rats. Fd Chem.Toxic. **24**, 17–22.

26. Varady, J. & Szanto, G. (1940) Klin.Wschr. **19**, 200.

27. Muramatsu, K., Maruyama, S. & Nishizawa, S. (1979) J.Fd Hyg.Soc. Japan, **20**, 106.

28. Stephany, R.W. & Schuller, P.L. (1978) The intake of nitrate, nitrite and volatile *N*-nitrosamines and the occurrence of volatile nitrosamines in human urine and veal calves. IARC Scientific Publ. No.19, 443–460.

29. Harada, M., Ishiwata, H., Nakamura, Y., Tanimura, A. & Ishidate, M. (1975) J.Fd Hyg.Soc. Japan **16**, 11.

30. Benjamin, N., O'Driscoll, F., Dougall, H., Duncan, C., Smith, L. & McKenzie, H. (1994) Nature, **368**, 502.

31. Mirvish, S.S., Patil, K., Ghadirian, P. & Kommineni, V.R.C. (1975) Disappearance of nitrate from the rat stomach; Contribution of emptying and other factors. J.Natl.Cancer Inst., **54**, 869–875.

32. Mueller, R.L., Hagel, H.J., Wild, H., Ruppin, H. & Domschke, W. (1986) Nitrate and nitrite in normal gastric juice. Oncology, **43**, 50–53.

33. Dolby, J.M., Webster, A.D., Borriello, S.P., Barclay, F.E., Bartholomew, B.A. & Hill, M.J. (1984) Bacterial colonization and nitrite concentration in the achlor-hydric stomachs of pateints with primary hypogamma-globulinaemia or classical perinicious anaemia. Scand J.Gastroenterol. **19**, 105–110.

34. Ruddell, W.S., Bone, E.S., Hill, M.J., Blendis, L.M. & Walters, C.L. (1976) Gastric juice nitrite. A risk factor for cancer in the hypochlorhydric stomach. Lancet 2, 1037–1039.

35. Speijers, G.J.A., van Went, G.F., van Apeldoorn, M.E., Montizaan, G.K., Janus, J.A., Canton J.H., van Gestel, C.A.M., van der Heijden, C.A., Heijna-Merkus, E., Knaap, A.G.A.C, Luttick, R. & de Zwart, D. (1987) Integrated Criteria Document Nitrate Effects. Report No.758473007, RIVM, Bilthoven, The Netherlands.

36. Agunod, M, Yamaguchi, M., Lopez, R., Luhby, A.L. & Glass, G.B.J. (1969) Correlative study of hydrochloric acid, pepsin and intrinsic factor secretion in newborn and infants. Am.J.Digest.Disord. **14**, 400–414.

37. Shuval, H.I. & Gruener, N. (1972) Epidemological and toxicological aspects of nitrates and nitrites in the environment. Am.J.Publ.Hlth. **62**, 1045–1052.

38. Fritsch, P., de Saint-Blanquat, G. & Klein, D. (1985) Excretion of nitrates and nitrites in saliva and bile in dog. Fd Chem.Toxic. **23**, 655–659.

39. Schneider, N.R. & Yeary, R.A. (1975) Nitrite and nitrate pharmokinetics in the dog, sheep and pony. Am.J.Vet.Res. **36**, 941–947.
40. Smith, J.E. & Beutler, E. (1966) Methaemoglobin formation and reduction in man and various animal species. Am.J.Physiol. **210**, 347–350
41. Thal, W., Lacchein, P.L. & Martinek, M. (1961) Welche Hamoglobinkonzentrationen sind bein Brunnenwasser-Methamoglobinamie noch mit Leben vereinbar. Arch.Toxicol. **19**, 25–33.
42. Kubler, W. (1958) The importance of the nitrate content of vegetables in child nutrition. Z. Kinderheilkunde **81**, 405–416
43. Wagner, D.A., Shuker, D.E., Bilmazes, C., Obiedzinski, M., Baker, I., Young V.R. & Tannenbaum, S.R. (1985) Effect of vitamins C and E on endogenous synthesis of N-nitrosamino acids in humans: precursor product studies with (^{14}N) nitrite. Cancer Res. **45**, 6519–6522
44. Tannenbaumn S.R. (1987) Endogenous formation of N-nitrosocompounds: a current perspective. The relevance of N-nitroso-compounds to human cancer, exposure and mechanisms. In IARC Scientific Publication No.84, pp 292–296, International Agency for Research on Cancer, Lyon
45. World Health Organisation (1996) Toxicological evaluation of certain food additives and contaminants: WHO Food Additives Series No.35 WHO, Geneva.

18

Identification of Nitrate Reducing Bacteria from the Oral Cavity of Rats and Pigs

Hong Li[1], Callum Duncan[1,2], Michael Golden[2] and Carlo Leifert[1]

[1] DEPARTMENT OF PLANT & SOIL SCIENCE, CRUICKSHANK BUILDING, UNIVERSITY OF ABERDEEN, ABERDEEN AB9 2UD, SCOTLAND, UK
[2] DEPARTMENT OF MEDICINE & THERAPEUTICS, POLWARTH BUILDING, UNIVERSITY OF ABERDEEN MEDICAL SCHOOL, FORESTERHILL, ABERDEEN AB9 2ZD, SCOTLAND, UK

Abstract

Within the oral cavity most nitrate reduction was detected on the dorsal surface of the tongue. Nitrite producing bacteria (NPB) were isolated from tongues of pigs obtained from a local abattoir and compared with previous results from laboratory rats (10).

The most commonly isolated genus was *Staphylococcus* in both pig and rat with *Staphylococcus sciuri* being the most common species. Other nitrate reducing bacteria included *Enterobacteriaciae*, *Micrococcus*, *Streptococcus* and *Pasteurella* spp. in the pig and *Pasteurella* and *Streptococcus* spp. in the rat.

Nitrate reduction and numbers of nitrate reducing bacteria were very similar over the surface of tongues from pigs. Conversely, in the rat the enumeration of culturable bacteria (cfu) showed an increase in the density of bacteria towards the posterior tongue. The proportion of culturable NPB in the total bacterial population increased from 6% on the anterior tongue to 65% on the posterior tongue. Nitrite production was sensitive to oxygen and significant nitrite generation was only detected on the posterior tongue of rats where the majority of bacteria are situated in deep clefts in the tongue surface. In the pig, deep clefts were found over the entire tongue surface.

1 Introduction

Human nitrate and nitrite intake and metabolism have received considerable interest because nitrite formation in saliva is suspected to be involved in N-nitrosamine carcinogenesis in the acid environment of the stomach (15, 17, 19, 20). Dietary nitrate mainly originates from vegetables (21) and is actively concentrated by the salivary glands in most mammals. As a result, salivary nitrate concentrations are approximately 10 to 20 times those found in plasma (6, 7). Saliva collected directly from salivary ducts contains nitrate but no nitrite. However, saliva collected at other locations in the oral cavity also contains nitrite (6, 13, 17).

A number of investigations have indicated that the oral microflora plays a major role in nitrate reduction (4, 7, 17). It has been demonstrated that the nitrite concentration increases after incubation of saliva at 37°C, but not if the saliva is filter sterilised (6). Several nitrite-producing organisms in human saliva have been identified and include *Veillonella* spp., *Staphylococcus aureus* and *Staphylococcus epidermidis*, *Nocardia* spp., *Corynebacterium* (probably *C. pseudodiphtheriticum*) and an anaerobic, filamentous organism tentatively identified as *Fusobacterium nucleatum* (17, 18).

By far the greatest nitrate reduction activity in the human oral cavity was demonstrated on the dorsum linguae (12). Studies with rats have shown that nitrate reduction occurs mainly on the posterior surface of the tongue (3) and could in principle be due to one or a combination of mechanisms (12). It could be caused by (i) the action of reducing substances in salivary secretions, (ii) reducing bacterial metabolites, (iii) mammalian nitrate reductase in the papilla linguae and/or (iv) nitrate reductase enzymes of the microorganisms colonising the tongue (12). Recent results have shown that nitrate reduction is virtually absent in germ-free rats, and that nitrite production on the tongue of adult man is greatly reduced after administration of broad spectrum antibacterial antibiotics (2, 3). These findings clearly indicate microbial nitrate reduction as the primary mechanism. Systematic studies on the population dynamics and the relative importance of individual species of nitrate reducing bacteria have so far only been carried out in laboratory rats (10).

It has now become clear that salivary nitrite can greatly augment the antimicrobial activity of gastric acid against swallowed pathogenic micro-organisms (1). Once swallowed, the acidic condition of the stomach protonates nitrite to form nitrous acid; which in turn dissociates to form oxides of nitrogen with strong antimicrobial activity (3). We have therefore postulated a novel host resistance mechanism which relies on a symbiotic relationship between mammals and nitrate reducing bacteria which live mainly on the tongue (3).

The aims of this study were: (i) map and identify the nitrite producing micro-organisms on the tongue surfaces of pigs obtained from a local abattoir (Scotch Premier Meat, Inverurie, Aberdeenshire, UK) and compare with those found on laboratory rats (10) and (ii) to compare nitrate reduction velocity on different tongue sections with the populations of nitrite producing bacteria.

2 Materials and Methods

2.1 Microbial Growth Media Used

The following media were used to isolate and enumerate microorganisms on pig tongues: nutrient broth (NB, Oxoid CM1, Oxoid Ltd, Basingstoke, Hampshire, UK), nutrient broth with (per litre) 10g KNO_3 (NBNO$_3$, BDH, Poole, UK), Columbia blood agar base (Oxoid CM331) with 5% laked horse blood (Oxoid SR48)(BA), Columbia blood agar base with 5% laked horse blood and *Streptococcus* selective supplement (Oxoid SR126E)(BAS), Columbia blood agar base with 5% laked horse blood and *Streptococcus/ Staphylococcus* selective supplement (Oxoid SR 070E)(BASS), Columbia blood agar base with *Pseudomonas* selective supplement (Oxoid SR 103 E)(BAP), De Man, Rogosa & Sharp's agar (MRS, Oxoid CM 3361), MacConkey agar (MC, Oxoid CM115), Sabouraud dextrose agar (SDA, Oxoid CM 41), nutrient agar (NA, Oxoid CM3) and Schaedler anaerobe agar (AA, Oxoid CM437).

2.2 Enumeration of Bacterial and Fungal cfu

Pig tongues (n = 5) were obtained from a local abattoir (Scotch Premier Meat, Inverurie, Aberdeenshire, UK). Five circular sections were removed from each tongue using a number 10 cork borer (1.5 cm diameter). Sections of the tongue were then homogenised separately in sterile 1/4 strength Ringer's solution (1/4 RS). The total number of bacteria and the number of nitrite producing bacteria were determined for each section by (i) a most probable number method (MPN) (11, 14) and (ii) plating of diluted tongue homogenates on to general and semi-selective agar-solidified microbial growth media.

For the MPN determinations, the tongue homogenate of each section was serially diluted in NBNO$_3$ (2.1 above) in 10 ml test tubes with screw top lids. Each homogenate was serial ten fold diluted five times. Test tubes were sealed with airtight screw tops, incubated for 72 h at 37°C, and then assessed for microbial growth and nitrite production in the different dilution steps. After serial dilution tubes contained 9 ml of medium, leaving only 1 ml of headspace gas. This results in a rapid reduction in oxygen tension, within the tube, due to microbial respiration (14). Nitrite production was detected by NO_2-Merckoquant strips (detection limit 5 ppm nitrite, Merck, Darmstadt, Germany). The number of positive tubes in each dilution step was recorded and the most probable numbers obtained by referring to Cochran's statistical table (11, 14).

For enumeration by plating, homogenates of each tongue section were serial diluted in 1/4 RS and 10 µl aliquots of different dilution steps transferred in triplicate on to the range of general and semi-selective media described above. BAS, MRS and AA agar plates were inoculated in duplicate. One plate was incubated aerobically and the other anaerobically using an anaerobic jar (AMSTA, Sweden) and a Gas Generating Kit (Oxoid Ltd, Basingstoke, Hampshire, UK) to create anaerobic conditions. The number of cfu appearing on aerobic incubating plates was counted after 16 and 28h incubation at 37°C. The number of cfu appearing on anaerobic incubating plates was counted after 28h incubation at 37°C. Results were then compared with those obtained previously from laboratory rats (10).

2.3   Isolation and Identification of Denitrifying Bacteria

Thirty well separated colonies were selected from all the agar solidified isolation media listed in Section 2.1 and subcultured on to BA plates. All isolates were tested for growth, gas and nitrite production in $NBNO_3$ using the method described for the MPN-determination. Eighty three strains of nitrite producing organisms were then identified using conventional bacteriological methods (Gram stain, catalase, coagulase, motility, oxidase, shape, heat resistance) and API test strips according to an identification scheme described previously (9) and compared with published results obtained for laboratory rats (10).

2.4   Histology

To visualise the microbial niche and microorganisms of the dorsal surface of the tongues, sections from each of 5 additional pig tongues were removed as described above and placed in 10% neutral buffered formalin (4% formalde-hyde) at room temperature for 24 hours, dehydrated and processed by standard means to glycol methacrylate resin (JB4, Polysciences, Warrington, Pennsylvania, USA), sectioned at 2μm and stained for micro-organisms by an adaptation of the Gram method (16). Sections were then examined using an Olympus CH2 optical microscope (Olympus Optical Co. Ltd., London UK) and compared with published results obtained for laboratory rats (10).

2.5   Measurement of Nitrite Production on Intact Tongue Sections

5 sections of equal thickness were cut from each of 5 additional pig tongues. Two circular plugs were removed from each section with a number ten cork borer (1.5 cm diameter) and quartered into 4 equal segments. Individual sections were placed into Bijou bottles, containing 2 ml of 1/4 strength Ringer's solution with potassium nitrate (KNO_3) to produce final concentrations of 100, 200, 400, 800, 1000, 2000, 10,000 or 20,000 μM NO_3^-, and incubated for 60 minutes at 37°C. The nitrite concentration in the medium was then assayed colorimetrically using a microplate based method (3) and compared with published results obtained for laboratory rats (10).

2.6   Statistical Analysis

Statistical analyses were carried out using the Minitab software (Minitab Ltd. USA). Data from the most probable number determinations and those from colony counts on agar solidified isolation media (BA, BAS, BASS, AA, NA) were log transformed and compared for different tongue sections by analysis of variance and by calculating the least significant difference. Colony counts for AA, BAS and BASS were also analysed for differences between aerobic and anaerobic incubation. Regression analyses were carried out between the calculated maximum reaction velocity of nitrite production and the density of culturable nitrite producing bacteria on different tongue sections.

3 Results

3.1 Enumeration of Culturable Bacteria (cfu).

There was a >30 fold increase in the density of culturable bacteria between the anterior and posterior end of the rat tongue ($p<0.05$), but no significant difference between the numbers found on different isolation media (NA, BA, BASS, BAS and AA) on the same section (10). There was no significant difference in culturable bacterial numbers between aerobically and anaerobically incubated isolation media AA, BAS and BASS. Numbers of cfu on SDA (used to estimate the numbers of fungi/yeasts on the tongue) remained at or below the detection limit of the dilution method used (5×10^1 cfu cm^{-2}). No colonies developed on MRS (selective for *Lactobacillus* spp.), MCA (selective for coliform bacteria) and BAP agar (selective for *Pseudomonas* spp.).

Conversely, there was no significant difference in the density of culturable bacteria between the anterior and posterior end of the pig tongue (Figure 1). Again no significant difference was found between the bacterial population densities determined using different isolation media (NA, BA, BASS, BAS and AA). There was no significant difference in culturable bacterial numbers between aerobically and anaerobically incubated isolation media AA, BAS and BASS. Numbers of cfu on SDA remained at detection limit of the dilution method used. No colonies developed on MRS, MCA and BAP agar.

3.2 Nitrate Reduction from Intact Tongue Sections

Nitrite production was virtually absent (below 7.8 nmoles cm^{-2} h^{-1}) from the anterior 3 sections of the rat tongue and increased very little with increasing nitrate substrate concentrations. On the two posterior tongue sections, nitrite production increased with increasing nitrate substrate levels to between 400 and 500 moles cm^{-2} h^{-1} at the highest (10 and 20 mM) nitrate concentrations.

Nitrite production was similar on all five sections of pig tongues and increased with increasing substrate levels to between 70 and 90 nmoles cm^{-2} h^{-1} at the highest (10 and 20 mM) nitrate concentrations.

3.3 Density of Nitrite Producing Bacteria and Nitrite Production on Tongue Sections

There was a significant ($p<0.001$) increase in the density of culturable nitrite producing bacteria and in nitrite production (Figure 1) between the anterior

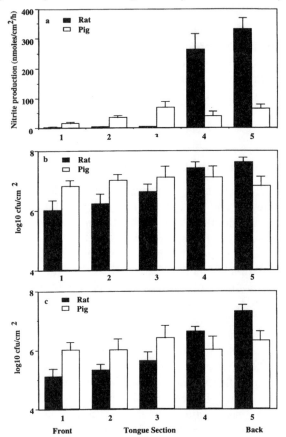

Figure 1   (a) *Nitrite production (nmoles/cm²/h), measured from intact tongue sections taken at roughly equal distances from front to back (1 to 5) and incubated in potassium nitrate (1mM), in laboratory rats (solid bars) and abattoir pigs (open bars). Compared with numbers of* (b) *total and* (c) *nitrite producing culturable cfu (MPN method). Means and SEM were from 5 determinations*

and posterior end of the rat tongue. The density of culturable nitrite producing bacteria and nitrite production were similar for all pig tongue sections (Figure 1). A significant correlation ($r = 0.93$, $p<0.05$) between the density of culturable nitrite producing bacteria and the maximum velocity of nitrite production was found for tongue sections from both rat and pig tongues.

3.4   Observation of the Histological Structure of Tongue Surface

The area of epithelial surface of a macroscopic centimetre square of rat tongue on the anterior surface was found to be approximately 1.6cm², having a relatively smooth surface. The posterior tongue, however, had a surface with regular clefts of 200 μm depth and 45 μm width (10) giving a macroscopic centimetre square of tongue of approximately 3.3 cm². The anterior and posterior surfaces of the pig tongue, on the other hand, had relatively similar numbers and depths of cleft.

3.5 Identification of Nitrite Producing Bacteria

The main nitrite producing genera isolated from rat tongues were *Staphylococcus* (65% of strains isolated), *Pasteurella* (20%) and *Streptococcus* (10%) and *Listonella* (5%) (Figure 2).

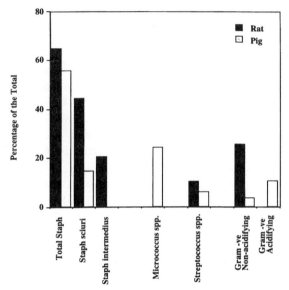

Figure 2 *The main nitrite producing genera and species isolated from laboratory rat (solid bars) and abattoir pig tongues (open bar). Percentages were calculated on 83 identified bacterial strains isolated from pig tongues and 63 bacterial strains from rat tongues*

The most frequently isolated bacterial species were *Staphylococcus sciuri* (40% of strains isolated) and *Staphylococcus intermedius* (25% of strains). Populations of *Streptococcus* consisted of a wider range of species including 5 streptococcus strains which could not be identified by the API computer software. Gram negatives isolated included three different species of *Pasteurella* (*P. pneumotropica, P. multocide, P. haemolytica*), *Listonella demsela, Enterococcus faecalis* and *Aerococcus viridans.*

In abattoir pigs there was a much greater diversity both at genus and species level (Figure 2). *Staphylococcus* (55% of isolates), *Streptococcus* (6%), *Pasteurella* (6%), *Micrococcus* (24%) and various genera (*Achromobacter, Aeromonas, Escherichia* and *Flavobacterium*) of Gram negative bacteria were also isolated.

Staphylococcus sciuri (15% of isolates) and *Micrococcus lylae* (20%) were the most frequently isolated species.

4   Discussion and Conclusion

The finding of similar numbers of bacteria on the general growth media (BA and NA) and blood agar containing *Streptococcus* or *Streptococcus/Staphylococcus* selective supplements in both the rat and the pig indicates that *Staphylococcus* and *Streptococcus* species are a main component of the microbial flora on the rat tongue.

For rats no *Micrococcus* spp. could be isolated from this media but relatively large proportion of isolates from pigs were identified as *Micrococcus* spp. (24%), indicating that *Micrococcus* spp. are a main inhabitant of pig tongues.

The results for rats and pigs are similar to those found in previous surveys on the microbial flora of human tongues. For example, Gordon & Gibbons (5) found that 70% of human tongue isolates were Gram-positive (mainly *Staphylococcus*, *Micrococcus* and *Streptococcus* species) and 30% Gram-negative.

The finding of similar numbers of cfu on aerobic and anaerobically incubated media both in rats and pigs indicates that the tongue microflora comprise few obligately aerobic species and that the majority of tongue bacteria are facultative anaerobes.

The populations of nitrate reducing bacteria inhabiting animal tongues have to date received very little interest. A large proportion of culturable bacteria the surface of the tongue (up to 65% of strains on the posterior tongue) were able to reduce nitrate to nitrite (10). A large proportion of culturable bacteria (up to 31%) from the surface of pig tongues were also found to reduce nitrate to nitrite. There was a significant positive correlation between the density of culturable nitrite producing bacteria and nitrite production, which again supports the hypothesis that microbial nitrate reduction is responsible for nitrite production.

A prominent feature of many animal tongues (including the two posterior sections of the rat tongue and all sections of the pig tongue), is the presence of deep clefts which are filled with bacteria (3). On the posterior rat tongue 58% of bacteria are present in these clefts (21% in the deep part of the cleft) which are constantly filled with saliva *in situ*. Their depths and the density of facultative-anaerobic bacteria present is likely to result in a steep gradient in oxygen tension due to bacterial respiration. Many bacteria which can use nitrate as an alternative electron acceptor are known to produce respiratory nitrate reductase enzymes only under low oxygen tensions (8). Given the apparent sensitivity of nitrate reduction on the tongue to oxygen (10), it is therefore reasonable to hypothesise that the increase in nitrite production on the posterior tongue is due to increased expression of respiratory nitrate reductase enzymes in conditions of reduced oxygen tension in the deep clefts on the posterior tongue. A relatively similar number and depth of clefts were found on all five sections of pig tongue, explaining the uniformity of nitrite production over the tongue surface.

Staphylococcus, Pasteurella and *Streptococcus* spp. were the dominant nitrite producing micro-organisms of the rat tongue and one species, *Staphylococcus sciuri* clearly dominated (40% of strains isolated) on the posterior tongue where most of the nitrite was produced. A large proportion of the nitrite production might therefore be due to just one bacterial species. Apart from *Staphylococcus*, a relatively large proportion of *Micrococcus* spp. and a number of other genera were found in abattoir pigs, indicating a greater diversity both at the genus and species level on pig tongues. Additional investigations into the kinetics of the nitrate reductase activity on the tongue surface and of nitrate reductase enzymes produced by different tongue bacteria would have to be carried out to identify the genera, species or strains of bacteria which contribute most towards nitrite production.

Staphylococcus sciuri and most of the other nitrite producing bacterial species identified are known commensal inhabitants of animals. Such organisms may provide the basis for probiotic therapy in domestic animals susceptible (*e.g.* neonatal animals which lack nitrate producing bacteria and adult animals after antibiotic therapy) to gastroenteritis (2, 3). An in-depth understanding of this new animal defence mechanism may not only reduce one of the most important problems in intensive animal husbandry, but, since the mechanism relies on nitrate intake by animals, may also change the currently negative perception of high nitrate/nitrite levels in foods such as green vegetables, cured meats and drinking water.

5 Acknowledgement

We would like to thank Mr A.D. McKinnon for his help with the preparation of tongue sections.

References

1. Benjamin, N., F. O'Driscoll, H. Dougall, C. Duncan, L. Smith, M. Golden and H. McKenzie *Nature*, 1994, **368**, 502.
2. Dougall, H.T., L. Smith, C. Duncan and N. Benjamin, *Br. J. Clin. Pharm.*, 1995, **39**, 460
3. Duncan, C., H. Dougall, P. Johnston, S. Green, R. Brogan, C. Leifert, L. Smith, M. Golden, and N. Benjamin, *Nature Med.*, 1995, **1**:546
4. Goaz, P.W. and H. A. Biswell, *J. Dent. Res.*, 1961, **40**, 355
5. Gordon, R.J. and R.J. Gibbons, *Arch Oral Biol.*, 1966, **11**, 627
6. Ishiwata, H., P. Boriboon, M. Harada, A. Tanimura, and M. Ishidate, *J. Food Hyg. Soc. Japan.*, 1975, **16**, 93
7. Ishiwata, H., A. Tanimura, and M. Ishidate. *J. Food Hyg. Soc. Japan.*, 1975, **16**, 89
8. Killham, K. 'Soil Ecology', Cambridge University Press, Cambridge, UK. 1994.
9. Leifert, C., W. M. Waites, and J. R. Nicholas, *J. Appl. Bacteriol.*, 1989, **67**, 353

10 Li, H., C. Duncan, J. Townend, K. Killham, L. Smith, P. Johnston, R. Dykhuizen, D. Kelly, M. Golden, N. Benjamin and C. Leifert, *Appl. Environ. Microbiol.*, 1997, **63**, 924

11. Page, A. L., R. H. Miller, and D. R. Keeney, 'Chemical and Microbiological Properties-Agronomy Monograph no. 9', Soil Science Society of America, Madison, Wis,1982 (2nd Edition), 815

12. Sasaki, T. and K. Matano, *J. Food Hyg. Soc. Japan.*, 1979, **20**, 363

13. Savostianov, G. M. *Fiziol. Zh. SSSR*, 1937, **23**, 159

14. Schinner, F., R. Öhlinger, E. Kandeler and R. Margesin 'Bodenbiologische Arbeitsmethoden', Springer-Verlag. Berlin, Germany 1993. (2. auflage)

15. Sen, N. P., D. C. Smith and L. Schwinghamer, *Food Cosmet. Toxicol.*, 1969, **7**, 301

16. Stevens, A. 'Theory and Practice of Histological Techniques', Churchill Livingstone Inc., New York, 1990, 3rd ed., 290

17. Tannenbaum, S. R., A. J.Sinskey, M. Weisman, and W. Bishop, *J. Natl. Cancer Inst.*, 1974, **53**, 79

18. Tannenbaum, S.R., M. Weisman, and D. Fett. Fd Cosmet. Toxicol., 1976, **14**, 549

19. Tannenbaum, S. R., D. Fett, V. R. Young , P. D. Land, and W. R. Bruce. *Science*, 1978, **200**, 1487

20. Tannenbaum, S.R., S. W. John, and D. L. Cynthia, *Am. J. Clin. Nutr.*, 1991, **53**, 247

21. White, J. W., Jr. *J. Agric. Food Chem.*, 1975, **23**, 886

19

Potential Risks and Benefits of Dietary Nitrate

Michael Golden[1] and Carlo Leifert[2]

[1] DEPARTMENT OF MEDICINE & THERAPEUTICS, POLWARTH BUILDING, UNIVERSITY OF ABERDEEN MEDICAL SCHOOL, FORESTERHILL, ABERDEEN AB9 2ZD, SCOTLAND, UK
[2] ABERDEEN UNIVERSITY CENTRE FOR ORGANIC AGRICULTURE (AUCOA), CRUICKSHANK BUILDING, ABERDEEN AB24 3UU, SCOTLAND, UK

1 Introduction

Governmental concern and public fear of nitrate has, over the last 40 years, resulted in increasingly stringent regulations to lower the levels of nitrate in food and water (WHO, 1970, 1984; DOE 1984). Vegetables and fruit are the main sources of nitrate, but drinking water and certain cured meat products may also contribute significantly (MAFF, 1987 & 1992). In cooler temperate zones where vegetables are grown with mineral nitrogen fertilisation at low light intensity and high temperatures in greenhouses, nitrate levels in vegetables may be particularly high. Moreover, the nitrate concentration of ground water is gradually increasing in many areas (Addiscott 1996). This is thought to be mainly due to the leaching of nitrate applied to agricultural soils as organic and mineral N-fertilisers. As a result waterboards struggle to maintain nitrate concentrations in drinking water below the allowable limit (50 mg/l) in many areas of intensive agricultural production. This and proposed legislation to limit food nitrate concentrations threaten conventional intensive agricultural and horticultural production systems in Northern Europe.

Legislation is currently mainly based upon the premise that dietary nitrate has detrimental effects such as an increased risk of gastric cancer (Anonymous, 1981; Cuello et al. 1976; Armijo and Coulson, 1975; Ohshima & Bartsch, 1981; Sen et al. 1969; Xu et al. 1992; Correa et al. 1990; Rademacher et al. 1992) and infantile methaemoglobinaemia (Comly, 1945; Donahoe, 1949). However, epidemiological and nutritional studies could not provide evidence for an increased risk of gastric and intestinal cancer in human population groups with high dietary nitrate intake (Boeing, 1991; Armijo *et al.* 1981, Pobel *et al.* 1995; Knight *et al.* 1990; Forman *et al.* 1985; Sobala *et al.* 1989).

More recent physiological studies indicated that nitrate may have beneficial effects on the physiology of the intestinal tract (McKnight *et al.* 1997, Lundberg *et al.* 1994, Bilski *et al.* 1994) and the cardiovascular system (Key *et al.* 1996) and may protect us against gastrointestinal diseases, tooth decay and oral infections (Duncan *et al.* 1997).

2   Potential Risks of Dietary Nitrate

2.1   Infantile Methaemoglobinaemia

The first concerns about nitrate in drinking water arose in the 1940s when it was recognised that infantile methaemoglobinaemia ('blue baby syndrome') was associated with the use of reconstituted baby foods in areas where water was drawn from local wells with high nitrate concentrations (Comly 1945, Donahoe 1949, Lecks 1950). It has since been realised that infantile methaemoglobinaemia is associated with bacterially contaminated well water, in which nitrate is reduced to nitrite by bacterial nitrate reductase enzymes.

$$NO_3^- \rightarrow NO_2^- \quad NR = \text{nitrate reductase}$$
$$NR$$

As young infants have low activity methaemoglobin reductase when nitrite oxidises the ferrous iron of the oxyhaemoglobin to ferric iron, the formation of methaemoglobin in blood can rise:

$$NO_2^- + oxyHb(Fe^{2+}) \rightarrow metHb(Fe^{3+}) + NO_3^-$$

In the UK this has led to the current *Water Supply (Water Quality) Regulations* (Ministry of Agriculture, Fisheries and Food 1987), which set a maximum nitrate concentration of 50 mg/L and a maximum nitrite concentration of 0.1 mg/L. The condition is now very rare in Europe and in the UK the last reported case of infantile methaemoglobinaemia from mains water was reported in 1972, and in the last 35 years there have been only 14 suspected cases, all associated with domestic shallow wells contaminated with nitrate reducing bacteria (Cottrell 1987).

No cases of methaemoglobinaemia associated with ingestion of well water with high nitrate concentrations have been reported for (i) breast fed infants (thought to be not at risk as nitrate is not transported into human milk) and in (ii) older children or adults where methaemoglobin is converted back to haemoglobin by the enzyme methaemoglobin reductase:

$$oxyHb(Fe^{3+}) \rightarrow metHb(Fe^{2+}) \quad MR = \text{methaemoglobin reductase}$$
$$MR$$

The enzyme system is only induced during the physiological post-weaning period and keeps methaemoglobin levels between 1 and 2% (Committee on

Nutrition 1970). However, accidental and occasional fatal poisoning has occurred where nitrite salts have been ingested (Gowans 1990; McQuiston & Belf, 1936).

2.2   Gastrointestinal Carcinogenesis

The other major public health concern about high dietary nitrate intake is that it may be causally associated with gastrointestinal cancer. It has been postulated that nitrous acid formed from swallowed dietary or salivary nitrite may nitrosate primary and secondary amines ingested in food to form nitrosamines (Sen *et al.* 1969, Tannenbaum *et al.* 1974, Oshima and Bartsch 1981) some of which have been shown to be carcinogenic in animal studies (Low 1974). Nitrosamines may also be formed during food storage or processing, between nitrite formed from nitrate by bacteria and amines and amides present in the food. The frequent consumption of certain foods containing preformed nitrosamines (*e.g.* the Korean Kimchi, a fermented cabbage dish) was linked to the persistently high incidence of gastric cancer in the Far East (Boring *et al.* 1993). These findings have led to calls for restrictions of nitrate levels in food products (Tannenbaum *et al.* 1976 & 1978, Tannenbaum 1983).

Many epidemiological studies have been undertaken in an attempt to establish whether nitrate intake is directly linked to gastrointestinal cancer. They have, however, produced conflicting and often contradictory results. On balance there seems to be no correlation between gastric cancer and dietary nitrate intake, but clear positive correlations between hygiene levels and food preparation methods and gastric cancer. Indeed several nutritional and epidemiological studies indicate that high vegetable based nitrate intake protects against cancer (see Duncan *et al.* 1997 for a recent literature review).

Most of these studies concentrated on comparisons between vegetarians and omnivores in the developed world. Vegetarians have a 3 times higher nitrate intake than non-vegetarians, 185–194 mg/person/day compared with 61 mg/person/day (Ministry of Agriculture, Fisheries and Food 1992), because the vast majority (80–90%) of nitrate taken in with the diet comes from vegetables, particularly green leafy vegetables. A number of recent studies have investigated the standardised mortality ratio (SMR) of vegetarians compared to non-vegetarians in relation to deaths from cancer (Thorogood 1995). A 20–40% reduction in death from cancer was shown for the vegetarian groups in the majority of studies. These studies, however, were based on volunteer subjects making them susceptible to the 'healthy volunteer' effect. One study that used randomly collected data was the Norwegian Adventist study (Fonnebo 1994), which showed very similar mortality ratios for both vegetarians and non-vegetarians. These studies suggest that a high vegetable based nitrate intake at least poses no greater risk for the development of gastric cancer and may even be protective.

It is important to stress the problems of using the absence of a clear correlation or negative correlations between vegetable based nitrate intake and gastric cancer as evidence against a potential causative role of nitrate in gastric cancer. For example, reducing agents such as ascorbic acid (which are commonly found in vegetables and fruit) are known to inhibit nitrosamine formation (Tannenbaum *et al.* 1991). This may explain why high vegetable nitrate uptake shows no negative effects in vegetarians, but cannot exclude the potential risk of high nitrate levels in water and meat products. However, Boeing (1991) found not only a consistently negative relationship between raw vegetables and gastric cancer in four cohort and 16 case-control studies, but also no difference in the development of cancer between groups consuming raw and cooked vegetables (during cooking most of the protective antioxidants such as vitamin C are lost).

Although there is no evidence for direct nitrate or nitrite carcinogenicity from animal studies, toxicological studies unequivocally demonstrated that preformed nitrosamines cause gastrointestinal cancer in animals (Magee & Barnes 1967). The most important question to answer must therefore be whether carcinogenic nitrosamines can be formed in the human stomach from nitrite (formed from dietary nitrate) and amines in doses sufficient to induce cancer.

McKnight *et al.* (1997) recently showed that following the ingestion of a 2 mM nitrate solution in 10 healthy volunteers the concentration of nitric oxide (NO) in the stomach rose to a mean of about 90 ppm at 60 minutes while the concentration of nitrite remained at very low levels ($\approx$ 20 μmoles/L) despite salivary nitrite levels reaching 1.5 moles/L at 20–40 minutes. Nitrite entering the normal acidic stomach appears to be rapidly converted to reduced N-forms which do not nitrosate secondary amines.

Pepsin, the major gastric proteinase, only attacks peptide bonds where a tyrosine or phenylalanine provide the amino group. If these free amino groups were nitrosated it would not pose a risk, since animal experiments indicate that nitrosamines based on tyrosine or phenylalanine are not carcinogenic (Magee & Barnes 1967, MAFF 1987).

Finally, whilst the intake of nitrate has increased, due to rising nitrate levels in drinking water and increased consumption of fruit and vegetables, the incidence of gastric cancer in Western Europe and North America has declined (Howson *et al.* 1986). This decrease in the western developed world is now believed to be linked to increased hygiene and better food preparation methods, and particularly the eating of raw vegetables which contain high levels of antioxidants (Hwang *et al.* 1994), but also to the reduction in the incidence of *Helicobacter pylori* infection (Hwang *et al.* 1994, Bolin *et al.* 1995, Correa 1995).

3   Potential Beneficial Effects of Dietary Nitrate

Potential beneficial effects of nitrite were first hypothesised when the metabolism and fate of dietary nitrate was studied (Tannenbaum *et al.* 1974, Grandli *et al.* 1989). It was found that when a dietary nitrate load is ingested, the nitrate is rapidly absorbed into the blood stream through the intestinal mucosa, primarily in the upper small intestine and to a lesser extent in the stomach (Witter *et al.* 1979, McKnight *et al.* 1997). Nitrate is then circulated in the blood and actively assimilated into the salivary glands (Spiegelhalder *et al.* 1976, Stephany and Schuller 1980) by an active transport mechanism, with salivary concentrations increasing to approximately ten times those found in the plasma. It is thought that the most likely anion transporter for the nitrate ion is that which also transports iodide and thiocyanate since there is competition between these anions for uptake. Nitrate inhibits iodide uptake, although it is less effective than perchlorate and thiocyanate and equally thiocyanate inhibits nitrate uptake (Tenouvo 1986). However, the exact nature of the transporter (*e.g.* a Ca^{2+} dependent channel) remains to be established. Approximately 1.5 litres of saliva are produced and swallowed each day containing nitrate and nitrite along with the other constituents, among which are reducing equivalents such as iodide, thiocyanate and ascorbic acid (Marsh & Martin 1992, Florin *et al.* 1990). Up to 25% of circulating nitrate from exogenous sources is taken up by the salivary glands (Spiegelhalder *et al.* 1976, Stephany and Schuller 1980).

In evolutionary terms, the finding that nitrate was recycled with the blood, actively concentrated in the saliva and repeatedly recirculated through the stomach rather than secreted immediately, strongly suggested a net beneficial physiological function for nitrate in animals.

3.1 Protection against intestinal pathogens

The presence of both nitrate and nitrite in the saliva has been known since the first half of this century (Savostianov, 1937; Varady and Szanto, 1940). Saliva collected directly from salivary ducts contains nitrate but no nitrite. However, saliva collected at other locations in the oral cavity also contains nitrite (Tannenbaum *et al.* 1974, Ishiwata *et al.* 1975). A number of investigations have shown that nitrate reductase enzymes of the oral microflora are responsible for the reduction of nitrate to nitrite. (Goaz and Biswell 1961, Ishiwata *et al.* 1975, Tannenbaum *et al.* 1974, Li *et al.* 1997, Li *et al.* this volume, pp. 259–268).

Acidification of ingested food in the stomach was until recently believed to be the primary resistance mechanism preventing food borne pathogens from entering the more distal intestine. However, re-examination of this mechanism revealed that two hour exposure (the maximum likely period of time food remains in the stomach) to pH values of between 2 and 4 (the pH values commonly found in the stomach) only had a bacteriostatic effect on food borne pathogens such as *Campylobacter*, *Salmonella*, *Shigella* and *E. coli* (Frazer *et al.* 1995; Dykhuizen, R, Benjamin, N. & Leifert, C. unpublished). Sufficient numbers of the pathogens would therefore be expected to survive the acid treatment in the stomach and re-grow and cause disease when reaching the neutral environment of the more distal intestine.

Addition of nitrite (at concentrations commonly found in the saliva) to hydrochloric acid solutions with pH values of between 2 and 4 increased its antimicrobial activity by up to 100 fold and bacteriocidal activity of acidified nitrite was established against a range of gastrointestinal pathogens including *Salmonella enteriditis, Salmonella typhimurium, Yersinia enterocolitica, Shigella sonnei,* and *Escherichia coli* 0157 (Dykhuizen *et al.* 1996 and this issue, pp. 295–316). They also demonstrated that the antimicrobial activities of nitrite and protons are synergistic within the physiological range of their concentrations in saliva and gastric juice respectively. Food pathogens differed in sensitivity to acidified nitrite (*Y. enterocolitica* > *S. enteriditis* > *S. typhimurium* = *Shig. sonnei* > *E. coli* 0157) and in their relative sensitivity to nitrite and protons. Addition of thiocyanate (SCN^-) and iodide, which are also concentrated by the salivary glands sharing the same transport mechanism as nitrate, increased the antibacterial activity of acidified nitrite (Dykhuizen *et al.* 1996; Fite, A., Golden, M. & Leifert, C. unpublished).

The active chemical species causing microbial kill in acidified nitrite solutions are, as yet unclear. The possible chemical species involved in kill have been reviewed by Duncan *et al.* (1997) and the most frequently proposed candidate chemical species are described below. Nitrite itself has been described as an antimicrobial (Wodzinski *et al.* 1977, 1978), but since the antibacterial activity of nitrite increases with decreasing pH other workers suggested nitrous acid (HNO_2) as the main antimicrobial species (Duncan *et al.* 1997):

$$NO_2^- + H^+ \rightarrow HNO_2 \text{ (pK}_a \text{ 3.2–3.4)}$$

The known reaction of nitrite at low pH to produce nitric oxide (NO) suggests that NO may have an important role in microbial killing (Duncan *et al.* 1995):

$$3HNO_2 \rightarrow H_2O + 2NO + NO_3^-$$

Although NO is implicated in microbial kill, reactive products, such as peroxynitrite ($ONOO^-$) have been described as likelier candidates for this role than NO. This is supported by results which show that nitric oxide under anaerobic conditions has virtually no effect on *Pseudomonas fluorescence* or *Staphylococcus aureus* (Shank *et al.* 1962). Investigations by Zhu *et al.* (1992) and Brunelli *et al.* (1995) indicate that peroxynitrite has a much higher antimicrobial activity compared to NO. Although microbiological evidence based on the relative rates of kill shown by NO, 'zermalite' and peroxynitrite, suggests that the rate determining step is the production of peroxynitrite some doubt still remains. Further recent evidence for the role of NO as a precursor and not the active agent is provided by Yoshida *et al.* (1993), who argue that the anti-microbial action of NO is enhanced by addition of an oxidising agent – PTIO and by low pH. A possible active agent in such a system is NO or equivalents in activity: N_2O_3 and N_2O_4, consistent with known mutagenic activity of these compounds. They further argue that these mechanisms are sufficient to explain the toxic effect of NO without invoking the production of peroxynitrite. Until the exact chemical species responsible for the antimicrobial activity are identified some concern must remain about their cytotoxic and mutagenic potential and there is clearly now a very practical need to initiate research to gain a better understanding of the chemistry of acidified nitrite solutions.

Preliminary *in vivo* experiments have confirmed *in vitro* results. In these studies bacterial suspensions of *Salmonella typhi* live attenuated oral vaccine strain Ty21 were enclosed in a short segment of dialysis tubing, impermeable to high molecular weight molecules such as immunoglobulins and interferon, and capsules then swallowed into the stomach on a length of soft dental floss by healthy volunteers. Survival rates of bacteria exposed to the stomach environment of volunteers which had been administered a drink containing the amount of nitrate commonly found in half a lettuce were significantly lower than that of bacteria which were exposed to the stomach of volunteers which had a nitrate free drink (Ndebele 1996).

This evidence suggests, that while there is no epidemiological evidence for the theory that dietary nitrate causes N-nitrosamine induced carcinogenesis, there is now a building body of evidence indicating that dietary nitrate plays an important beneficial role, particularly in protecting the intestinal tract from common pathogens.

3.2 Prevention of *Helicobacter pylori* Infection

Helicobacter pylori causes gastritis, MALT lymphoma, and duodenal ulcer, and is associated with gastric ulceration and gastric malignancy (Parsonnet 1993, Dykhuizen *et al.* 1998). Peptic ulceration affects at least 10% of adults and gastric cancer is the fourth commonest cause of cancer death. *H. pylori* represents a particular problem because the bacterium is adapted to grow in the strongly acid environment of the stomach and can persist for many years in the host. Recent *in vitro* results demonstrated the bacteriocidal activity of acidified nitrite against *Helicobacter pylori* (Dykhuizen *et al.* 1998) and the possibility that nitrite in the acidic stomach may contribute to the natural defence against *H. pylori* is under investigation *in vivo*. If a protective activity of dietary nitrate intake could be confirmed it would provide an explanation for the epidemiological studies which showed a negative correlation between dietary nitrate intake and gastric cancer incidence (see section 1.2.).

3.3 Prevention of Tooth Decay and Oral Infections

Conversion of nitrite into antimicrobial nitrogen compounds also occurs in acid zones of the oral cavity itself. Bacteria involved in tooth decay (*e.g.* *Streptococcus* and *Lactobacillus* spp.) produce acid, in fact acid production is one of the main reasons for tooth decay (Marsh & Martin 1992). Similarly yeasts such as *Candida albicans* which cause oral thrush are known to produce acid during growth. It is, therefore, likely that the production of nitrite in areas of the tongue with neutral pH will inhibit acid producing dental pathogens (Duncan *et al.* 1995). Although there is no direct evidence that salivary nitrite gives significant protection against tooth decay, some indirect evidence is available. For example, it has been known for some time that medications which inhibit saliva production result in an acceleration of tooth decay. Furthermore, antibiotic treatments, which inhibit nitrite production in the oral cavity are known to increases the risk of oral thrush (Marsh & Martin 1992).

4   Conclusions

There is now a growing body of evidence to suggest a beneficial role for dietary nitrate intake facilitated by the bacterial reduction of nitrate to nitrite in the oral cavity and the enterosalivary circulation of nitrate. Nitrite formed from nitrate is thought to protect both the oral cavity and the gastrointestinal tract from common pathogens. On the other hand, there is no conclusive epidemiological evidence that dietary nitrates are causally linked to carcinogenesis. Conversely, some epidemiological studies show a reduced rate of gastric and intestinal cancer in groups with a high vegetable based nitrate intake and this may be explained by a protective effect of acidified nitrite against *Helicobacter pylori* infection which is increasingly recognised to be associated with gastric malignancy (Parsonnet 1993).

However, although the apparent beneficial effects of nitrite in the prevention of infectious diseases are likely to change the current negative perception of vegetable based dietary nitrate intake, the authors feel that a sensible limit for levels of nitrate in drinking water should be enforced. This is for several reasons:

- First, although controversial and experimentally weak, a correlation between diabetes and nitrate levels in drinking water has been proposed by McKinney (this issue, pp. 327–339). These data should be evaluated and confirmed by independent experts.
- Second, although very rare, infantile methaemoglobinaemia may be caused by reduction of nitrate to nitrite in contaminated wells.
- Finally, and most importantly, rising nitrate levels are a symptom of improper use of nitrogen fertilisers and poor agricultural management practice, which can result in environmental damage especially through eutrophication and algal blooms in water reservoirs, rivers and costal waters.

5   References

Addiscott, T.M *Fertilisers and nitrate leaching. In: Agricultural Chemicals and the Environment [R.E. Hester and R.M. Harrison, editors] Issues in Environmental Science and Technology, No. 5*, The Royal Society of Chemistry, Cambridge, UK., 1996, pp. 1–26

Anonymous. *The health effects of nitrate, nitrite and N-nitroso compounds. Part 1 of a 2 part study by the committee on nitrite and alternative curing agents in food.* National Academy Press. Washington D.C. 1981.

Armijo, R. and Coulson, A.H. Epidemiology of stomach cancer in Chile: the role of nitrogen fertilizers. *International J. of Epidemiology,* 1975, **4,** 301.

Armijo, R., Gonzales, A., Orellana, M., Coulson, A.H., Sayre, J.W. and Detels, R. Epidemiology of gastric cancer in Chile II. Nitrate exposures and stomach cancer frequency. *International J. of Epidemiology*, 1981, **10**, 57.

Benjamin, N., O'Driscoll, F., Dougall, H., Duncan, C., Smith, L., Golden, M., and McKenzie, H. Stomach NO synthesis. *Nature*, 1994, **368**, 502.

Bilski, J., Konturek, P.C., Konturek, S.J., Cieszkowski, M., and Czarnobilski, K. Role of endogenous nitric oxide in the control of gastric acid secretion, blood flow and gastrin release in conscious dogs. *Regulatory peptides*, 1994, **54**, 175.

Boeing, H. Epidemiological research in stomach cancer: progress over the last ten years. *J. Cancer Re. Clin. Oncol.*, 1991, **117**, 133.

Bolin, T.D., Hunt, R.H., Korman, M.G., Lambert, J.R., Lee, A., Talley, N.J. *Heliobacter pylori* and gastric neoplasia: Evolving concepts. *Med. J. Australia.* 1995, **163**, 253.

Boring, C.C., Squires, T.S., Tong, T. Cancer statistics 1993. *CA Cancer J. Clin.*, 1993, **43**, 7.

Brunelli, L., Crow, J.P. and Beckman, J.S. The comparative toxicity of nitric oxide and peroxynitrite to *Escherichia coli*. *Arch. of Biochem. and Biophysics.* 1995, **316**, 327.

Comly, H.H. Cyanosis in infants caused by nitrates in well water. *J.A.M.A.* 1945, **129**, 112.

Committee on Nutrition, *Paediatrics*, 1970, 46, 475.

Correa, P. *Helicobacter pylori* and gastric carcinogenesis. *Am. J. Surg. Pathol.* 1995, **(Suppl)**, S37.

Correa, P., Haenszel, W., Cuello, C., Zavala, D., Fontham, E., Zarama, G., Tannenbaum, S., Collazoa, T. and Ruiz, B. Gastric precancerous process in a high risk population: cross-sectional studies. *Cancer Research*, 1990, **50**, 4731.

Cottrell, R. Nitrate in Water. *Nutrition in food Science*, 1987, May/June, 20.

Cuello, C., Correa, P, Haenszel, W., Gordillo, G., Brown, C., Archer, M. & Tannenbaum, S. Gastric cancer in Colombia. 1. Cancer Risk and Suspected Environmental Agents. *Journal of the National Cancer Institute*, 1976, **57**, 1015.

Department of the Environment, Nitrates in water supplies, *DOE Press Notice 180*, 9 April, 1984.

Donahoe, W.E. Cyanosis in infants with drinking water as a cause. *Paediatrics.*, 1949, **3**, 308.

Duncan, C., Dougall, H., Johnston, P., Green, S., Brogan, R., Leifert, C., Smith, L., Golden, M., Benjamin, N. Chemical generation of nitric oxide in the mouth from the enterosalivary circulation of dietary nitrate. *Nature Medicine*, 1995, **1**, 546.

Duncan, C., Li, H., Dykhuizen, R., Frazer, R., Johnston, P.,MacKnight, G., MacKenzie, H., Batt, L. Golden, M., Benjamin, N. & Leifert, C. Protection against oral and gastrointestinal diseases: Importance of dietary nitrate intake, oral nitrate reduction and enterosalivary nitrate circulation. *Comp. Biochem. Physiol.*, 1997, **118A**, 939.

Dykhuizen, R.S., Duncan, C., Frazer, R., Smith, C.C., Golden, M., Benjamin, N., Leifert, C. Dietary nitrate might protect against infective gastroenteritis: The antimicrobial effect on acidified nitrite on gut pathogens. *Antimicrob. Agents Chemother.* 1996, **40**, 1422.

Dykhuizen, R.S., Fraser, A., McKenzie, H., Golden, M., Leifert, C., and Benjamin, N. *Heliobacter pylori* is killed by nitrite under acidic conditions. *Gut* 1998, **42**, 334.

Florin, T.H., Neale, G. and Cummings, J.H. The effect of dietary nitrate on nitrate and nitrite excretion in man. *Brit. J. Nutrition*, 1990, **64**, 387.

Fonnebo, V. The Healthy Seventh-day Adventist lifestyle: What is the Norwegian experience? *Am. J. Clin. Nut.*, 1994, **59**, 1124S.

Forman, D., Al-Dabbagh, S.A., and Doll, R. Nitrates, nitrites and gastric cancer in Great Britain. *Nature*, 1985, **313**, 620.

Frazer, R., Duncan, C., Li, H., White, D., Benjamin, N., Leifert, C. Sensitivity of Fungi and bacteria to acidified nitrite. *Society of General Microbiology 132nd Ordinary Meeting, Abstract Booklet*, 1995, 109.

Goaz, P.W. and Biswell, H.A. Nitrite reduction in whole saliva. *J. Dent. Res.*, 1961, **40**, 355.

Gowans, W.J. Fatal methaemoglobinaemia in a dental nurse. A case study of sodium nitrite poisoning. *Brit. J. General Practice*, 1990, **40**, 470.

Grandli., T., Dahl, R., Brodin and P., Bockman, O.C. Nitrate and nitrite concentration in human saliva: variations with salivary flow rate. *Fd. Chem. Toxical.*, 1989, **27**, 675.

Howson, C.P., Hiyama, T., Wynder, E.L. The decline in gastric cancer: Epidemiology of an unplanned triumph. *Epidemiol. Rev.*, 1986, **8**, 1.

Hwang, H., Dwyer, J., Russell, R.M. Diet, *Helicobacter pylori* infection, food preservation and gastric cancer risk: Are there new roles for preventative factors? *Nutr. Rev.*, 1994, **52**, 75.

Ishiwata, H., Tanimura, A. and Ishidate, M. Nitrite and nitrate concentrations in human saliva collected from salivary ducts. *J. Food Hyg. Soc. Japan.*, 1975, **16**, 89.

Key, T.J.A., Thorogood, M., Appleby, P.N. and Burr, M.L. Dietary habits and mortality in 11000 vegetarians and health conscious people: results of a 17 year follow up. *British Medical J.*, 1996, **313**, 775.

Knight, T.M., Forman, D., Piratsu, R., Comba, P., Iannarilli, R., Cocco, P.L. Angotzi, G., Ninu, E., and Schierano, S. Nitrate and nitrite exposure in Italian populations with different gastric cancer rates. *International J. of Epidemiology*, 1990, **19**, 510

Lecks, H.I. Methemoglobinemia in infancy. *Am. J. Dis. Child.*, 1950, **79**, 117.

Li, H., Duncan, C., Frazer, R., Townend, J., Killham, K., Smith, L.M., Johnston, P., Dykhuisen, R., Golden, M., Benjamin, N., Leifert, C. Nitrate reducing bacteria on rat tongues. *Appl. Environ. Microbiol.* 1997, **63**, 924.

Low, H. Nitroso compounds: Safety and Public Health. *Archives of Environmental Health*, 1974, **29**, 256.

Lundberg, J.O.N., Weitzberg, E., Lundberg, J.M. and Alving, K. Intragastric nitric oxide production in humans: measurements in expelled air. *Gut*, 1994, **35**, 1543.

Magee, P.N. and Barnes, J.M. Carcinogenic nitroso compounds. *Advances in Cancer Research*, 1967, **10**, 163.

Marsh, P. and Martin, M. Oral Microbiology, 1992, Chapman & Hall, London, UK.

McKnight, G., Smith, L.M., Drummond, R.S. Duncan, C.W., Golden, M.N.H. and Benjamin, N. The Chemical Synthesis of Nitric Oxide in the Stomach from Dietary Nitrate in Man. *Gut*, 1997, **40**, 211.

McQuiston, T.A.C. and Belf, M.B. Fatal poisoning by sodium nitrite. *Lancet*, 1936, **ii**, 1153.

Ministry of Agriculture, Fisheries and Food. *Food surveillance paper No. 20: Nitrate, nitrite and N-nitroso compounds in food.* 1987, HMSO, London.

Ministry of Agriculture, Fisheries and Food. *Food surveillance paper No. 32: Nitrate, nitrite and N-nitroso compounds in food: Second Report.* 1992, HMSO, London.

Ndebele, N., The in vivo sensitivity of *Salmonella typhi* TY21A to acidified nitrite in the stomach. 1996. M.Sc. Thesis, Aberdeen.

Oshima, H and Bartsch, H. Quantitative estimation of endogenous nitrosation in humans by monitoring N-nitosoproline excreted in urine. *Cancer Res.*, 1981, **41**, 3658.

Parsonnet, J. *Helicobacter pylori* and gastric cancer. *Gastroenterol Clin. North Am.*, 1993, **22**, 89.

Pobel, D., Riboli, E., Cornee, J., Hemon, B., and Guyader, M. Nitrosamine, nitrate and nitrite in relation to gastric cancer: a case control study in Marseilles, France. *European J. of Epidemiology*, 1995, **11**, 67.

Rademacher, J.J., Young, T.B., and Kanarek, M.S. Gastric cancer mortality and nitrate levels in Wisconsin drinking water. *Archives of Environmental Health*, 1992, **47**, 292.

Savostianov, G.M. On the question of nitrites in saliva. *Fiziol. Zh. SSSR*, 1937, **23**, 159.

Sen, N.P., Smith, D.C. and Schwinghamer, L. Formation of N-nitrosamines from secondary amines and nitrite in human and animal gastric juice. *Food Cosmet. Toxicol.*, 1969, **7**, 301.

Shank, J.L., Silliker, J.H., and Harper, R.H. The effect of nitric oxide on bacteria. *Appl. Microbiol.*, 1962, **10**, 185.

Sobala, G.M., Schorah, C.J., Sanderson, M., Dixon, M.F., Tompkins, D.S., Godwin, P. and Axon, A.T. *Gastroenterology*, 1989, **97**, 257.

Spiegelhalder, B., Eisenbrand, G. Preussmann. R. Influence of dietary nitrate on nitrite content of human saliva: Possible relevance to in vivo formation of N-nitroso compounds. *Fd. Cosmet. Toxicol.*, 1976, **14**, 545.

Stephany, R.W., and Schuller, P.L. Daily dietary intakes of nitrate, nitrite and volatile N-nitrosamines in the Netherlands using the duplicate portion sampling technique. *Oncology*, 1980, **37**, 203.

Tannenbaum, S.R. N-nitroso compounds: a perspective on human exposure. *Lancet*, 1983, **1**, 629.

Tannenbaum, S.R., Sinskey, A.J., Weisman, M., Bishop, W. Nitrite in human saliva. Its possible relationship to nitrosamine formation. *J. Natl. Cancer Inst.* 1974, **53**, 79.

Tannenbaum, S.R., Weisman, M. and Fett, D. The effect of nitrate on nitrite formation in human saliva. *Food Cosmet. Toxicol.* 1976, **14**, 549.

Tannenbaum, S.R., Fett, D., Young., V.R., Land, P.D., Bruce, W.R. Nitrite and nitrate are formed by endogenous synthesis in the human intestine. *Science*, 1978, **200**, 1487.

Tannenbaum, S.R., John, S.W., Cynthia, D.L. Inhibition of nitrosamine formation by ascorbic acid. *Am. J. Clin. Nutr.*, 1991, **53**, 247.

Tenouvo, J. The biochemistry of nitrates, nitrites, nitrosamines and other potential carcinogens in human saliva. *J. Oral Pathol.*, 1986, **15**, 303.

Thorogood, M. Vegetarianism and health. *Nutr. Res. Rev.* 1995, **8**, 179.

Varady, J. and Szanto, G. Untersuchungen über den Nitritegehalt des Speichels, des Magensaftes und des Harnes. *Klin. Wochenschr.*, 1940, **19**, 200.

Witter, J.P., Balish, E., and Gatley, S.J. Distribution of nitrogen 13 from labelled nitrate and nitrite in germfree and conventional-flora rats. *Appl. Environ. Microbiol.*, 1979, **38**, 870.

Wodzinski, R.S., Labeda, D.P. Alexander, M. Toxicity of SO_2 and NO_x: Selective inhibition of blue-green algae by bisulfite and nitrite. *J. Air Pollut. Control Assoc.*, 1977, **27**, 891.

Wodzinski, R.S. Labeda, D.P. Alexander, M. Effects of low concentrations of bisulfite and nitrite on microorganisms. *Appl. Environ. Microbiol.*, 1978, **35**, 718.

World Health Organisation. *European Standards for Drinking-Water*. 1970.

World Health Organisation. *Guidelines for Drinking-Water Quality. Vol. 1 Recommendations*. 1984.

Xu, G., Song, P. and Reed, P.I. The relationship between gastric mucosal changes and nitrate intake *via* drinking water in a high-risk population for gastric cancer in Moping County, China. *European J. of Cancer Prevention*, 1992, **1**, 437.

Yoshida, K., Akaike, T., Doi, T., Sato, K., Ijiri, S., Suga, M., Ando, M. and Maeda, H. Pronounced enhancement of NO-dependent antimicrobial action by an NO-oxidising agent, imidazolineozyl N-oxide. *Infect. Immun.*, 1993, **61**, 3552.

Zhu, L., Gunn, C., Beckman, J.S. Bactericidal activity of per-oxynitrite. *Arch. Biochem. Biophys.*, 1992, **298**, 452.

20

Metabolism of Nitrate in Humans – Implications for Nitrate Intake

Nigel Benjamin and Gillian McKnight

DEPARTMENT OF CLINICAL PHARMACOLOGY,
ST BARTHOLOMEW'S AND THE ROYAL LONDON SCHOOL OF
MEDICINE AND DENTISTRY, LONDON EC1, UK

1   Introduction

Inorganic nitrates (NO_3^-) and nitrite (NO_2^-) have been used as a food preservative for centuries. As well as its beneficial effect to limit the growth of serious pathogens such as *Clostridium botulinum* [1], nitrite has also the dubious benefit of rendering muscle tissue a bright pink colour, by the formation of nitrosomyoglobin. Despite their long use, there have been considerable concerns about the content of these ions in vegetables and drinking water. There are two main reasons for this:

- There is the theoretical possibility of forming carcinogenic N-nitroso compounds in food to which these ions are added and in humans in-vivo, due to nitrosation of secondary amines also present in the diet or ingested as drug therapy [2].
- Nitrosation of amines and other chemicals will occur rapidly as a result of reaction with nitrous acid which is formed under acidic conditions (such as in the human stomach) when nitrite is present [3]. It will also occur in the stomach under more neutral pH when there are bacteria present. The mechanism of this presumably enzymatic nitrosation is not understood.

Nitrous acid is an effective nitrosating agent due to its ability to donate an NO^+ group. These nitrosation reactions are catalysed by halide ions (such as chloride) and thiocyanate due to the formation of nitrosohalide (eg NOCl) and nitrosothiocyanate (NOSCN) respectively, due to the ability of these intermediates to more effectively donate NO^+ groups. Both these ions are present in gastric juice in high concentrations, the latter derives from the diet (particulary brassicas such as cabbage) and is concentrated in saliva. Nitrate in the diet is similarly concentrated in saliva following absorption and reduced to nitrite in the mouth. It therefore seems that the stomach is an ideal reaction chamber for the nitrosation of susceptible swallowed chemicals.

This scheme of human metabolism of nitrate and nitrite seems to be counterproductive for human health and has prompted reconsideration of exactly how these ions are utilised. A scheme was formulated whereby the formation of nitrogen oxides in the stomach and on the skin surface, derived from dietary nitrate and nitrite, may protect against bacterial, and possibly viral, pathogens.

A second potential mechanism for nitrite toxicity is the formation of methaemoglobin due to oxidation of the iron in haemoglobin. This is due to the reaction of nitrite with oxyhaemoglobin, where it is oxidised to form nitrate in addition to methaemoglobin.

It should also be noted that nitric oxide also reacts quickly with oxyhaemo-globin to produce methaemoglobin and nitrate – this may be more important than nitrite in causing this problem in infants fed on well water. Methaemoglo-binaemia is generally only a problem in young infants, but has been the main reason for statutory limitations on the concentrations of nitrate in drinking water. For nitrate to cause methaemoglobinaemia it has to be reduced to nitrite in food, water or in the body.

2   Mammalian Nitrate Synthesis

This interest in the metabolism of nitrate stimulated studies in humans which confirmed a discovery originally made in 1916 [4], that mammals, including humans, synthesise inorganic nitrate [5,6,7,8]. This means that even on a nitrate-free diet, there are considerable concentrations of nitrate in plasma (around 30 µM) and in the urine (around 800 µmoles/24 hours). It was also found that, although it is not protein bound, nitrate has a long half-life of 5–8 hours [9], which seems to be because it is about 80% reabsorbed from the renal tubule by an active transport mechanism [10]. It is now thought that endogenous nitrate synthesis derives from constitutive NOS enzymes acting on L-arginine [11]. The NO formed is rapidly oxidised to nitrate when it encounters superoxide or oxidised haemoglobin. It is still not clear whether all endogenous nitrate synthesis derives from this route as, following prolonged infusion of ^{15}N-labelled arginine, the enrichment of urinary nitrate with this heavy isotope is only about one half of the steady state of ^{15}N arginine enrichment [12]. This may mean that nitrate also derives from another source, or that the intracellular enrichment of labelled arginine is less than that in the plasma due to transamination reactions.

This peculiar metabolism of nitrate – renal salvage, salivary concentration and conversion to nitrite in the mouth made us consider that this may be a purposeful mechanism to provide oxides of nitrogen in the mouth and stomach to provide host defence against swallowed pathogens [13]. The first studies performed were to investigate the mechanism of nitrate reduction to nitrite in the mouth.

3 Oral Nitrate Reduction

Although Tannenbaum and his colleagues [8] had considered that salivary bacteria may be reducing nitrate to nitrite, Sasaki and colleagues [14] showed that in humans this activity is present almost entirely on the surface of the tongue. They contested that the nitrate reductase enzyme was most likely to be a mammalian nitrate reductase. Using a rat tongue preparation, we also found that the dorsal surface of the tongue in this animal had very high nitrate reductase activity, which was confined to the posterior two-thirds [15]. Microscopic analysis of the tongue surface revealed a dense population of gram negative and gram-positive bacteria, 80% of which, in-vitro, showed marked nitrate reducing activity.

Our suspicion that the nitrate reduction was being accomplished by bacteria was strengthened by the observation that the tongues of rats bred in a germ-free environment, which had no colonisation of bacteria, demonstrated no nitrate reducing activity on the tongue. Furthermore, treatment of healthy volunteers with the broad spectrum antibiotic amoxycillin results in reduced salivary nitrite concentrations [16].

Although it has not been able to characterise the organisms in normal human tongues (this would require a deep biopsy as the majority of the bacteria are at the bottom of the papillary clefts of the tongue surface), the most commonly found nitrite-producing organisms in the rat [17] were *Staphylococcus sciuri,* followed by *Staphylococcus intermedius, Pasteurella spp.* and finally *Streptococcus spp.* Both morphometric quantification of bacteria on tongue sections and enumeration of culturable bacteria showed an increase in the density of bacteria towards the posterior part of the tongue.

We now believe that these organisms are true symbionts, and that the mammalian host actively encourages the growth of nitrite-forming organisms on the surface of the tongue. The bacteria are facultative anaerobes which, under hypoxic conditions, use nitrate instead of oxygen as an electron acceptor for oxidation of carbon compounds to derive energy.

For the bacteria, nitrite is an undesirable waste product of this process, but it is, likely, used by the mammalian host for its antimicrobial potential elsewhere.

4 Acidification of Nitrite – Production of NO in the Mouth and Stomach

Nitrite formed on the tongue surface can be acidified in two ways. It can be swallowed into the acidic stomach, or it may encounter the acid environment around the teeth provided by organisms such as *Lactobacillus* or *Streptococcus mutans* which are thought to be important in caries production by virtue of their ability to produce large amounts of acid when metabolising sugars.

Acidification of nitrite produces nitrous acid (HNO_2) which has an acid dissociation constant of 3.2, so that in the normal fasting stomach (pH1–2) complete conversion will occur.

$$NO_2^- + H^+ \leftrightarrow HNO_2$$
$$2HNO \leftrightarrow N_2O_3 + H_2O \leftrightarrow NO + NO_2$$
$$3HNO_2 \leftrightarrow 2NO + NO_3 + H^+ + H_2O$$

Nitrous acid is unstable and will spontaneously decompose to NO and nitrogen dioxide (NO_2). Under reducing conditions more NO will be formed than NO_2. Lundberg and his colleagues [18] were the first to show that there was a very high concentration of NO in gas expelled from the stomach in healthy volunteers, which increased when nitrate intake was increased and reduced when gastric juice acidification was impaired with the proton pump inhibitor omeprazole. Further studies on the amount of NO produced following ingestion of inorganic nitrate, measured more directly using nasogastric intubation of healthy human volunteers. After swallowing 1 millimole of inorganic nitrate – the amount of nitrate found in a large helping of lettuce, there follows a pronounced increase in stomach headspace gas NO which peaks at about 1 hour and continues to be elevated above the control day for at least 6 hours [19]. The concentration of NO measured in the headspace gas of the stomach in these experiments would be lethal after about 20 minutes if breathed continuously.

The concentration of NO in the stomach is much higher than would be expected from the concentration of nitrite in saliva and the measured pH in the gastric lumen. *In vitro* studies suggest that these concentrations of nitrite and acid would generate about one tenth of the NO that is actually measured (McKnight, Smith & Benjamin, unpublished data). It is likely that a reducing substance such as ascorbic acid [20,21,22] (which is actively secreted into the stomach) or reduced thiols (which are in high concentrations in the gastric mucosa) are responsible for the enhanced NO production.

We were surprised to find that NO is also generated in the oral cavity from salivary nitrite [15] as saliva is generally neutral or slightly alkaline. The most likely mechanism for this production is acidification at the gingival margins as noted above. It will be important to determine if this is the case as the NO formed in this way may be able to inhibit the growth of organisms which generate acid. Such a mechanism for local NO synthesis from nitrite may in part explain the importance of saliva in protection from caries. As in the stomach, acidification of saliva results in larger amounts of NO production than would be expected from the concentration of nitrite present. Saliva contains vitamin C [23] and treatment with ascorbic acid oxidase partly reduces the NO to the expected value. It is again likely that there are other reducing agents present in normal saliva which augment NO production.

5   Nitric Oxide Synthesis from the Skin

Again using a chemiluminescence NO detector, it was possible to show generation of NO from normal human skin [24]. As NO has the ability to diffuse readily across membranes, it was considered most likely that NO which had escaped from vascular endothelium to the skin surface, manufactured by constitutive NOS was being measured. However, when the NOS antagonist monomethyl arginine was infused into the brachial artery of healthy volunteers in amounts sufficient to maximally reduce forearm blood flow, it was found that the release of NO from the hand was not affected. Futhermore, application of inorganic nitrite substantially elevated skin NO synthesis.

This coupled with the observations that NO release was enhanced by acidity, and reduced by antibiotic therapy makes it likely that again NO is being formed by nitrite reduction. Normal human sweat contains nitrite at a concentration of about 5 μM, and this concentration is precisely in line with that which one would predict would be necessary to generate the amount of NO release observed from the skin. The source of nitrite is not clear, but is likely to be from bacterial reduction of sweat nitrate by skin commensal organisms which are known to elaborate the nitrate reductase enzyme.

This observation has led to the hypothesis that skin NO synthesis may also be designed as a host defence mechanism to protect against pathogenic skin infections, especially against fungi. The release of NO is inevitably increased following licking of the skin (due to the large amount of nitrite in saliva), which may explain why animals and humans have an instinctive urge to lick their wounds [25]. We have also shown that the application of inorganic nitrite and an organic acid is effective in the treatment of patients with tinea pedis (athlete's foot; Weller, Ormerod & Benjamin, submitted for publication).

6 Antimicrobial activity of acidified nitrite

Nitrate is generally non-reactive with organic molecules and has to be chemically or enzymatically reduced to nitrite (NO_2^-) to be effective as an antimicrobial agent [26], which is much potentiated in an acidic environment.

It is not at all clear which is the exact species responsible for microbial killing in mammalian cells which synthesise NO [27]. Indeed, different organisms may be susceptible to different reactive nitrogen species. Acidification of nitrite results in a complex mixture of nitrogen oxides as well as nitrous acid. It may also be important to provide the additional stress of acidification to make the organisms more susceptible to these nitrogen oxides. Nitrous acid, dinitrogen trioxide and nitrogen dioxide are all effective nitrosating agents (NO^+ donors) [3]. Nitrosation may occur on the microbial cell surface or due to intermediates such as nitrosothiols, which are also good NO^+ donors. Reduced thiols are in high concentration in gastric mucosa and will inevitably be nitrosated in the presence of nitrite and acid. Thiocyanate is also concentrated in saliva and chloride ions are in high concentration in the stomach. Both of these anions will catalyse nitrosation reactions by the formation of more reactive intermediates and may add to the toxicity of acidified nitrite [3].

Many human pathogens which cause gastrointestinal disease are remarkably resistant to acid alone. Incubation of *Candida albicans* at pH1 for one hour has no detectable effect on its future growth. Addition of nitrite to the acid incubation medium at concentrations found in saliva results in nearly complete sterilisation.

Similarly, *E.coli* viability is markedly affected by addition of nitrite to an incubation medium buffered to pH3. As little as 10 µM nitrite will slow the subsequent growth of this organism. The concentration of nitrite in saliva ranges between 100 µM and 1mM, depending on dietary nitrate intake.

The common enteric pathogens such as *S. typhimurium, Y. enterocolitica, S. sonnei and E. coli 0157* are also similarly sensitive to the combination of acid and nitrite [28] at pH3; most of these organisms would not be killed following exposure for one hour, but would be susceptible to addition of nitrite in the concentration normally found in saliva.

7 Conclusions

We have described a novel and potentially important mechanism for host defence of epithelial surfaces – the production of reactive nitrogen oxides by the reduction of inorganic nitrate to nitrite and subsequent acidification. Whereas it is clear that acidified nitrite is effective in killing a variety of human gut and skin pathogens, we have no definite evidence as yet that this mechanism is truly protective in humans exposed to a contaminated environment. Studies currently underway will determine whether augmenting this system by increasing dietary nitrate intake helps prevent gastroenteritis will be important in this respect. If nitrate supplementation is useful in preventing infection from contaminated food, it will be necessary then to consider the possible risks of this augmentation in comparison with the benefits, although as yet, it has been difficult to convincingly demonstrate significant risk with moderate amounts of this anion which is normally found in green vegetables. Understanding the system of enterosalivary circulation of nitrate and subsequent production of nitrogen oxides may also help develop new antimicrobial therapies based on augmenting what seems to be a simple and effective epithelial host defence system.

References

1. Reddy, D., Lancaster, J.R. and Cornforth, D.P. (1983). Nitrite inhibition of Clostridium botulinum: electron spin resonance detection of iron-nitric oxide complexes. *Science.* **221**, 769–770.
2. Tannenbaum SR, Sinskey AJ, Weisman M & Bishop W (1974) Nitrite in human saliva. Its possible relationship to nitrosamine formation. *J.Nat.Cancer.inst.* **53** 79.
3. Williams, D.H.L. Nitrosation (1988). Cambridge University Press. Cambridge.
4. Mitchell HH, Shonle HA, Grindley HS. (1916) The origin of the nitrates in the urine. *J Biol Chem* **24**:461–490.
5. Green LC. Tannenbaum SR. Goldman P. (1981) Nitrate synthesis in the germfree and conventional rat. *Science.* **212**(4490):56–8.
6. Green LC. Ruiz de Luzuriaga K. Wagner DA. Rand W. Istfan N. Young VR. Tannenbaum SR. Nitrate biosynthesis in man. *Proceedings of the National Academy of Sciences of the United States of America.* **78**(12):7764–8.
7. Tannenbaum SR. Nitrate and nitrite: origin in humans (1979) I. B(4413):1332, 1334–7.

8. Tannenbaum SR. Fett D. Young VR. Land PD. Bruce WR. Nitrite and nitrate are formed by endogenous synthesis in the human intestine. I. 200(4349):1487–9.

9. Wagner DA. Schultz DS. Deen WM. Young VR. Tannenbaum SR. (1983) Metabolic fate of an oral dose of 15N-labeled nitrate in humans: effect of diet supplementation with ascorbic acid. *Cancer Research.* **43**(4):1921–5.

10. Kahn T. Bosch J. Levitt MF. Goldstein MH. Effect of sodium nitrate loading on electrolyte transport by the renal tubule. (1975) *American Journal of Physiology.* **229**(3):746–53.

11. Hibbs JB Jr. Westenfelder C. Taintor R. Vavrin Z. Kablitz C. Baranowski RL. Ward JH. Menlove RL. McMurry MP. Kushner JP. *et al.*, Evidence for cytokine-inducible nitric oxide synthesis from L-arginine in patients receiving interleukin-2 therapy (1992) [published erratum appears in *J Clin Invest* 1992 Jul;**90**(1):295] *J Clin Invest.* **89**(3):867–77.

12. Macallan DC. Smith LM. Ferber J. Milne E. Griffin GE. Benjamin N. McNurlan MA. (1997)Measurement of NO synthesis in humans by L-[15N2]arginine: application to the response to vaccination. *Am.J Physiol.* **272**(6 Pt 2):R1888–96.

13. Benjamin N. O'Driscoll F. Dougall H. Duncan C. Smith L. Golden M. McKenzie H.(1994) Stomach NO synthesis. *Nature.* **368**(6471):502.

14. Sasaki, T. and Matano, K. Formation of nitrite from nitrate at the dorsum linguae.(1979) *J. Food. Hyg. Soc. Japan.* **20**, 363–369.

15. Duncan C. Dougall H. Johnston P. Green S. Brogan R. Leifert C. Smith L. Golden M. Benjamin N. (1995) Chemical generation of nitric oxide in the mouth from the enterosalivary circulation of dietary nitrate.*Nature Medicine.* **1**(6):546–51.

16. Dougall HT. Smith L. Duncan C. Benjamin N. The effect of amoxycillin on salivary nitrite concentrations: an important mechanism of adverse reactions? (1995). *B.J.Clin.Pharm.* **39**(4):460–2.

17. Li H. Duncan C. Townend J. Killham K. Smith LM. Johnston P. Dykhuizen R. Kelly D. Golden M. Benjamin N. Leifert C. (1997) Nitrate-reducing bacteria on rat tongues.*Applied & Environmental Microbiology.* **63**(3):924–30.

18. Lundberg JON, Weitzberg E, Lundberg JM, Alving K. (1994) Intragastric nitric oxide production in humans: measurements in expelled air. *Gut* **35**: 1543–1546.

19. McKnight GM. Smith LM. Drummond RS. Duncan CW. Golden M. Benjamin N. (1997) Chemical synthesis of nitric oxide in the stomach from dietary nitrate in humans. *Gut.* **40**(2):211–4.

20. Schorah CJ, Sobala GM, Sanderson M, Collis N, Primrose JM. (1991) Gastric juice ascorbic acid: effects of disease and implications for gastric carcinogenesis.*Am J Clin Nutr* **53**: 287S–293S.

21. Sobala GM, Pignatelli B, Schorah CJ, Bartsch H, Sanderson M, Dixon MF, King RFG, Axon ATR. (1991) Levels of nitrite, nitrate, *N*-nitroso compounds, ascorbic acid and total bile acids in gastric juice of patients with and without precancerous conditions of the stomach. *Carcinogenesis* **12**: 193–198.

22. Sobala GM, Schorah CJ, Sanderson M, Dixon MF, Tompkins DS, Godwin P, Axon ATR. (1989) Ascorbic acid in the human stomach. *Gastroenterology* **97**: 357–363.

23. Leggott PJ. Robertson PB. Rothman DL. Murray PA. Jacob RA. Response of lingual ascorbic acid test and salivary ascorbate levels to changes in ascorbic acid intake(1986)] *Journal of Dental Research.* **65**(2):131–4.

24. Weller R. Pattullo S. Smith L. Golden M. Ormerod A. Benjamin N. Nitric oxide is generated on the skin surface by reduction of sweat nitrate.(1996) *Journal of Investigative Dermatology.* **107**(3):327–31.

25. Benjamin N. Pattullo S. Weller R. Smith L. Ormerod A. Wound licking and nitric oxide (1997)*Lancet.***349**(9067):1776.
26. The function of nitrate, nitrite and bacteria in curing bacon and hams. Department of Scientific and Industrial Research. Food Investigation Board Special Report No. 49. His Majesty's Stationery, London, United Kingdom.
27. Fang FC. Perspectives series: host/pathogen interactions. Mechanisms of nitric oxide-related antimicrobial activity (1997). *Journal of Clinical Investigation.* **99**(12):2818–25.
28. Dykhuizen RS. Frazer R. Duncan C. Smith CC. Golden M. Benjamin N. Leifert C. Antimicrobial effect of acidifiednitrite on gut pathogens: importance of dietary nitrate in host defense.(1996) *Antimicrobial Agents & Chemotherapy.***40**(6):1422–5.

21

Bacterial Nitrate Reductase Activity is Induced in the Oral Cavity by Dietary Nitrate

Callum Duncan[1,2,3], Hong Li[1], Denise Kelly[3], Michael Golden[2] and Carlo Leifert[1]

[1] DEPARTMENT OF PLANT & SOIL SCIENCE, CRUICKSHANK BUILDING, UNIVERSITY OF ABERDEEN, ABERDEEN AB9 2UT, SCOTLAND, UK
[2] DEPARTMENT OF MEDICINE & THERAPEUTICS, POLWARTH BUILDING, UNIVERSITY OF ABERDEEN MEDICAL SCHOOL, FORESTERHILL, ABERDEEN AB9 2ZD, SCOTLAND, UK
[3] NUTRITION DIVISION, ROWETT RESEARCH INSTITUTE, GREENBURN ROAD, BUCKSBURN, ABERDEEN AB2 9SB, SCOTLAND, UK

Abstract

Dietary nitrate forms the basis for a non-immune defence mechanism against gastrointestinal and oral pathogens in animals and man. Facultatively anaerobic bacteria residing in specific low oxygen niches in the oral cavity produce nitrite, from salivary nitrate, which then forms antimicrobial compounds in the stomach acid. The expression of nitrate reductase has mainly been studied in soil bacteria where, in most bacteria, it is controlled by oxygen and nitrate concentrations at the transcriptional level. Since the concentration of nitrite secreted into the oral cavity *via* the saliva is heavily influenced by dietary nitrate intake, an increased dietary nitrate intake is likely to have a significant effect on nitrite production in the oral cavity. Nitrite production on the dorsal tongue was significantly greater in rats fed a nitrate supplemented diet for 14 days than that of rats fed a non-supplemented diet ($p<0.001$), while no statistical difference in bacterial density was found. These results suggest that a high nitrate diet increases the expression of bacterial nitrate reductase by oral bacteria, augmenting the rate of reduction of nitrate to nitrite and, thus, the enterosalivary circulation of nitrate.

1 Introduction

Dietary nitrate and nitrite intake and metabolism have received considerable interest over the past 20 years. High dietary nitrate intake has been suggested to pose a health risk, causing infantile methaemoglobinaemia and gastric/intestinal cancer through the production of carcinogenic N-nitrosamines (1). In recent studies, however, it has been shown that dietary nitrate forms the basis for a non-immune defence mechanism against gastrointestinal and oral pathogens in animals and man (2–4,6,7).

Dietary nitrate is absorbed from the stomach and proximal small intestine into the plasma. Approximately 25% of the daily dietary nitrate intake is actively concentrated (10 fold) from the plasma into the salivary glands (3, 8, 9) and re-secreted into the mouth resulting in an entero-salivary circulation of nitrate (8, 9). The dorsal surface of the tongue harbours large numbers of symbiotic nitrate reducing facultatively-anaerobic bacteria which rapidly reduce nitrate to nitrite (3,6). By this mechanism a high concentration of nitrite is generated in the mouth from dietary and salivary nitrate (7). When the nitrite is swallowed it increases the antimicrobial activity of the stomach acid by upto 100 fold (4).

Li *et al.* (6) demonstrated a significant positive correlation between the density of culturable nitrite producing bacteria and nitrite production on the posterior part of the tongue surface. They, also, demonstrated that nitrate reductase activity on tongue sections was sensitive to oxygen, indicating that respiratory nitrate reductase enzymes are responsible for nitrite production. Deep clefts, filled with bacteria, are a prominent feature of the dorsal tongue. Their depths and the density of facultative-anaerobic bacteria present are thought to result in reduced oxygen tension or anaerobic conditions due to bacterial respiration (6).

Nitrate respiration has mainly been studied in soil bacteria where the expression of nitrate reductase, in most bacteria studied, was controlled by oxygen and nitrate concentrations at the transcriptional level (5,10). The concentration of nitrite secreted into the oral cavity *via* the saliva is heavily influenced by dietary nitrate intake (7, 8, 9) and is thus likely to have a significant effect on nitrite production in the oral cavity.

The aim of this study was, therefore, to determine the effect of a high dietary nitrate intake, over 14 days, on the nitrite producing capacity and the density of nitrite producing bacteria on the tongue dorsum.

2 Materials and Methods

Twenty adult male Sprague Dawely rats were provided with distilled water supplemented with 10mM potassium nitrate and a standard CRM diet (Labsure, Lavender Mill, Manea, Cambridgeshire: < 3mg/kg nitrate) supplemented with 1g/kg potassium nitrate for 14 days. Twenty age and weight matched control rats were provided with distilled water and non-supplemented diet. Water and food intake and the health of animals was monitored. Equal volumes of water and food were consumed by both groups and no deterioration in health was noted.

After 14 days the rats were anaesthetised in 100% CO_2 and killed by cervical dislocation. Their tongues were excised aseptically as far as the larynx. Two circular sections were removed from the back of each tongue using a size 1 cork borer. Each section was quartered into equal segments. Individual segments were placed into microwell plates, containing 250 1 of 1/4 strength Ringer's solution with potassium nitrate (KNO_3) to produce final concentrations of 100, 200, 400, 800, 1000, 2000, 10,000 or 20,000 μM NO_3^-, and incubated at 37°C for 15, 30, 45 and 60 minutes to establish a time course of nitrite production (n = 5 per time point). All tongues were handled under low oxygen conditions as previously described (6). Four tongue sections, from four different rats, were used per replicate to allow both assessment of nitrite production and microbiological analysis. Nitrite production was assayed colorimetrically using a microplate based method (3) and the density of bacteria on tongue segments was determined using a standard most probable number method (6).

Statistical analysis was carried out using a two-way analysis of variance. Differences in the density of bacteria between the two groups were analysed using paired T-tests.

3 Results and Discussion

Concentrations of nitrate in the saliva are known closely to correlate with dietary nitrate intake (7, 8, 9). Following a bolus oral dose of 2 mmoles potassium nitrate, salivary nitrate levels in healthy volunteers rose to a maximum level of 1516.7±280.5 μmol/l after 20–40 minutes, rapidly followed by an increase in salivary nitrite levels to 761.5±187.7 μmol/l (7).

Nitrate concentrations in the drinking water and CRM diet, of laboratory rats, were supplemented to levels corresponding to the maximum levels permitted under EU legislation in drinking water (50mg/l) and those found in green vegetables (1). Significantly higher nitrate reduction (p<0.001) was measured in tongue sections of rats consuming the nitrate supplemented diet (Figure 1) compared to the control rats on a nitrate free diet (<3mg/kg food, <1mg/l water). This could indicate that nitrate supplementation resulted in (i) an increase in the population of nitrate reducing bacteria and/or (ii) an increased expression of nitrate reductase enzymes in oral bacteria.

Experiments were carried out in low oxygen conditions because nitrite production by the tongue microflora was previously shown to be sensitive to atmospheric oxygen levels (6). Substantially more nitrite was produced from posterior dorsal tongue sections that were maintained in a low oxygen environment, than those exposed to atmospheric oxygen concentrations. In addition, measurement of nitrite production at 15, 30, 45 and 60 minutes clearly demonstrated a lag phase of approximately 20 minutes prior to nitrite production on tongue sections exposed to atmospheric oxygen. Tongue sections maintained in a low oxygen environment, on the other hand, had a

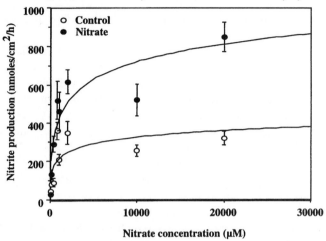

Figure 1 *Nitrite production (nmoles/cm²/h) from the posterior rat tongue following 14 days of a nitrate supplemented diet (closed circles), compared with those maintained on an non-supplemented diet (open circles). Means and SEM are from five determinations*

relatively short lag phase of approximately 5 to 10 minutes. In experiments described here tongue sections were maintained in a low oxygen environment, and no difference in lag phase was found between tongue sections of rats fed nitrate supplemented and non-supplemented diets (Figure 2).

The density of both total and nitrite producing bacteria on tongue sections from rats fed supplemented and non-supplemented diets were not significantly different, indicating that over the 14 day period there was no increase in the density of either total or nitrate reducing bacteria on the tongue dorsum. The increase in nitrite producing capacity observed in rats fed the nitrate supplemented diet is thus likely to be due to an increased expression of nitrate reductase enzymes in the bacterial population rather than an increase in the nitrite producing microflora of the tongue dorsum.

These results strongly suggest that a high nitrate diet augments the ability of the oral microflora to reduce nitrate to nitrite in the oral cavity, thus enhancing its ability to increase the antimicrobial activity in the stomach.

References

1. Anonymous,'Nitrate, nitrite and N-nitroso compounds in food: Second report', Food Surveillance Paper No. 32, Ministry of Agriculture, Fisheries and Food, HMSO, London, 1992.
2. Duncan, C., H. Li, R. Dykhuizen, R. Frazer, P. Johnston, G. McKnight, L. Smith, K. Lamza, H. McKenzie, L. Batt, D. Kelly, M. Golden, N. Benjamin and C. Leifert, Protection against oral and gastrointestinal diseases: The importance of dietary nitrate intake, oral nitrate reduction and enterosalivary nitrate circulation, Comp. Biochem. Physiol., 1997, **118A**, 939–948.

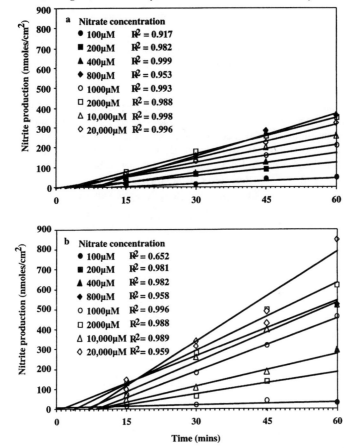

Figure 2 *Nitrite production (nmoles/cm²), measured at four time points (15–60 minutes), from intact tongue sections (4 and 5) incubated in increasing concentrations of potassium nitrate (100-20,000 μM). Tongue sections were incubated anaerobically (a) following maintenance of a virtually nitrate free diet and (b) following nitrate supplementation for 14 days. Linear regressions are plotted through the linear proportions of each graph, the x-intercept indicating the approximate starting point of nitrite production. Means and SEM are from five determinations*

3. Duncan, C., H. Dougall, P. Johnston, S. Green, R. Brogan, C. Leifert, L. Smith, M. Golden and N. Benjamin, Chemical generation of nitric oxide in the mouth from the enterosalivary circulation of dietary nitrate, Nat. Med., 1995, **1**, 546–551.

4. Dykhuizen, R.S., C. Duncan, R. Frazer, C.C. Smith, M. Golden, N. Benjamin and C. Leifert, Dietary nitrate might protect against infective gastroentcritis: the antimicrobial effect of acidified nitrite on gut pathogens. Antimicrob. Agents Chemother., 1996, **40**, 1422–1425.

5. Gunsalus, R.P., Control of electron flow in *Escherichia coli*: Coordinated transcription of respiratory pathway genes, J. Bacteriol., 1992, **174**, 7069–7074.

6. Li, H., C. Duncan, J. Townend, K. Killham, L.M. Smith, P. Johnston, R. Dykhuizen, D. Kelly, M. Golden, N. Benjamin and C. Leifert, Nitrate reducing bacteria on rat tongues, Appl. Environ. Microbiol., 1997, **63**, 924–930.

7. McKnight, G., L.M. Smith, R.S. Drummond, C.W. Duncan, M. Golden and N. Benjamin, Chemical synthesis of nitric oxide in the stomach from dietary nitrate in man, Gut, 1997, **40**, 211–214.

8. Spiegelhalder, B., G. Eisenbrand and R. Preussmann, Influence of dietary nitrate on nitrite content of human saliva: Possible relevance to in vivo formation of n-nitroso compounds, Fd. Cosmet. Toxicol., 1976, **14**, 545–548.

9. Tannenbaum, S.R., M. Weisman and D. Fett, The effect of nitrate on nitrite formation in human saliva, Fd. Cosmet. Toxicol., 1976, **14**, 549–552.

10. Unden, G., S. Becker, J. Bongaerts, J. Schirawski, S. Six, Oxygen regulated gene expression in facultative anaerobic bacteria, Antonie van Leeuwenhoek, 1994, **66**, 3–23.

22

Antimicrobial Effect of Acidified Nitrite on Gut Pathogens: the Importance of Dietary Nitrate in Host Defence

R. S. Dykhuizen, A. Fraser, C. Duncan, M. Golden,
N. Benjamin and C. Leifert[1]

DEPARTMENT OF MEDICINE AND THERAPEUTICS, MEDICAL
SCHOOL, POLWARTH BUILDING, FORESTERHILL, ABERDEEN,
AB25 2ZD, UK
[1] DEPARTMENT OF PLANT AND SOIL SCIENCE, CRUICKSHANK
BUILDING, UNIVERSITY OF ABERDEEN, ABERDEEN AB9 2UD, UK

Abstract

Addition of nitrite achieves kill of micro-organisms where acid alone allows growth to continue. The synergism in antimicrobial action of acid and nitrite is evident against common gut pathogens such as the *Enterobacteriaceae*, including *Escherichia coli 0157*, but also against the stomach pathogen *Helicobacter pylori,* normally very resistant to acid alone. The antibacterial action of acidified nitrite becomes apparent at physiological concentrations of acid and nitrite after exposure times that are within the passage time of a food bolus through the stomach. The antimicrobial activity of acidified nitrite is enhanced by thiocyanate, also present in gastric juice. Ascorbic acid provides protection against the antibacterial action of acidified nitrite, suggesting that $NO^{.}$ is not the antibacterial agent.

The mechanism of chemical host defence *via* acidification of salivary nitrite may be of fundamental importance. A conclusive demonstration of the antimicrobial effect *in vivo* would require a major re-interpretation of the role of dietary nitrate in human health and animal husbandry.

1 Introduction

Nitrogen oxides and nitrous acid are known in organic chemistry as noxious compounds and atmospheric pollutants, which represent a significant health risk[1]. They have been implicated in the generation of the potentially carcinogenic N-nitrosamines in the mammalian stomach[2]. In humans, ingested nitrate (NO_3^-) is absorbed from the gastrointestinal tract into the bloodstream and concentrated in the salivary glands by an active transport system shared with iodide and thiocyanate, increasing concentrations up to ten times that of plasma[3,4]. Salivary nitrate is then rapidly reduced to nitrite (NO_2^-) by nitrate reductase expressed by micro-organisms in the mouth[5]. N-nitrosamines are formed from nitrite and secondary amines in the stomach[6]. Concerns about this endogenous formation of N-nitroso compounds has led to calls for restrictions of nitrate and nitrite in food products and drinking water[7].

We have recently suggested that the production of salivary nitrite may serve a useful purpose as a host defence mechanism against swallowed pathogens *via* the formation of bacteriocidal compounds in the acid conditions of the stomach[8]. Indeed it has been shown that expelled stomach air contains high concentrations of the antimicrobial gas nitric oxide (NO·) which are enhanced by a high nitrate intake[9]. We proposed that the salivary generation of nitrite is accomplished by a symbiotic relationship involving nitrate reducing bacteria on the tongue surface, which is designed to provide host defence against microbial pathogens in the mouth and lower gut *via* chemical NO· production[10].

Patients with infective gastroenteritis have increased plasma nitrate levels compared to healthy controls[11], septicaemic patients[12] and patients with inflammatory bowel disease[13]. During infective gastroenteritis salivary generation of nitrite might be greatly enhanced, resulting in increased gastric NO· production. To investigate the role of salivary nitrite in the bactericidal function of the stomach, we studied the effect of acidified nitrite against micro-organisms that are known to cause infective gastroenteritis and against *Helicobacter pylori*. The micro-organisms were exposed to acidified solutions and various concentrations of nitrite characteristic of concentrations found in saliva. We also studied the effects of thiocyanate and ascorbic acid on the antibacterial activity of acidified nitrite.

2   Materials and Methods

Patient isolates of *Salmonella enteriditis*, *Salmonella typhimurium*, *Shigella sonnei*, *Yersinia enterocolitica*, and *Escherichia coli 0157* were tested. All experiments used early log phase cultures and were carried out in triplicate. The following procedure was applied to obtain standardised inocula of 2×10^7 colony forming units per milliliter (cfu/ml). Organisms were grown on nutrient agar plates for 24 hours at 37 °C. A 75ml volume of nutrient broth (Oxoid CM1) was prepared in 125 ml Erlenmeyer flasks, sealed with aluminium foil, autoclaved, and cooled. A visible quantity of a single bacterial colony was transferred to the flask using a flamed loop. The flask was then resealed and incubated on a shaker (New Brunswick Scientific Company, U.S.) for 18 hours at 37 °C. After incubation, the optical density was measured at 570nm using a spectrophotometer (Cecil Instruments, CE303 grating spectrophotometer). A dilution series was made using nutrient broth ($1, 5 \times 10^{-1}, 2 \times 10^{-1}, 1 \times 10^{-1}$) and the corresponding optical densities were determined. Two further dilution series were made starting from the 5×10^{-1} and the 1×10^{-1} suspensions by serial transfers of 1 ml of suspension into 9 ml of nutrient broth in universal containers down to a dilution factor of 10^{-7}. Nutrient agar plates were marked in quadrants representing the dilution factors, and three 20 μl drops were pipetted into each segment from the corresponding dilution. The plates were incubated for 24 hours. The number of colonies in each segment were counted and the most probable number of organisms in the original suspensions calculated. A calibration curve was produced for each organism relating the number of cfu/ml to the optical density at 570nm. This curve was then used for all the further experiments to produce standardised inocula of 2×10^7 cfu/ml by dilution of early log phase cultures with fresh nutrient broth.

2.1 Determination of the Bacteriostatic Activity of Acidified Nitrite

The experiments were carried out on disposable, flat bottomed plates (96 wells of 300μl). The reaction plate was prepared by combining three separately prepared nutrient broth solutions in the microwells by pipette transfer with multichannel pipettes (Anachem, Luton, UK) with sterile tips (Gilson).

The following solutions were used:

1. potassium nitrite amended nutrient broth   60μl
2. nutrient broth with bacterial inoculum (8×10^7 cfu/ml)   60μl
3. pH amended nutrient broth   <u>120μl</u>

                      Total volume of microwell: 240μl

(1). Potassium nitrite solutions to give final concentrations of 0, 50, 100, 200, 500, 1000, 2000 and 10,000 μmol/l of nitrite in the microwells were prepared.

(3). The pH of the final microwell contents was extrapolated from a titration curve produced by measuring the pH (Jenway 3020 pH Meter) of 500 ml of nutrient broth adding 1M hydrochloric acid. The effect of autoclaving, mixing with different concentrations of nitrite and addition of bacterial inoculum on the pH of the cell contents was tested experimentally by scaling up the cell volume from 240 μl to 24 ml. This enabled determination of the actual pH in the microwells avoiding the difficulty of contamination and physical limitations imposed by the small size of the microwell cell. The final gradients of pH were: 5.4, 4.8, 4.2, 3.7, 3.0, and 2.1 (see Figure 1).

Potassium amended broth (1) and nutrient broth with bacterial inoculum (2) were first added to the microwells. The experiment was then initiated with the addition of pH amended nutrient broth (3).

The plates were sealed using a sterile adhesive film (Sealplate[R] Anachem, Luton, UK), and incubated for 24 hours at 37 °C on the shaker. The inhibitory effect of acidified nitrite on bacterial growth was determined by measurement of the optical density (570nm) of the wells using a microwell plate reader (MRX Microplate Reader, Dynatech Products Limited, Guernsey, Channel Islands, Great Britain). All experiments were carried out in triplicate. The

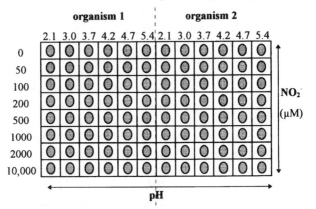

Figure 1   *pH and nitrite gradients on the microwell plates*

minimum inhibitory concentration (MIC) of nitrite at the different pH settings was defined as the lowest nitrite concentration whereby no growth of micro-organisms had taken place after 24 hours.

2.2   Determination of the Bacteriocidal Activity of Acidified Nitrite

From the microwell plate setup in Figure 1, 20 μl of the microwell suspensions were sampled after 30 minutes, two hours, and 24 hours exposure to acidified nitrite, and transferred to 180μl of a recovery medium (nutrient broth; pH 7.0). The dilution was repeated with a further 20μl transferred into 180 μl recovery medium. The two transfers accomplished neutralisation of acid (by two points on the pH-scale), dilution of nitrite concentration (one hundred fold) and reduction of the original inoculum size (to a final number of 10,000 micro-organisms).

Recovery media were incubated on the shaker for 24 hours at 37 °C to allow growth of the bacteria that had survived the exposure to the test solutions in the original microwell plate. After incubation, the bacterial growth was assessed on the microwell plate reader. All experiments were carried out in triplicate. The minimum bacteriocidal concentration (MBC) was defined as the lowest nitrite concentration whereby no growth was detected after transfer into recovery media.

2.3   Statistical Analysis of MIC and MBC

MIC and MBC values (mol/l) for nitrite were log transformed for statistical analysis. The relationship between pH and nitrite for the various antibacterial activities (MIC, $MBC_{30\,min}$, MBC_{2h}, $MBC_{24\,h}$) against the various micro-organisms was plotted and the slopes assessed using regression analysis.

Multiple regression analysis was applied to determine differences between sensitivity of micro-organisms to acidified nitrite.

2.4 Antibacterial Activity of Acidified Nitrite Against *Helicobacter pylori*

H.pylori is a fastidious organism that is dependent on an anaerobic environment and is very sensitive to overgrowth by contaminants. The organism synthesises urease which in the presence of urea generates an alkaline environment to protect itself from the bacteriocidal effect of acid. The enzyme mediates the hydrolysis of urea to ammonia and bicarbonate, neutralising any hydrogen ion penetrating the cell wall[14]:

$$NH_2$$
$$|$$
$$C = O + 2H_2O + H^+ \xrightarrow{\text{urease}} 2NH_4^+ + HCO_3^-$$
$$|$$
$$NH_2$$

Attempts to grow *H.pylori* in reproducible numbers on microwell plates failed and therefore a different procedure from the one used for the *Enterobacteriaceae* was followed to determine the sensitivity of the organism to acidified nitrite.

H.pylori, isolated from human gastric biopsy specimens, were cultured on horse-blood agar plates incubated at $37\,^{\circ}C$ in an atmosphere of 10% CO_2, 5% oxygen, and 85% nitrogen ['campygas']. After 3 days incubation, the bacteria were harvested and suspended in normal saline at pH7 to give a final concentration of approximately 10^9 cells/ml [turbidity = McFarland's no.6].

Inoculate (1ml) was added to 4.5ml of of $0.2M$ HCl/KCl buffer at pH2 with or without urea (5mM) in the solution. Immediately thereafter, normal saline (1ml) or potassium nitrite (1ml) to reach a final concentration of 1mmol/l was added in universal containers. As a control, the experiment was repeated with 4.5ml normal saline at pH7 instead of 0.2M HCl/KCl buffer.

The samples were incubated at $37\,^{\circ}C$. After 30 minutes, aliquots of each sample were diluted with normal saline in serial 10-fold dilutions for determination of the number of colony forming units (cfu). The diluted suspensions (10μl) were inoculated onto horse-blood agar plates and incubated in an anaerobic incubator with 5% CO_2 (95% N_2) for up to 5 days. Colony counts per plate were calculated as {no. of colonies $\times$ [1/dilution] $\times$ [1/0.01]} per ml. The lower limit of detection was 10^2 organisms/ml.

2.5 Dose Relationship of the Antimicrobial Effect of Nitrite on *H.pylori* at pH2

Inoculate (1ml) was added to 4.5ml of 0.2M HCl/KCl buffer at pH2 with 5mM urea in the solution. Immediately thereafter, potassium nitrite solution (1ml) was added to reach final concentrations of 0, 50, 100, 200, 500, and 1000 μmol/

l. As a control, the experiment was repeated with 4.5ml normal saline at pH7 instead of 0.2M HCl/KCl buffer at pH2.

The containers were incubated at 37 °C. After 30 minutes, samples of each of the containers were taken and diluted with normal saline in serial 10-fold dilutions for cfu determination. The diluted suspensions were inoculated onto horse-blood agar plates and incubated in an anaerobic incubator with 5% CO_2 for up to 5 days. Colony counts per plate were calculated as {no. of colonies × [1/dilution] × [1/0.01]} per ml. The lower limit of detection was 10^2 organisms/ml.

The pH in the universal containers was measured with a glass pH electrode before and after each 30 minute incubation of *H.pylori* in acidified nitrite. The experiment was carried out in triplicate.

2.6   The Effect of Vitamin C and Thiocyanate on the Antibacterial Activity

Thiocyanate is secreted by the salivary glands[15] and vitamin C is secreted by the gastric mucosa into gastric juice[16]. The former is an oxidant and known to facilitate the formation of N-nitroso compounds[17]; the latter is a reductant, and potently enhances the production of nitric oxide from other oxides of nitrogen[18]. In this section, the effect of physiological concentrations of thiocyanate and vitamin C on the bacteriocidal effect of acidified nitrite is tested against *Yersinia enterocolitica.*

Experiments were carried out on microwell plates as described in section 2.1. Solutions were transferred into the microwells in the following sequence:

1.	nutrient broth	50μl
2.	KCl/HCl buffer	30μl
3.	vitamin C or thiocyanate solution	50μl
4.	potassium nitrite solution	50μl
5.	nutrient broth with bacterial inoculum (1×10^9 cfu/ml)	20μl
	Total volume of microwell:	200ml

(2). The KCl/HCl buffer was prepared by mixing 0.2M KCl with varying volumes of 0.2M HCl. The volume of HCl required to obtain pH2 in the microwells under test conditions was determined by mimicking the exposure medium in proportionally scaled up volumes for each combination of potassium nitrite and vitamin C or thiocyanate solutions.

(3). Solutions of ascorbic acid were prepared to give final concentrations of 125, 250, 500, 1000 μM in the microwells. Solutions of ascorbic acid were freshly prepared for each experiment and not autoclaved but micro-filtered to obtain sterile conditions. Solutions of potassium thiocyanate were prepared to give final concentrations of 25, 50, 125, 250, and 500 μM.

(4). Potassium nitrite solutions were prepared to give final concentrations of 125, 250, 375, 500, 625, 750, and 1000 μM (see Figure 2).

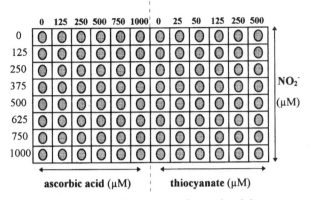

Figure 2   *Concentration gradients of nitrite, ascorbic acid and thiocyanate*

After 5, 10, 15, and 20 minutes of exposure of *Y. enterocolitica* to the test solutions, samples of 20 µl were serial diluted into recovery media and survival of the organisms was determined as described in section 2.2.

2.7   Statistical Analysis

Nitrite, ascorbic acid, and thiocyanate concentrations were log transformed for statistical analysis. Multiple regression analysis was applied to determine the influence of exposure times, ascorbic acid, and thiocyanate on the antibacterial activity of nitrite at pH2.

3   Results

The nitrite concentrations showing antimicrobial activity against gut pathogens at pH 2.1, 3, 3.7, 4.2, 4.8, and 5.4 are tabulated in Table 1.

The synergistic antibacterial action of nitrite and acid is clearly demonstrated. At pH 2.1 *Shig. sonnei* and *E.coli* 0157 could survive for 30 minutes or even 2 hours, unless nitrite was present in the solution. At pH 5.4 no bacteriocidal action within 2 hours exposure time could be demonstrated for any of the organisms tested.

Figures 3–6 show the relationship between pH and log transformed nitrite concentrations for the four antibacterial activities (MBC after 30 minutes, 2 hours, 24 hours exposure times, and MIC). The points in each of the curves represent the means of triplicate experiments. Each individual replicate was used for the regression, but replicates showing no antibacterial activity with $\leq 10.000 \mu mol/l$ nitrite were omitted from the analysis.

Y. enterocolitica is the most sensitive to acid of the five organisms tested. Without the addition of nitrite, complete inhibition of growth is observed at pH 4.0. However the organism is able survive two hours exposure to the same pH, unless $>1000 \mu mol/l$ nitrite is present in the solution. There is significant negative correlation of acid and nitrite to achieve antibacterial activity ($p<0.001$).

Table 1 *Antibacterial activity of nitrite (mmol/l) against common gut pathogens at six different pH settings. Values for MBC and MIC are the mean of three experiments (total n = 288). The maximum difference between replicates is two steps away (see Figure 1)*

| Organism | Antibacterial activity[†] | Exposure time | ←—pH—→ | | | | | |
			2.1	3.0	3.7	4.2	4.8	5.4
Y.enterocolitica	MBC	30 minutes	0	0.02	1.33	6.67	>10[‡]	>10
	MBC	2 hours	0	0	0.07	1.67	10.0	>10
	MBC	24 hours	0	0	0	0.05	2.0	10.0
	MIC	24 hours	0	0	0	0	0.5	6.67
S.enteriditis	MBC	30 minutes	0	0.13	1.33	10.0	>10	>10
	MBC	2 hours	0	0	0.4	2.0	10.0	>10
	MBC	24 hours	0	0	0	0.05	1.0	10.0
	MIC	24 hours	0	0	0	0.02	0.67	6.67
Shig.sonnei	MBC	30 minutes	0.20	1.67	10.0	>10	>10	>10
	MBC	2 hours	0.07	0.67	3.67	10.0	>10	>10
	MBC	24 hours	0	0	0.20	1.0	6.67	>10
	MIC	24 hours	0	0	0	0.40	3.0	10.0
E.coli 0157	MBC	30 minutes	1.0	6.67	>10	>10	>10	>10
	MBC	2 hours	0.5	1.33	3.0	10.0	>10	>10
	MBC	24 hours	0	0.02	0.3	0.83	6.67	>10
	MIC	24 hours	0	0	0	0	1.0	10.0

[†] MBC = minimum bactericidal concentration; MIC = minimum inhibitory concentration
[‡] No antibacterial activity at nitrite concentrations of >10 μmol/l.

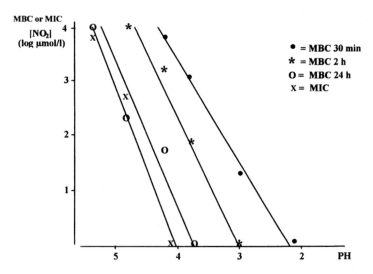

Figure 3 *Synergistic relationship of the antibacterial activity of acid (pH) and nitrite (log μmol/l) for Yersinia enterocolitica*

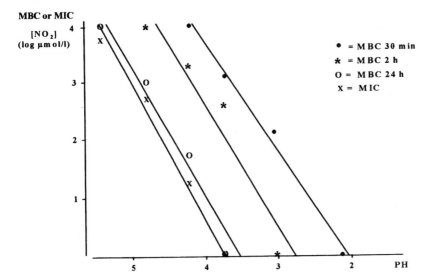

Figure 4   *Synergistic relationship of the antibacterial activity of acid (pH) and nitrite (log μmol/l) for Salmonella enteriditis*

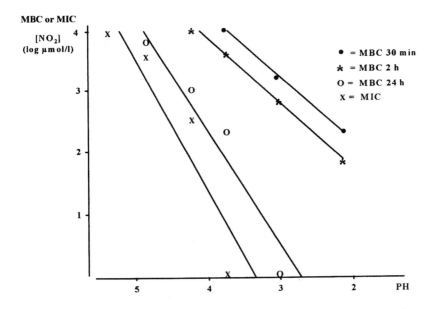

Figure 5   *Synergistic relationship of the antibacterial activity of acid (pH) and nitrite (log μmol/l) for Shigella sonnei*

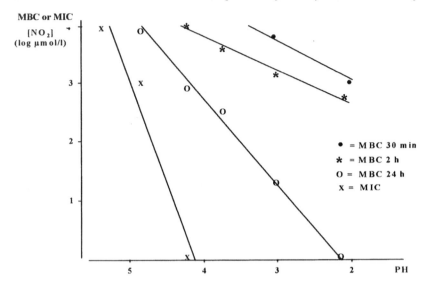

Figure 6   *Synergistic relationship of the antibacterial activity of acid (pH) and nitrite (log μmol/l) for Escherichia coli 0157*

The growth of *S. enteriditis* is inhibited at pH 3.7. To achieve kill of the organism at two hours exposure time at the same pH, addition of >125μmol nitrite/l is needed. Kill is achieved within 30 minutes if >2000 μmol/l nitrite is available. The synergism between acid and nitrite is significant for all four regression curves ($p<0.001$).

The growth of *Shig sonnei* is inhibited by acid at pH 3.5, but acid is relatively ineffective in killing the organism at exposure times of 30 minutes and two hours. To achieve these antibacterial activities, addition of 1500 and 6500 μmol/l nitrite respectively is needed. The synergism of acid and nitrite is significant but less convincing at short exposure times ($p = 0.027$ for MBC_{30min}, $p = 0.035$ for MBC_{2h}, <0.001 for MBC_{24h} and MIC).

Growth of *E. coli 0157* is inhibited at pH 4.2 but the organism is only killed after prolonged exposure (24 hours) at pH 2. Addition of nitrite achieves kill even at quite high pH and short exposure times. The synergism is highly significant ($p = 0.015$ for MBC_{30min}, $p<0.001$ for the others), and from the flat slopes of the MBC_{30min} and MBC_{2h} curves it seems that increasing amounts of nitrite are highly effective in achieving eradication of the organism.

The differences between organisms can be demonstrated by plotting the regression curves of MBC after 2 hours exposure time (Figure 7). The slopes of the curves clearly demonstrate that the increase in antibacterial activity due to the addition of nitrite to acid is most pronounced in *E. coli 0157* ($p<0.001$).

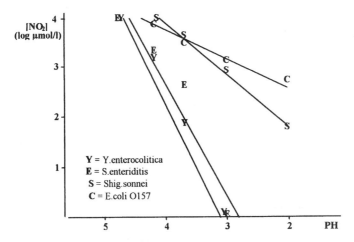

Figure 7   *Regression curves for MBC after 2 hours exposure time against individual micro-organisms* [$R_Y = 0.90$, df = 11, p < 0.001, int. = −4.45, x = 1.94 ± 0.58; $R_{SE} = 0.96$, df = 11, p < 0.001, int. = −6.14, x = 2.19 ± 0.47; $R_{Shig} = 0.88$, df = 11, p < 0.001, int. = −1.29, x = 1.57 ± 0.62; $R_{E0157} = 0.95$, df = 11, p < 0.001, int. = 1.40, x** = 0.59 ± 0.14 (**p<0.001 compared with other micro-organisms)]

3.1   The Effect of Vitamin C and Thiocyanate

Significant correlation is demonstrated between the concentrations of ascorbic acid or thiocyanate in the reaction medium and the minimum bacteriocidal activity of nitrite at pH2. The correlation is positive for vitamin C and negative for thiocyanate (Table 2).

Table 2   *Effect of vitamin C and thiocyanate on the MBC of nitrite (mol/l) against Y.enterocolitica at pH2. Experiments in triplicate*

		Ascorbic acid (μmol/l)					
	Time (min)	0	125	250	500	750	1000
Y.enterocolitica	5 #	1000	>1000	>1000	>1000	>1000	1000
	10	917	875	833	833	1000	>1000
	15	458	542	542	583	833	>1000
	20	292	333	500	500	792	>1000

		Thiocyanate (μmol/l)					
		0	25	50	125	250	500
Y.enterocolitica	5	>1000	>1000	>1000	750	542	0
	10	667	292	167	0	0	0
	15	292	167	0	0	0	0
	20	167	0	0	0	0	0

Because the majority of experiments needed >1000 μmol/l of nitrite to achieve bacterial kill, the exposure time of 5 minutes is not used for statistical analysis.

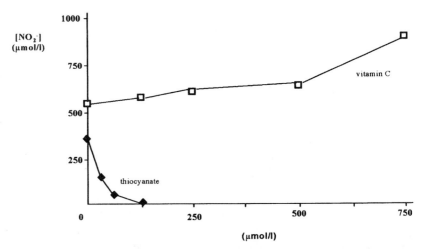

Figure 8   *Dose related effect of thiocyanate and ascorbic acid (vitamin C) on the
bacteriocidal activity of nitrite at pH2*

Figure 8 shows the effect of thiocyanate and ascorbic acid on the bacteriocidal activity of nitrite at pH2. The points in the graphs represent the
arithmetic mean MBC of the exposure times. As anticipated, regression
analysis (Table 3) of log transformed data showed a strong influence of
exposure time on the MBC (p<0.001).

Table 3   *Effect of vitamin C and thiocyanate on the Minimum Bacteriocidal
Concentration of nitrite against Y.enterocolitica at pH2: regression
statistics for the individual exposure times.*

Asc. acid	R	df	Intercept	Slope	p
5 min #	–	–	–	–	–
10 min	0.19	14	860	0.10 ± 0.10	0.49
15 min	0.80	14	450	0.44 ± 0.20	<0.001
20 min	0.84	14	282	0.62 ± 0.24	<0.001
overall	0.44	44	530	0.39 ± 0.25	0.003
SCN^-	R	df	Intercept	Slope	p
5 min	0.99	9	1020	-2.02 ± 0.24	<0.001
10 min	0.63	18	341	-0.93 ± 0.60	0.005
15 min	0.52	18	136	-0.38 ± 0.33	0.03
20 min	0.32	18	51	-0.15 ± 0.22	0.19
overall	0.33	63	226	-0.45 ± 0.33	0.007

\# Because 15 out 18 experiments needed >1000 (mol/l of nitrite to achieve bacterial kill, the
exposure time of 5 minutes was not used for statistical analysis.

3.2 Antibacterial Activity of Acidified Nitrite Against Helicobacter pylori

The effects of exposure of *H.pylori* to acid and nitrite are presented in Table 4. The control experiments show some 1.5×10^6 cfu/ml after 30 minutes exposure at pH7 in the universal containers and a similar survival rate was observed when 1mM potassium nitrite is present. The organisms are also able to survive exposure to pH2 if urea is present. However, the combination of acid and nitrite resulted in complete kill with or without the presence of urea (p<0.001).

Table 4 *Survival of* H.pylori *(log cfu/ml ± SD) after exposure to acid (pH2) and potassium nitrite (1mmol/l)*

	Without nitrite (n° of experiments)	1mM nitrite (n° of experiments)
Control (pH7)	6.14 ± 0.98 (5)	6.32 ± 1.62 (4)
0.2M HCl/KCl (pH2) plus urea (5mmol/L)	5.06 ± 0.91*(5)	† (5)
0.2M HCl/KCl (pH2) no urea added	† (5)	† (5)

† No detectable survival of micro-organisms
‡ p<0.001 versus 0.2M HCl/KCl (pH2) plus urea with 1 mmol/l nitrite; number of replicates are given in parentheses.

The dose relationship of the antibacterial effect of nitrite towards *H. pylori* at pH2 shows a dose dependent kill of micro-organisms after exposure for 30 minutes (Figure 9). Increased kill was seen after addition of 50 and 100 μM of nitrite in all three experiments. Although the same isolate was used in all three experiments, the inoculum of the first series seemed markedly more sensitive to the mechanism than those in the other two series. No survival was detected in the first series at concentrations >200μmol/l, and in the other two at >500μmol/l. The difference in sensitivity of the inoculates is most likely due to differences in viability of the inoculates used which was also apparent when they were exposed to acid alone for 30 minutes. Whereas 1×10^6 cfu/ml were

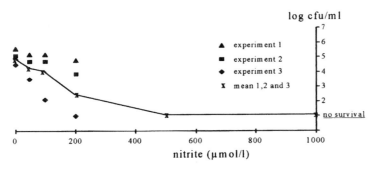

Figure 9   *Dose relationship between survival of* H.pylori *(log cfu/ml) and nitrite (mmol/l) at pH2 in the presence of urea (5mmol/l)*

recovered after inoculum one had been exposed to normal saline at pH7, only 3×10^4 cfu/ml survived after exposure to HCl/KCl buffer at pH2. These figures were 3×10^5 at pH7 versus 1×10^5 at pH2 for the second, and 1×10^6 at pH7 versus 4×10^5 at pH2 for the third inoculum.

The pH was 2 in all universals at the start of exposure. A rise in pH during the exposure was seen in those containers where survival of *H.pylori* could be demonstrated, and the pH at the end of the exposure time varied from 2.8 in the containers without nitrite, to 2.1 in those with 200μmol/l nitrite.

4   Discussion

4.1   Pathogens Involved in the Aetiology of Infective Gastroenteritis

The antibacterial potential of nitrite at low pH is clearly demonstrated by the data from this study. A highly significant contribution ($p<0.001$) of nitrite to MIC, MBC_{30min}, MBC_{2h}, and MBC_{24h} was observed for all micro-organisms tested. The synergistic relationship between nitrite and acid in the antibacterial activity against the *Enterobacteriaceae* is demonstrated in Figures 3–7. Multiple regression analysis identifies several co-variables as determinants of the amount of nitrite needed to bring about the required antibacterial activity: the type of antibacterial activity to be achieved (MIC, MBC_{30min}, MBC_{2h}, or MBC_{24h}), the organism in question, and the pH (all $p<0.001$).

In view of the lower MBC_{2h} values compared to MBC_{24h} it would appear that the antibacterial action of acidified nitrite is prolonged beyond two hours. Differences between MBC_{24h} and MIC show that even after 24 hours some organisms that were inhibited in their growth still proved viable after transfer to a recovery medium.

Susceptibilities to the acidified nitrite solutions ranked as follows: *Y.enterocolitica* > *S.enteriditis* > *Shig.sonnei* > *E. coli 0157* ($p<0.05$). *E. coli 0157* and *Shig. sonnei* are most resistant to acid; they survive exposure at pH 2.1 for 30 minutes which kills the other micro-organisms. However, *E. coli 0157* shows inhibition of growth at pH 4.2 when the other organisms, apart from *Y. enterocolitica,* manage to maintain growth unless nitrite is present in the solution (Table 1). It seems that *E. coli 0157* manages to survive a relatively acid environment by slowing down growth activity. Its ability to survive this way is overcome by the addition of nitrite to the medium.

The acidity in the lumen of the human stomach is dependent on physiological variables such as previous food intake, anxiety, age, medication like antacids, and previous gastric surgery. Under fasting conditions the median of the luminal pH in healthy volunteers is around 2.0, ranging from 1.5–5.5[19]. Ingestion of a meal characteristic of the main meal of a western diet produces an immediate rise in the median gastric pH to about 6.0, which will return over the following 2–3 hours to premeal values of about 2.0[20,21]. Human salivary nitrate and nitrite concentrations are greatly influenced by the amount of nitrate ingested. Values between 50 and 10,000 μmol/l have been reported[22].

Gastric nitrite concentrations are significantly lower than salivary concentrations because of the formation of nitrous acid (HNO_2) which reacts to generate other oxides of nitrogen. Nitric oxide and nitrogen dioxide escape into the gaseous phase. After a nitrate drink (1 m.mole NO_3^-), the gastric headspace gas nitric oxide concentration rose from less than 20 parts per million to a maximum of 291 parts per million in human volunteers[23]. The depletion of nitrite due to the formation of other oxides of nitrogen is the main cause of the discrepancy between salivary and gastric nitrite concentrations, which cannot be explained by dilution of saliva with gastric contents alone.

The antimicrobial effect of acidified nitrite is accomplished by the reactive oxides of nitrogen that are formed from nitrite *via* nitrous acid. Figure 10 illustrates the potential for the effect of a high dietary nitrate intake on host defence against ingested gut pathogens. Because the concentrations of the reactive oxides of nitrogen that are formed in the stomach from salivary nitrite cannot be easily quantified, the physiological range of salivary nitrite has been extrapolated to the stomach. During the first two hours after a major meal, the average pH might be between 3 and 4[20,21]. If the meal had been contaminated with *S. enteriditis*, it could depend on the amount of nitrite delivered by saliva to the stomach whether the organism will still be viable after passage through the stomach to colonise the lower gut. If the host has low salivary nitrite concentrations (<50μmol/l), the organism may continue to grow (white area in Figure 10). A moderate amount of nitrate in the saliva might inhibit growth, but the organism may still be viable after passage through the stomach (dotted area Figure 10). Higher salivary nitrite concentrations may kill the organism during its passage through the stomach provided the transit time of the food bolus is long enough (grey area Figure 10). Salivary nitrite concentrations close to 10,000 μmol/l may lead to kill of the organism within as little time as 30 minutes.

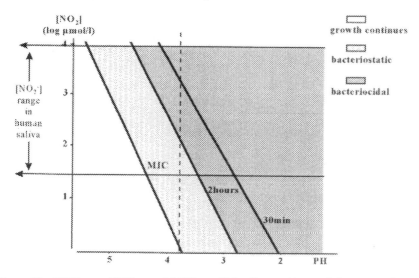

Figure 10 *MBC_{30min}, MBC_{2h}, and MIC μmol/l for S. enteriditis and the range of nitrite concentration in human saliva*

There is epidemiological evidence that exposure to antimicrobials not only increases the risk of infection with antimicrobial-resistant strains of *Salmonella*, but also with antimicrobial sensitive strains[24,25]. Dougall *et al*[26] studied the effect of amoxycillin on the salivary nitrite concentration, following a 200 mg potassium nitrate oral load in 10 healthy volunteers. They showed an almost 50% reduction in nitrite production in the antibiotic treated group compared with controls. The authors suggested that destruction of nitrate reductase containing bacteria in the mouth was the cause of the decreased production. In the context of the data presented in this report, it could be speculated that reduction in salivary nitrite concentration is to blame for the increased susceptibility to enteric infections following antibiotic usage.

The very high serum nitrate levels observed in patients that are suffering from infective gastroenteritis may protect against the faecal-oral route of reinfection *via* increased generation of salivary nitrite. This mechanism could limit the impact of outbreaks of gastroenteritis, which would be relevant in humans, but also be of particular importance in other mammalian species and animal husbandry.

Health conscious individuals and government authorities have advocated restriction of dietary nitrates for the last 20 years after ingestion of amines and nitrates had been associated with gastric cancer in animal models. Although the harmful and potentially carcinogenic activity of N-nitroso amines cannot be dismissed, epidemiological evidence for this association has not been forthcoming. Review of the literature learns that of the 32 published studies on the relationship between exposure to nitrates and cancer, 11 report a positive relationship[27-37], 13 papers could not demonstrate any relationship[38-50], and 8 papers report a negative relationship[51-58].

Addition of nitrite to acidic solutions achieves kill of gut pathogens where acid alone allows growth to continue. Physiological concentrations of nitrite accomplish kill after exposure times that are comparable with the transfer time of a food bolus through the stomach. The mechanism of chemical host defence that seems to take place in symbiosis with nitrate reducing bacteria that live on the dorsum of the tongue may be of fundamental importance. A conclusive demonstration of the antimicrobial effect *in vivo* would require a major re-interpretation of the role of dietary nitrate in human health and animal husbandry.

4.2   Helicobacter pylori

H.pylori is killed at pH2 unless urea is present in the solution (Table 4). However, even in the presence of urea, the organism seems to be unable to survive when 1mM of nitrite is added to the medium. Figure 9 demonstrates the (negative) dose dependent relationship between nitrite and the number of surviving micro-organisms at pH2.

Table 4 shows no effect of nitrite at neutral pH on the survival of *H.pylori*. This indicates that reactive oxides of nitrogen rather than nitrite itself are responsible for the antibacterial action of acidified nitrite.

Ingestion of foods rich in nitrate may protect against colonisation of the stomach by *H.pylori*. There is no epidemiological evidence that people with a high nitrate intake might have a reduced prevalence of the organism. On the contrary, transmission of the infection has been related to consumption of uncooked vegetables[59], and the infection is acquired earlier with a high percentage of the adult population infected in developing countries where nitrate intake is expected to be relatively high[60]. However, no investigations have been conducted to specifically investigate the relation between dietary nitrate intake and survival of *H. pylori*.

4.3   The Influence of Ascorbic Acid and Thiocyanate

Thiocyanate is secreted into saliva by the salivary duct cells by the same transport mechanism that is responsible for the concentration of nitrate in saliva[61]. The concentration in the gastric lumen is about 250 μmol/l[62], but can be up to four times higher in smokers[63]. In the presence of hydrogen peroxide the hypothiocyanate anion is formed $(OSCN^-)$[64], which has a strong bactericidal action. This oxidation reaction does occur in saliva with the help of peroxide produced by bacteria and leucocytes[65,66]. Thiocyanate catalyses the formation of nitrosamines in an acidic environment with an optimal pH of 2.3[67].

Table 2 and Figure 8 show the effect of the addition of various concentrations in the range from 25 and 500 μmol/l thiocyanate on the antibacterial activity of nitrite against *Y. enterocolitica* at pH2. The intrinsic antibacterial activity of thiocyanate is evident from the Table; in the presence of (125μmol/l thiocyanate at pH2, no nitrite is needed to achieve kill after (10 minutes exposure time. The effect of the addition of thiocyanate was highly significant at short exposure times, but less evident after longer exposure times when the acidic environment itself was sufficient to kill the bacterium even without nitrite or thiocyanate.

According to the data in this study, the kill of *Y. enterocolitica* takes place at physiological concentrations of acid, nitrite, and thiocyanate after exposure times that are well within the passage time of a food bolus through the stomach. The three antimicrobial agents are highly synergistic in their antimicrobial action.

Ascorbic acid is actively secreted into the gastric lumen[18] and reaches concentrations between 50 and 500 μmol/l in healthy individuals[68]. It is a powerful antioxidant and can react with nitrous acid converting it to nitric oxide, whereas it is oxidised itself to dehydroascorbic acid (Figure 11). Ascorbic acid thus potently enhances nitric oxide formation under acidic conditions and has potential importance as an *in vivo* nitrite scavenger in the stomach[68]. Nitrosamine formation is inhibited by ascorbic acid[69] and low ascorbic concentrations are associated with *H. pylori* infection and intestinal metaplasia, a precancerous condition of the stomach[70].

Table 2 and Figure 8 show the effect of the addition of various concentrations in the range from 125 and 1000 μmol/l ascorbic on the antibacterial activity of nitrite against *Y. enterocolitica* at pH2. The micro-organism is

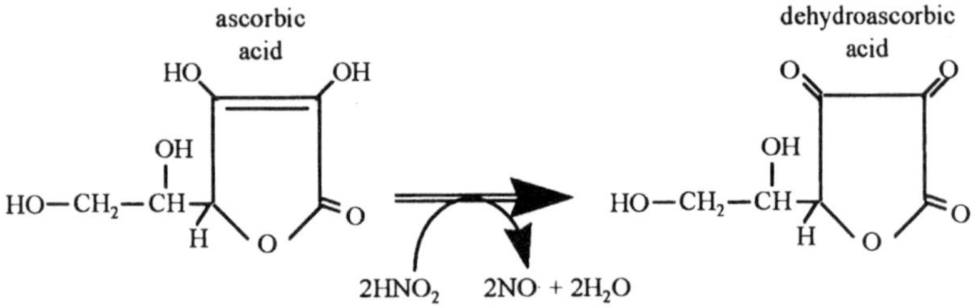

ascorbic
acid

dehydroascorbic
acid

Figure 11 *Oxidation of ascorbic acid to dehydroascorbic acid by nitrous acid.*

protected by high concentrations of ascorbic acid and 1000 µmol/l prevents kill completely. Regression analysis shows a significant correlation after 15 and 20 minutes exposure (Table 3); at shorter exposure times no kill of the micro-organism is achieved even without the protection of ascorbic acid. From the data in this study it can be concluded that the presence of ascorbic acid reduces the antimicrobial activity of acidified nitrite against *Y. enterocolitica in vitro*.

Yoshida *et al.*[71] studied the antifungal activity of acidified nitrite (1mM $NaNO_2$ at pH4) in the presence of various concentrations of an $NO^{\cdot}$-oxidising agent, imidazolineoxyl N-oxide. Reacting from 2-phenyl-4,4,5,5-tetramethylimidazoline-3-oxide-1-oxyl (PTIO) to 2-phenyl-4,4,5,5-tetramethylimidazoline-1-oxyl (PTI), the agent acts as an oxygen donor, producing $NO_2^{\cdot}$ from $NO^{\cdot}$. They showed convincingly that imidazolineoxyl N-oxide enhanced the antifungal activity of the acidified nitrite solution. The protection of micro-organisms against the antimicrobial activity of acidified nitrite by ascorbic acid which drives reactions towards $NO^{\cdot}$ formation, and the enhancement of the antimicrobial activity by imidazolineoxyl N-oxide which drives the reaction towards other oxides of nitrogen suggest that $NO^{\cdot}$ is not responsible for the

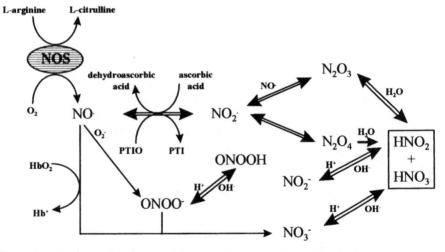

Figure 12 *Biochemical pathways of the formation of reactive oxides of nitrogen*

antimicrobial activity of acidified nitrite. It is more likely that the antimicrobial activity is mediated by other oxides of nitrogen such as dinitrogentrioxide or dinitrogentetroxide or even peroxynitrite (Figure 12).

References

1. Pryor WA, Lightsey JW. Mechanisms of nitrogen dioxide reactions: initiation of lipid peroxidation and the production of nitrous acid. Science 1981;**214**:435–7.
2. Lijinski W. Nitrosamines and nitrosamides in the aetiology of gastrointestinal cancer. Cancer 1977;**40**:2446–9.
3. Spiegelhalder B, Eisenbrand G, Preussmann R. Influence of dietary nitrate on nitrite content of human saliva: possible relevance to in vivo formation of N-nitroso compounds. Food Cosmet Toxicol 1976;**14**: 545–8.
4. Tannenbaum SR, Weisman M,Fett D. The effect of nitrate intake on nitrite content in human saliva. Food Cosmet Toxicol 1976;**14**: 549–52.
5. Ishiwata H, Boriboon P, Harada M, Tanimura A. Ishidate M. Studies on in vivo formation of N-nitrosocompounds (IV). Changes of nitrite and nitrate concentration in incubated human saliva. J. Food Hyg Soc 1975;**16**:93–8.
6. Ohshima H, Bartsch H. Quantitative estimation of endogenous nitrosation in humans by monitoring N-nitrososoproline excreted in urine. Cancer Res 1981;**41**:3658–62.
7. Tannenbaum SR. N-nitroso compounds: a perspective on human exposure. Lancet 1983;**I**:629–31.
8. Benjamin N *et al.* Stomach NO synthesis. Nature 1994;**368**:502.
9. Lundberg JON, Weitzberg E, Lundberg JM, Alving K. Intragastric nitric oxide production in humans: Measurements in expelled air. Gut 1994;**35**:1543–6.
10. Duncan C *et al.* Chemical generation of nitric oxide in the mouth from the enterosalivary circulation of dietary nitrate. Nature Medicine 1995;**1**:546–51.
11. Dykhuizen RS, Copland M, Smith CC, Douglas G, Benjamin N. Plasma nitrate concentration and urinary nitrate excretion in patients with gastroenteritis. J Infect 1995;**31**:73–5.
12. Neilly IJ, Copland M, Haj M, Adey G, Benjamin N, Bennett B. Plasma nitrate concentrations in neutropenic and non-neutropenic patients with suspected septicaemia. Br. J. Haem. 1995;**89**:199–202.
13. Dykhuizen RS, Masson J, McKnight G, Mowat ANG, Smith CC, Smith LM, Benjamin N. Plasma nitrate concentration in infective gastroenteritis and inflammatory bowel disease. Gut 1996;**39**:393–395.
14. Marshall BJ, Barrett LJ, Prakash C, McCallum RW, Guerrant RL. (1990) Urea protects *Helicobacter (Campylobacter pylori)* from the bactericidal effect of acid. Gastroenterology **99**: 697–702.
15. Burgen ASV, Emmelin NG. (1961) Physiology of the salivary glands; Monographs of the Physiological Society, No 8. Edward Arnold Ltd., London, United Kingdom.
16. Schorah CJ, Sobala GM, Sanderson M, Collis N, Primrose JM. (1991) Gastric juice ascorbic acid: effects of disease and implications for gastric carcinogenesis. Am J Clin Nutr **53**: 287S-293S.
17. Ruddell WSJ, Blendis LM, Walters CL. (1976) Nitrite and thiocyanate in gastric juice. Gut **17**: 401.
18. Sobala GM, Schorah CJ, Sanderson M, Dixon MF, Tompkins DS, Godwin P,

Axon ATR. (1989) Ascorbic acid in the human stomach. Gastroenterology **97**: 357–363.

19. Verdu E, Viani F, Armstrong D, Fraser R, Siegrist HH, Pignatelli B, Idström JP, Cederberg C, Blum AL, Fried M. (1994) Effect of omeprazole on intragastric bacterial counts, nitrates, nitrites, and N-nitroso compounds. Gut **35**: 455–460.

20. Snepar R, Poporad GA, Romano JM, Kobasa WB, Kay D. (1982) Effect of cimetidine and antacid on gastric microbial flora. Infect Immun **36**: 518–524.

21. Konturek JW, Thor P, Maczka M, Stoll R, Domschke W, Konturek SJ. (1994) Role of cholecystokinin in the control of gastric emptying and secretory response to a fatty meal in normal subjects and duodenal ulcer patients. Scand J Gastroenterol **29**: 583–590.

22. Tannenbaum SR, Weisman M, Fett D. (1976) The effect of nitrate intake on nitrite formation in human saliva. Fd Cosmet Toxicol **14**: 549–552.

23. McKnight, Smith LM, Drummond RS, Duncan CW, Golden M, Benjamin N. (1997) Chemical synthesis of nitric oxide in the stomach from dietary nitrate in humans. Gut **40**: 211–214.

24. Pavia AT, Shipman LD, Wells JG, Puhr ND, Smith JD, McKinley TW, Tauxe RV. (1990) Epidemiologic evidence that prior antimicrobial exposure decreases resistance to infection by antimicrobial-sensitive *Salmonella.* J Inf Dis **161**: 255–260.

25. Neal KR, Brij SO, Slack RCB, Hawkey CJ, Logan RFA. (1994) Recent treatment with H₂-antagonists and antibiotics and gastric surgery as risk factors for salmonella infection. BMJ **308**: 176.

26. Dougall HT, Smith L, Duncan C, Benjamin N. (1995) The effect of amoxycillin on salivary nitrite concentrations: an important mechanism of adverse reactions?. Br J Pharmac **39**: 460–462.

27. Hill MJ, Hawksworth G, Tattersall G. (1973) Bacteria, nitrosamines and cancer of the stomach. Br J Cancer **28**: 562–567.

28. Zaldivar R, Robinson H. (1973) Epidemiological investigation on stomach cancer mortality in Chileans: association with nitrate fertilizer. Z Krebs-Forsch **80**: 289–295.

29. Armijo R, Coulson AH. (1975) Epidemiology of stomach cancer in Chile. The role of nitrogen fertilizers. Int J Epidemiol **4**: 301–309.

30. Haenszel W, Kurihara M, Locke FB, Shimuzu K, Segi M. (1976) Stomach cancer in Japan. J Natl Cancer Inst **56**: 265–278.

31. Cuello C, Correa P, Haenszel W, Gordillo G, Brown C, Archer M, Tannenbaum S. (1976) Gastric cancer in Colombia: I Cancer risk and suspect environmental agents. J Natl Cancer Inst **57**: 1015–1020.

32. Fraser P, Chilvers C. (1981) Health aspects of nitrate in drinking water. The Science of the Total Environment **18**: 103–116.

33. Jensen OM. (1982) Nitrate in drinking water and cancer in Northern Jutland, Denmark, with special reference to stomach cancer. Exotoxicology and Environmental Safety **6**: 258–267.

34. Clough PWL. (1983) Nitrates and gastric carcinogenesis. Minerals and the Environment **5**: 91–95.

35. Gilli G, Corrao C, Favilli S. (1984) Concentrations of nitrate in drinking water and incidence of gastric carcinomas: first descriptive study of the Piemonte Region, Italy. The Science of the Total Environment **34**: 35–48.

36. Dutt MC, Lim HY, Chew RKH. (1987) Nitrate consumption and the incidence of gastric cancer in Singapore. Fd Chem Toxic **25**: 515–520.

37. Zandjani F, Høgsaet B, Anderson A, Langård S. (1994) Incidence of cancer among nitrate fertilizer workers. Int Arch Occup Environ Health **66**: 189–193.

38. Geleperin A, Moses VJ, Fox G. (1976) Nitrate in water supplies and cancer. Illinois Med J **149**: 251–253.

39. Vincent P, Dubois G, Leclerc H. (1983) Nitrates dans l'eau de boisson et mortalité par cancer. Etude épidémiologique dans le Nord de la France. Rev Epidém et Santé Publique **31**: 199–207.

40. Fontham E, Zavala D, Correa P, Rodriguez E, Hunter F, Haenszel W, Tannenbaum SR. (1986) Diet and chronic atrophic gastritis: a case control study. JNCI **76**: 621–627.

41. Al-Dabbagh S, Forman D, Bryson D, Stratton I, Doll R. (1986) Mortality of nitrate fertiliser workers. Brit J Ind Med **43**: 507–515.

42. Fraser P, Chilvers C, Day M, Goldblatt P. (1989) Further results from a census based mortality study of fertiliser manufacturers. Brit J Ind Med **46**: 38–42.

43. Buiatti E, Palli D, Decarli A, Amadori D, Avellini C, Bianchi S, Bonaguri C, Cipriani F, Cocco P, Giacosa A, Marubini E, Minacci C, Puntoni R, Russo A, Vindigni C, Fraumeni JF, Blot WJ. (1990) A case-control study of gastric cancer and diet in Italy: II Association with nutrients. Int J Cancer **45**: 856–901.

44. Rafnsson V, Gunnarsdottir H. (1990) Mortality study of fertiliser manufacturers in Iceland. Brit J Ind Med **47**: 721–725.

45. Boeing H, Frentzel-Beyme R, Berger M, Berndt V, Göres W, Körner M, Lohmeier R, Menarcher A, Männl HFK, Meinhardt M, Müller R, Ostermeier H, Paul F, Schwemmle K, Wagner KH, Wahrendorf J. (1991) Case-control study of stomach cancer in Germany. Int J Cancer **47**: 858–864.

46. Leclerc H, Vincent P, Vandevenne P. (1991) Nitrates de l'eau de boisson et cancer, Ann Gastroénterol Hépatol **27**: 326–332.

47. Forman D. (1991) Nitrate exposure and human cancer, in: Nitrate contamination exposure consequence and control (Bogardi I, Kuzelka RD, eds). NATO ASI series G, Ecological Sciences 30 Springer Verlag Berlin: 281–288.

48. Hagmar L, Bellander T, Andersson C, Linden K, Attewell R, Moller T. (1991) Cancer morbidity in fertiliser workers. Int Arch Occup Environ Health **63**: 63–67.

49. Rademacher JJ, Young TB, Kanarek MS. (1992) Gastric cancer mortality and nitrate levels in Wisconsin drinking water. Arch Environ Health **47**: 292–294.

50. Fandrem SI, Kjuus H, Anderson A, Amlie E. (1993) Incidence of cancer among workers in a Norwegian nitrate fertiliser plant. Br J Ind Med **50**: 647–652.

51. Armijo R, Gonzalez A, Orellana M, Coulson AH, Sayre JW, Detels R. (1981) Epidemiology of gastric cancer in Chile: II Nitrate exposures and stomach cancer frequency. Int J Epidemiol **10**: 57–62.

52. Beresford SAA. (1985) Is nitrate in the drinking water associated with the risk of gastric cancer in the UK?. Int J Epidemiol **14**: 57–63.

53. Risch HA, Jain M, Choi NW, Fodor JG, Pfeiffer CJ, Howe GR, Harrison RW, Craib KJP, Miller AB. (1985) Dietary factors and the incidence of cancer of the stomach. Am J Epidemiol **122**: 947–959.

54. Kamiyama J, Ohshima H, Shimada A, Saito N, Bourgade MC, Ziegler P, Bartsch H. (1987) Urinary excretion of N-nitrosamino acids and nitrate by inhabitants in high- and low-risk areas of stomach cancer in Northern Japan. IARC Sci Publ **84**: 497–502.

55. Palli D, Bianchi S, Decarli A, Cipriani F, Avellini C, Cocco P, Falcini F, Puntoni R, Russo A, Vindigni C, Fraumeni JF, Blot WJ, Buiatti E. (1992) A case-control study of cancers in the gastric cardia in Italy. Brit J Cancer **65**: 263–266.

56. Hansson LE, Nyren O, Bergström R, Wolk A, Lindgren A, Baron J, Adami HO. (1994) Nutrients and gastric cancer risk. A population-based case-control study in Sweden. Int J Cancer **57**: 638–644.

57. Forman D, Al-Dabbagh S, Doll R. (1985) Nitrate, nitrites and gastric cancer in Great Britain. Nature **313**: 620–625.

58. Knight TM, Forman D, Pirastu R, Comba P, Iannarilli R, Cocco PL, Angotzi G, Ninu E, Schierano S. (1990) Nitrate and nitrite exposure in Italian populations with different gastric cancer rates. Int J Epidemiol **19**: 510–515.

59. Hopkins RJ, Vial PA, Ferreccio C, Ovalle J, Prado P, Sotomayor V. (1993) Seroprevalence of *Helicobacter pylori in Chile*: vegetables may serve as one route of transmission. J Infect Dis **168**: 222–226.

60. Megraud F, Brassens-Rabbe MP, Denis F, Belbouri A, Hoa DQ. (1989) Seroepidemiology of *Campylobacter pylori* infection in various populations. J Clin Microbiol **27**: 1870–1873.

61. Edwards DAW, Fletcher K, Rowlands EN. (1954) Antagonism between perchlorate, iodide, thiocyanate, and nitrate for secretion in human saliva: Analogy with the iodide trap of the thyroid. Lancet **1**: 498–499.

62. Das D, De PK, Banerjee RK. (1995) Thiocyanate, a plausible physiological electron donor of gastric peroxidase. Biochem J **305**: 59–64.

63. Boyland E., Walker SA. (1975) Thiocyanate catalysis of nitrosamine formation and some dietary implications. Pp. 124–126 in P. Bogovski and EA Walker, eds. N-nitroso compounds in the environment, IARC Scientific Publication No. 9. International Agency for Research on Cancer, Lyon, France.

64. Pruit KM, Tenovua J. (1982) Kinetics of hypothiocyanide production during peroxidase catalysed oxidation of thiocyanate. Biochim Biophys Acta **704**: 204–214.

65. Kraus FW, Nickerson JF, Perry WI, Walker AP. (1957) Peroxide and peroxidogenic bacteria in human saliva. J Bacteriol **73**: 727–735.

66. Klinkhammer JM. (1963) Human oral leucocytes. J Am Soc Periodontists **1**: 109–117.

67. Fan TY, Tannenbaum SR. (1973) Factors influencing the rate of formation of nitrosomorpholine from morpholine and nitrite: Acceleration by thiocyanate and other anions. J Agric Food Chem **21**: 237–240.

68. Sobala GM, Pignatelli B, Schorah CJ, Bartsch H, Sanderson M, Dixon MF, King RFG, Axon ATR. (1991) Levels of nitrite, nitrate, *N*-nitroso compounds, ascorbic acid and total bile acids in gastric juice of patients with and without precancerous conditions of the stomach. Carcinogenesis **12**: 193–198.

69. Tannenbaum SR, Wishnok JS, Leaf CD. (1991) Inhibition of nitrosamine formation by ascorbic acid. Am J Clin Nutr **53**: 247S–250S.

70. Schorah CJ, Sobala GM, Sanderson M, Collis N, Primrose JM. (1991) Gastric juice ascorbic acid: effects of disease and implications for gastric carcinogenesis. Am J Clin Nutr **53**: 287S–293S.

71. Yoshida K, Akaike T, Doi T, Sato K, Ijiri S, Suga M, Ando M, Maeda H. (1993) Pronounced enhancement of NO·-dependent antimicrobial action by an NO·-oxidising agent, imidazolineoxyl N-oxide. Inf Immun **61**: 3552–3555.

23

Effects of Nitrates and Nitrites in Experimental Animals

A. B. T. J. Boink[1], J. A. M. A. Dormans[2], G. J. A. Speijers[3] and W. Vleeming[1]

[1] LABORATORY OF HEALTH EFFECT RESEARCH, [2] LABORATORY OF PATHOLOGY AND IMMUNOBIOLOGY, [3] CENTRE OF SUBSTANCES AND RISKS, NATIONAL INSTITUTE OF PUBLIC HEALTH AND THE ENVIRONMENT, PO BOX 1, NL-3720 BA BILTHOVEN, THE NETHERLANDS

1 Introduction

High dietary nitrate intake due to the high nitrate content of several drinking water resources and of certain vegetables, cultivated at low temperature and moderate sun light intensity, has arisen concern about the possible health effects. The toxicity of nitrate *per se* is low but in humans 5–10% of the ingested dose of nitrate is converted into the more toxic nitrite by salivary or gastrointestinal reduction. Nitrite may interact with secondary or tertiary amines to form N-nitroso compounds (NOC). Several NOCs have proven to be carcinogenic in animal studies but epidemiological studies revealed no equivocal relation between nitrate uptake and cancer risk.

2 Results and Discussion

2.1 Toxicity Data

Numerous toxicological and epidemiological studies on the health effects of nitrite, nitrate and NOCs have been published. Comprehensive reviews of these studies were prepared by the Joint FAO/WHO Expert Committee on Food Additives (JECFA) [1,2]. JECFA has evaluated these studies to allocate the acceptable daily intake (ADI) for nitrite and nitrate. A summary of the most relevant toxicity data for nitrite and nitrate of the various types of studies is given in Table 1.

From the data in Table 1 it can be concluded that in all types of studies

Table 1   *Review of toxicity data in rodents*

Study type		nitrate	nitrite
			(mg per kg bw per day)*
Acute	LD_{50}	>3100	120
Short term	NOEL	365	5.5
Chronic/carcinog.	NOEL	365	5.5
Reproduction	NOEL	365	145
Mutagenicity	equivocal		

* expressed as NO_3^- and NO_2^-.

nitrite is much more toxic than nitrate. Moreover, there was no difference between the no-observed-effect level (NOEL) of nitrite in the short-term toxicity studies and the chronic toxicity studies although the toxicological end points differed.

The ADI for nitrite (0–0.06 mg/kg bw expressed as nitrite ion) was based on:

1. A 2 year study in rats. The NOEL in this study was 6.7 mg/kg bw/day, expressed as nitrite ion. The effects observed in this study were: thin dilated (instead of thickened and narrowed) coronary arteries and dilated bronchi with infiltration of lymphocytes and alveolar hyper-inflation [3].
2. A 90 day study in rats. The NOEL was 5.4 mg/kg bw/day, expressed as nitrite ion. Hypertrophy of the adrenal zona glomerulosa was observed as the toxic effect [4].

A safety factor of 100 was used to calculate the ADI for nitrite from the NOEL of both studies.

The ADI for nitrate (0–3.7 mg/kg bw, expressed as nitrate ion) was based on:

1. A 2 year study in rats. The NOEL was 365 mg/kg bw /day. Growth inhibition of the rats was observed as the toxic effect [5].
2. The 'transposed' NOEL, *i.e.* the NOEL for nitrite times 20 (5% rate of conversion of nitrate into nitrite) [6].

A safety factor of 50 was used to calculate the ADI for nitrate from the 'transposed' NOEL, which was justified by the fact that partly human data (conversion figures) were used.

Data on the toxicity of nitrite and nitrate in humans originate from accidental expositions. Intoxication by nitrate is caused predominantly by consumption of well water containing high nitrate levels. Human lethal doses of 67–833 mg nitrate ion / kg bw have been reported. Toxic doses – with methaemoglobin formation as a criterion for toxicity – ranged from 33–350 mg nitrate ion / kg bw [2]. Accidental human intoxications by nitrite have been reported due to its presence in food. The oral lethal dose for humans was estimated to vary from 33 to 250 mg nitrite ion/kg bw, the lower doses applying to children and elderly people. Doses of 1 to 8.3 mg nitrite ion /kg bw gave rise to induction of methaemoglobinaemia [1]. Comparison of these data for humans with the data for experimental animals (Table 1) makes clear that nitrite toxicity is similar for humans and experimental animals. Humans are 10–100 times more sensitive to nitrate than rats because the latter species lacks the mechanisms to convert nitrate into nitrite. Consequently, nitrate studies with rats are not appropriate to evaluate the human health effects of nitrate [6]. However, it seems justified to extrapolate results of nitrite studies in rats to the effects of nitrate on human health. Therefore, this paper deals with the mechanisms of action of nitrite only.

2.2 Carcinogenicity

It is beyond doubt that several – but not all – N-nitroso compounds are carcinogenic. NOCs can be formed in the stomach from a reaction of nitrite with amides and amines. It has been reported that this endogenous nitrosation contributes substantially to the total NOC burden as shown in Table 2 where the total NOC content in gastric juice of fasting and non-fasting subjects is compared with the estimated daily intake of preformed NOCs [7].

Table 2 *NOC content in gastric juice of fasting and non-fasting subjects compared with the estimated daily intake of preformed NOCs*

	Total NOC in gastric juice mean		
Fasting:	420 nM	(range: 10–40000 nM)	(n=455)
Non fasting:	200–4000 nM	(peak 18000 nM)	(n=4 repeatedly over the day)
Dietary intake preformed NOCs:		1000 nM per day	

It is still a controversial question whether endogenous NOC formation poses a relevant risk to human health. The amino acid proline undergoes nitrosation readily to form the non carcinogenic compound nitrosoproline. Nitrosoproline formation has been used as a measure of the potential for endogenous formation of carcinogenic NOCs [8] but it might also be considered as a physiological scavenger preventing the formation of carcinogenic NOCs (personal communication, Prof. R. Walker). Mice and rats receiving orally nitrosatable compounds and nitrite at doses which were extremely high in comparison to human exposure, showed an increase in the number of tumours [1]. However, none of 21 studies in mice and rats concerning the possible carcinogenicity of solely nitrite which were summarized in a report by the US National Academy of Science, indicated any carcinogenic effect [9]. In at least three studies with F344 rats a decrease in tumour incidence was observed [1].

Epidemiological studies have not provided any evidence that there is an increased risk of cancer related to high nitrate intake from other sources than vegetables [8]. Case control studies based on food frequency questionnaires tend to show a protective effect of the estimated nitrate intake on gastric cancer, probably due to the protective effect of vegetables and fruits.

2.3   Methaemoglobinaemia

The best known effect of nitrite is its ability to react with haemoglobin to form nitrate and methaemoglobin. As a consequence of the formation of methaemoglobin the oxygen delivery to tissue is impaired. Moderate methaemoglobinaemia (< 30% of total haemoglobin oxidised) causes discomfort (nausea, head ache) but severe methaemoglobinaemia (> 50%) may be life threatening. Numerous cases of methaemoglobinaemia due to nitrite and nitrate poisoning by contaminated (well) drinking water or meat have been reported [1,2].

The stoichiometry of the reaction between nitrite and haemoglobin is not clear. Two different mechanisms have been proposed as shown in Table 3 [10]. Moreover, a difference was observed in the stoichiometry of the reaction of nitrite with haemoglobin of haemolysed blood on the one hand and with intact erythrocytes *in vitro* or *in vivo* on the other hand [11].

Table 3   *Proposed mechanisms and stoichiometry of the reaction between nitrite and haemoglobin*

Oxygenated haemoglobin:
$$2(Hb)Fe^{2+} - O_2 + 3NO_2^- + 2H^+ \rightarrow 2(Hb)Fe^{3+} + 3NO_3^- + H_2O$$

Deoxygenated haemoglobin:
$$(Hb)Fe^{2+} + NO_2^- + 2H^+ \rightarrow (Hb)Fe^{3+} - NO + H_2O$$

	MetHb formed : NO_2^- added (mol : mol)
Haemolysed erythrocytes:	1 : 1
Ex-vivo intact erythrocytes:	2 : 1
In vivo	2 : 1

Spontaneous formation of methaemoglobin from haemoglobin occurs always in vitro and in vivo at a slow rate. Two mechanisms reduce the risk involved in the formation of this non-functional haemoglobin derivative, namely i) protection against oxidation and ii) the methaemoglobin reductase systems. The capacity of the pathways capable of reducing methaemoglobin to haemoglobin is 250 times greater than the rate at which haemoglobin is normally oxidised [12]. The high reducing capacity implicates that methaemoglobinaemia due to nitrite will occur only when the nitrite dose is higher than the methaemoglobin-reductase capacity. This is illustrated in Figure 1 which shows the results of a study conducted in our institute of methaemoglobin concentration measurements in blood of two groups of rats exposed to nitrite

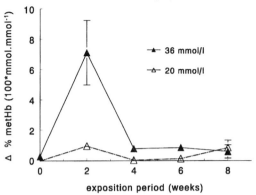

Figure 1 *Effect of nitrite concentration in drinking water and time on the concentration of methaemoglobin in blood of adult rats*

via the drinking water at a concentration of respectively 20 mmol/l and 36 mmol/l . Two weeks after the start of the exposition, blood obtained from rats drinking water with 20 mmol/l nitrite hardly showed methaemoglobinaemia while a 5 fold increase in methaemoglobin was observed in rats drinking the water with 36 mmol/l nitrite. Upon prolongation of the exposition, methaemoglobin was reduced remarkably. This decline of the extent of methaemoglobinaemia suggested that a metabolic adaptation to prolonged exposure to high nitrite doses has occurred.

2.4   Blood pressure

Nitrite and a series of organic nitrates have vasodilating properties [13]. Dilatation and thinning of intramuscular coronary blood vessels of rats given 100–300 mg $NaNO_2$ /l in the drinking water for 2 years were seen [3]. A fatal collapse of blood pressure following nitrate poisoning has been reported in cows [14]. Because conversion of nitrate into nitrite occurs very efficiently in ruminants, this fall of blood pressure was probably due to nitrite rather than to ingested nitrate.

Figure 2 shows that intravenous administration of nitrite to anaesthetised rats induced an immediate dose dependent decrease in blood pressure. The decrease in blood pressure preceeded the increase in methaemoglobin concentration (data not shown). This observation suggests that hypotension is the primary effect of nitrite. Intravenous dose of 30 µmol/kg bw caused a 10–20% decrease of blood pressure. The daily drinking water consumption of rats is approximately 50 ml/kg bw. When drinking water contains 36 mmol/l nitrite the total daily dose of nitrite will be 1800 µmol/kg bw. Thus, assuming 100% bioavailability, a single uptake of 1/60 of the total daily consumption is able to decrease blood pressure by 10–20%. From Figure 2 it is clear that restoration of normal blood pressure occurred within 30 minutes following a single dose of 30 µmol/kg bw nitrite. Therefore, continuous monitoring of blood pressure is

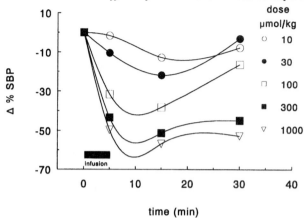

Figure 2 *Dose-effect response of intravenous NaNO$_2$ administration on systolic blood pressure in anaesthetised rats*

necessary to detect the effect of nitrite on blood pressure in a drinking-water study. Figure 3 represents typical registrations of blood pressure and heart rate by means of a telemetry device of a freely-moving rat drinking water containing KCl (upper panel; days 4 and 5, control period) or KNO$_2$ (lower panel; days 7 and 8, test period) both at a concentration of 36 mmol/l. From Figure 3 it is obvious that blood pressure and heart rate increased during the control period at night time, *i.e.* the activity phase of the rat during which rats drink. In contrast, at night time during the test period, repetitive episodes of decreased blood pressure and increased heart rate occurred. From this experiment it can be concluded that nitrite in drinking water induced transient hypotension in rats [15].

2.5 Hypertrophy Adrenal Zona Glomerulosa

In 1988 Til and coworkers reported that hypertrophy of the adrenal zona glomerulosa was observed in a subchronic oral toxicity study with potassium nitrite [4]. More recently the same authors showed that the no-observed-effect level of potassium nitrite is 50 mg/l in the drinking water, equivalent to about 5 mg/kg bw per day [16]. This nitrite-induced effect on the adrenals of the rats was subject of more detailed studies at the RIVM [17]. Adrenal morphometry and histopathological examination of the adrenals of rats exposed to nitrite *via* the drinking water (36 mmol/l potassium nitrite) showed that slight hypertrophy of the zona glomerulosa was present already after 28 days and continued without aggravation during a 90 day exposure period as illustrated in Figure 4. This nitrite-induced hypertrophy disappeared slowly after cessation of nitrite exposure. Slight focal and slight diffuse hypertrophy were still present after a 30 day recovery period but had disappeared after 60 days.

Enlargement of the zona glomerulosa of rats exposed to 12 or 3.6 mmol/l in the drinking water could not be detected by means of adrenal morphometry but on histopathological examination the adrenal zona glomerulosa of 8 (of

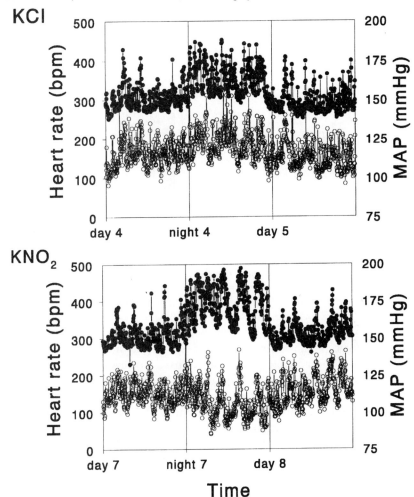

Figure 3 *Effect of potassium chloride (36 mmol/l, top) and potassium nitrite (36 mmol/ l, bottom panel) in the drinking-water on heart rate (●) and blood pressure (○) of a freely moving rat*

10) rats of the 12 mmol/l group and 3 (of 10) rats of the 3.6 mmol/l group was classified to be minimally hypertrophic. So, this effect of nitrite on the adrenal gland is dose related. The NOEL in this study was less than 3.6 mmol/l potassium nitrite in the drinking water corresponding to approximately 10 mg/ kg bw per day (expressed as nitrite ion).

As stated above in section 2.3 on methaemoglobinaemia, nitrite in blood is readily converted into nitrate. The plasma half-life of nitrite has been estimated to be less than 30 min in various species while the half-life of nitrate ranged from 4 h (sheep and horse) to 44 h (dog). This rapid conversion of nitrite into nitrate and the longer plasma half-life of the latter as compared to the half-life

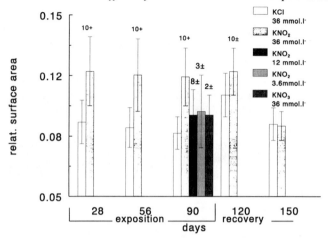

Figure 4   *The average relative adrenal zona glomerulosa fractional surface area of rats drinking water containing potassium chloride, potassium nitrite, or potassium nitrate is indicated by the vertical bars ($\pm$ s.d.). The numbers at the top of the bars indicate the number of rats in which histopathology revealed minimal ($\pm$) or slight (+) hypertrophy of the zona glomerulosa in a total of 10 rats per group*

of nitrite suggests that nitrate could play a role in the etiology of the hypertrophy of the adrenal zona glomerulosa. However, in rats exposed to potassium nitrate *via* the drinking water (36 mmol/l) during 90 days no hypertrophy could be observed by means of morphometry. On histopathological examination the adrenal zona glomerulosa of only 2 (of 10) rats of the 36 mmol/l nitrate group was classified to be minimally hypertrophic. The nitrate dose was 10 times higher than the lowest nitrite dose but the adrenal lesion was less severe. Thus, this lesion must be attributed to nitrite ion solely.

2.6   Role of the Adrenals in Blood Pressure Regulation

The adrenal gland plays a role in the regulation of blood pressure *via* the renin-angiotensin-aldosteron axis (RAA-axis) [18]. When blood pressure decreases, baroreceptors in the juxtaglomerular apparatus stimulate the release of renin. Renin catalyzes the conversion of angiotensinogen into angiotensin I which in turn is converted into angiotensin II by a converting enzyme. Angiotensin II effectively contributes to restoration of normal blood pressure by its vasoconstrictive effect on blood vessels and by stimulating the release of aldosteron from the zona glomerulosa of the adrenals. Aldosteron enhances the reabsorption of sodium in the renal tubules which leads to expansion of the extracellular volume. The rise in blood volume will also cause an increase in blood pressure.

It was shown above that exposition of rats to nitrite *via* the drinking water caused repeated dips of blood pressure. Therefore, each time the rats drank nitrite containing water the RAA-axis was activated. This over and over again activation of the RAA-axis may have been the cause of the hypertrophy of the adrenal zona glomerulosa of the rat. This hypothesis is currently under investigation in our laboratory. Rats are given ramipril, an inhibitor of angiotensin converting enzyme, in their feed. Ramipril inhibits the formation of angiotensin II and consequently the vasoconstriction and stimulation of the adrenal zona glomerulosa to release aldosteron. Indeed, in a pilot experiment with a limited number of rats, no hypertrophy of the zona glomerulosa could be detected. Although this observation has to be confirmed in a definitive study, the absence of nitrite-induced hypertrophy in rats given ramipril indicated that the RAA-axis was involved in the etiology of the hypertrophy of the adrenal zona glomerulosa. This suggests that hypertrophy of the adrenal zona glomerulosa can be considered to be a physiological adaptation to repeated episodes of hypotension caused by nitrite.

2.7 Summarizing Conclusions

The toxicity of nitrate itself is low but in humans salivary or gastrointestinal conversion of nitrate into nitrite occurs. Nitrite is much more toxic than nitrate. Its role in the induction of tumours is equivocal. The primary effect of nitrite is vasodilatation which results in a decrease of blood pressure. The RAA-axis is stimulated by this decrease in blood pressure resulting in the hypertrophy of the adrenal zona glomerulosa. This slight hypertrophy can be expected to occur in humans as well. On the basis of the experimental data it is to be considered as a physiological adaptation rather than a harmful lesion.

The restoration of blood pressure by the RAA-axis takes time. The formation of methaemoglobin and concurrent conversion of nitrite into nitrate acts to a certain extent as an efficient detoxifying mechanism to prevent a fatal collapse of blood pressure. Only if the nitrite dose is so high that severe methaemoglobinaemia occurs, *i.e.* when the capacity of the methaemoglobin-reductase system is exceeded, oxygen transport will become impaired which can eventually be life threatening.

References

1. Speijers G.J.A. *et al.*, Nitrite (and potential endogenous formation of N-nitroso compounds). In: 'Toxicological evaluation of certain food additives and contaminants in food'. WHO Food Additives Series: 35, WHO, Geneva, 1996, 269–323.
2. Speijers G.J.A. *et al.*, Nitrate. In: 'Toxicological evaluation of certain food additives and contaminants in food'. WHO Food Additives Series: 35, WHO, Geneva, 1996, 325–360.
3. Shuval H.I., N. Gruener. Epidemiological and toxicological aspects of nitrates and nitrites in the environment. *Am. J. Public Health*, 1972;**62**:1045–1052.
4. Til H.P., H.E. Falke, C.F. Kuper. Evaluation of the oral toxicity of potassium nitrite in a 13 week drinking-water study in rats. *Food Chem Toxic*, 1988;**26**:851–859.
5. Lehman A.J. Quarterly report to the editor on topics of current interest : Nitrates and nitrites in meat products. *Quart Bull As Food Drug Off*, 1958;**22**: 136–138.

6. Speijers G.J.A. Different approaches of establishing safe levels for nitrate and nitrite. In: 'Health aspects of nitrates and its metabolites (particularly nitrite)'. Council of Europe Press, Strasbourg, 1995, 287–298.

7. Shephard S.E. Endogenous formation of N-nitroso compounds in relation to the intake of nitrate or nitrite. In: 'Health aspects of nitrates and its metabolites (particularly nitrite)'. Council of Europe Press, Strasbourg, 1995, 137–149.

8. Møller H. Adverse health effects of nitrate and its metabolites: epidemiological studies in humans. In: 'Health aspects of nitrates and its metabolites (particularly nitrite)'. Council of Europe Press, Strasbourg, 1995, 255–268.

9. National Academy of Sciences. The health effects of nitrate, nitrite and N-nitroso compounds. Part 1 of a 2-part study by the committee on nitrite and alternative curing agents in food. National Academy Press, Washington, 1981.

10. Blaauboer B.J. Formation and reduction of ferrihemoglobin in red cells. Thesis. University of Utrecht, 1978.

11. Jaffé E.R. Methaemoglobinaemia. *Clinics in Haematology*, 1981;**10**:99-122.

12. Kortboyer J.M., A.B.T.J. Boink, M.J. Zeilmaker, W. Slob, J. Meulenbelt. Methemoglobin formation due to nitrite: dose-effect relationship in vitro. RIVM report 235802 006, August 1997, RIVM, Bilthoven, The Netherlands.

13. Nickerson M. Vasodilator drugs. In: L.S. Goodman, A. Gilman A eds, 'Pharmacological basis of therapeutics'. 5th ed. Macmillan, London, 1975, pp 727–743.

14. Dinkla E.T. Enkele gevallen van nitraatvergiftiging bij runderen in de provincie Groningen. *Tijdschrift voor Diergeneeskunde* 1976;**101**: 1096–1099.

15. Vleeming W., A. Van de Kuil, J.G. te Biesebeek, J. Meulenbelt, A.B.T.J. Boink. Effect of nitrite on blood pressure in anaesthetized and free-moving rats. *Food Chem Toxic*, 1997;**35**:615–619.

16. Til H.P., C.F. Kuper, H.E. Falke. Nitrite-induced adrenal effects in rats and the consequences for the no-observed-effect level. *Food Chem Toxic*, 1997;**35**:349–355.

17. Boink A.B.T.J., J.A.M.A. Dormans, G.J.A. Speijers. The role of nitrite and/or nitrate in the etiology of the hypertrophy of the adrenal zona glomerulosa of rats. In: 'Health aspects of nitrates and its metabolites (particularly nitrite)'. Council of Europe Press, Strasbourg, 1995, 213–227.

18. Garrison J.C., M.J. Peach. Renin and Angiotensin. In: A. Goodman Gilman, T.W. Rall, A.S. Nies, P. Taylor. eds. : Goodman and Gilman's The pharmacological basis of therapeutics. 8th ed. McGraw-Hill, New York, 1992, 749–763.

24

Nitrate Exposure and Childhood Diabetes

P. A. McKinney, R. Parslow and H. J. Bodansky

THE PAEDIATRIC EPIDEMIOLOGY GROUP, CENTRE FOR
HEALTH SERVICES RESEARCH, LEEDS UNIVERSITY

1 Introduction

Insulin dependent diabetes mellitus (IDDM) is a condition resulting from an autoimmune process in which insulin producing pancreatic β–cells are selectively destroyed. In the UK 20,000 young people under the age of 20 have IDDM. The causes are largely unknown but genetic susceptibility may account for up to 10% clearly indicating the key role environmental factors play in the aetiology of the disease. However, the aetiology of the condition is at present unclear.

Various environmental exposures have been studied in relation to IDDM, the majority concentrating on diet or viral infections (Bach, 1994). Specific experimental evidence and case reports show that chemical toxins such as alloxan, chlorotocin and certain pesticides can poison pancreatic β cells directly or trigger an immune response which damages these cells (Assan and Larger, 1993). Particular attention has focused on nitrosamines and β cell destruction as streptozotocin (STZ), a wide spectrum antibiotic, is also a nitrosamine specifically used as a toxin to induce IDDM in rats. STZ causes DNA strand breaks, selectively destroying β cells either following single high doses or multiple low doses (Dulin and Soret, 1978). In humans, the reduction of ingested nitrate to nitrite and amination in the stomach to nitrosamines is well documented (IARC 1987) and a limited number of previous studies have investigated the possible links between dietary intake of nitrate and nitrite and the incidence of IDDM. Helgason *et. al,* (1982) have shown that mice fed with a nitrosamine rich diet of smoked and cured meat developed diabetes and sustained morphological damage to islet β cells. In Scandinavia conflicting evidence has emerged from dietary studies of nitrate and nitrite (Dahlquist *et al.*, 1990; Virtanen *et al.*, 1994) and a case-control study in Australia found no association between estimated intake of nitrosamines from food and IDDM (Verge *et al.*, 1984). However, an ecological study in the USA (Kostraba *et al.*, 1992) has demonstrated a significant positive correlation between levels of nitrate in drinking water and IDDM.

Estimates of daily nitrate intake in the UK vary between 64 and 297 mg/person/day (Chilvers *et al.*, 1984). EC Drinking Water Directive 80/778/EEC (EEC 1980) defines a maximum acceptable concentration of nitrate in drinking water as 50 mg l^{-1} and recommends that levels do not exceed 25 mg l^{-1}. In the UK, a nitrate level of 50 mg l^{-1} in drinking water is estimated to provide over 50% of total dietary nitrate intake (Chilvers *et al.*, 1984).

The Paediatric Epidemiology Group at Leeds University is engaged on a series of studies investigating the environmental causes of childhood diabetes. Descriptive epidemiological work has shown a clear excess of childhood IDDM in rural areas (McKinney *et al.*, 1996). Nitrate levels in drinking water tend to be higher in rural compared to urban areas in association with fertiliser use. So based on the experimental and epidemiological studies described above, the present study set out to test the hypothesis that a raised incidence of childhood IDDM is present in areas with higher levels of nitrate in domestic drinking water. The analysis of small area variation in levels of nitrate in drinking water in relation to the incidence of IDDM is described (Parslow *et al.*, 1997).

2   Materials and Methods

2.1   Study Population

The Yorkshire Register of Childhood Diabetes (YRCD) is population-based and contains demographic details of 1797 children and young adults aged 0–16 years diagnosed with IDDM while resident within the geographical area of the former Yorkshire Regional Health Authority. The register covers the period 1978–1994 and is estimated to be 97% complete (Staines *et al.*, 1993). Registrations are ascertained from three independent sources (hospital clinics, general practitioners and hospital episode statistics) and diagnosis of IDDM confirmed by hospital note abstraction. A postcode is identified for the residential address at diagnosis of each child on the register using the Postcode Address File and located in the smallest areal unit of the 1991 decennial census, an Enumeration District (ED), using the Central Postcode Directory.

2.2   Exposure to Nitrate in Drinking Water

Water supplies in England are provided by privately owned utilities which are regulated under the Water Industry Act 1991 (HMSO, 1991). Regulation 2 of the Water Supply (Water Quality) Regulations under that Act, requires supply companies to create distribution networks comprising 'water supply zones' (WSZ) supplying a maximum of 50 000 people. The chemical composition of water supplied to every household within each WSZ is considered to be the same at any one time. Nitrate levels in drinking water for the study area were obtained from Yorkshire Water plc and the York Water Company. Complete data were only available from 1990 onwards for the entire study area and Figure 1 depicts its location and the WSZ boundaries. Sampling of treated

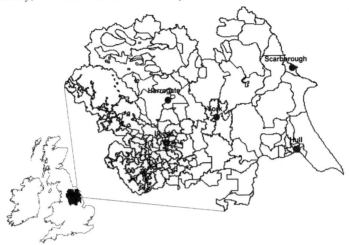

Figure 1 *Location of the study area and detail of the WSZ boundaries*

water in the distribution network is carried out at both regular (usually monthly) and random intervals to ascertain biological and chemical composition. A sample taken in a WSZ will represent the water quality throughout the zone at that time. A total of 9330 water samples were analysed for nitrate content in 148 WSZs in Yorkshire over the period 1990–1995. Over the sampling period more than 81% of WSZs showed no significant linear change over time using linear regression. Of the remaining 19%, 16 WSZs showed a decrease and 12 an increase in nitrate levels over time. A mean of the monthly mean nitrate level was taken to estimate an average level of nitrate load to account for short-term seasonal variation.

2.3 Linkage of WSZ to Cases of IDDM

The geographical boundaries defining the WSZs and EDs (The 1991 Census, Crown Copyright. ESRC purchase) were obtained in digital form. The 1991 census was the first census in which ED boundaries have been digitised allowing geographical analyses at such a fine scale. EDs were assigned to the 148 WSZs using a Geographical Information System (GIS – ARC/INFO v 7.0). The two sets of boundary data were not entirely coterminous and where an ED lay in more than one WSZ it was assigned to the WSZ which contained the largest proportion of its area. Systematic visual checking of the overlay confirmed that this method compensated for minor digitising errors in the original boundary data. In total, 1797 observed cases were assigned to 148 WSZs.

2.4 Population Census Data

Certain characteristics of populations were known *a priori* to be associated with the patterns of distribution of IDDM. In order to examine these in the analysis, census data was obtained for calculating indices of ethnicity, population density and social class for each WSZ. Thus data from the 1991 Census (The 1991 Census, Crown Copyright. ESRC purchase) were aggregated to provide a population structure for each WSZ. Population totals by age and sex, ethnicity (white or non-white), levels of car ownership, unemployment, owner-occupation and overcrowding were available for each WSZ. The 1991 Census is the first decennial census to provide information on ethnicity (OPCS, 1996). The latter four parameters were used to calculate the Townsend score, a measure of socio-economic status with high indices indicating lower socio-economic status (Phillimore *et al.*, 1994). Area-based population densities were calculated for children (0–16 years old) by dividing the respective populations in each ED by the area in hectares. The population weighted average of the area-based population density for each of the EDs was aggregated to WSZs to provide person weighted population densities. This measure more accurately reflects the density at which the average person lives (Dorling and Atkins, 1995).

2.4   Statistical Analysis

Statistical analyses were carried out using STATA (StataCorp, 1995). Expected numbers of cases were derived using the age-sex stratum specific rates and census population totals. Standardised incidence ratios (SIRs) were calculated for IDDM in each WSZ as the ratio of observed to expected cases multiplied by 100. Univariate analyses examined the effect of mean nitrate levels, ethnicity, person-based population density and Townsend Score and its four components on SIR using the Poisson trend statistic (Breslow and Day, 1987).

The incidence of IDDM was assumed to be Poisson distributed and the data were fitted to a Poisson regression model together with significant risk factors from the univariate analysis – person-based childhood population density, childhood ethnicity and the Townsend Score. The number of cases in each WSZ was used as the dependent variable and the log of the expected number of cases used as the offset. The Townsend score and ethnicity were divided into two categories and population density and mean nitrate levels divided into three categories, all with approximately equal populations. A base model was created which included childhood person-based population density, childhood ethnicity and Townsend score. The effect of adding mean nitrate levels to the model was calculated using the likelihood ratio test.

3   Results

3.1   Water Supply Zone Nitrate Levels and Population Characteristics

The distribution of the 148 WSZs by nitrate levels averaged for each WSZ, is shown in Figure 2. The majority (85%) of WSZs receive water with an

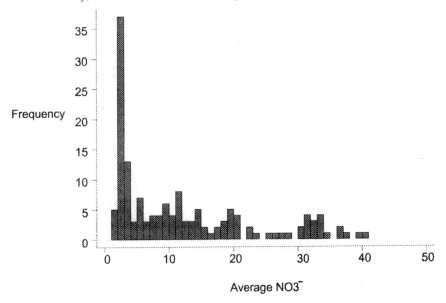

Figure 2   *Distribution of average nitrate levels for water supply zones in the Yorkshire Study Region (n = 148)*

average nitrate level below 25 mg l^{-1}. This is based on 9330 individual water samples taken from 1990–1995 across the entire area, with an average of 63 samples per WSZ. From the total number of samples, 2775 (30%) exceeded 25 mg l^{-1} nitrate of which 31 (0.11%) exceeded 50mg l^{-1}. Sixty-seven WSZs had at least one nitrate sample value greater than 25 mg l^{-1} and 12 WSZs had at least one sample value exceeding 50 mg l^{-1}. Table 1 details the underlying characteristics of the WSZs indicating the extent of variation in the data set in terms of population, geographical area and incidence of childhood IDDM.

Table 1   *Characteristics of water supply zones (WSZ) in Yorkshire*

WSZ (n = 148)	Mean	Min	Max	Total
Area hectares	6780	91	61252	1003400
Population (0–16 years)	4703	12	12641	696032
Area-based population density (0–16 years per hectare)	2.72	0.01	13.5	–
Person-based population density[a]	9.16	0.01	38.49	–
Observed number of cases	12.14	0	43	1797
Incidence rate[b] / 100, 000 /yr	15.92	0	99.42	–

[a] Parameter taken for analyses (19)
[b] Incidence rates adjusted for age and sex

Table 2   *SIRs for IDDM by nitrate levels, ethnicity, population density and Townsend Score*

	Range		O^b	E	SIR	95% CI		Population (0–16)
Nitrate level (mg l^{-1}),	1.48	<3.22	498	584	85	78	93	225708
$\chi^2 = 26.81$, 1df, p<0.001^a	3.22	<14.85	591	598	99	91	107	232373
	14.85	40.01	708	615	115	107	124	237951
Ethnicity (proportion non-white)	0.00	<0.02	1026	903	114	107	121	347966
$\chi^2 = 33.57$, 1df, p<0.001^a	0.02	0.81	771	894	86	80	93	348066
Population density	0.1	<8.27	703	605	116	108	125	231269
(0–16 years per hectare)	8.27	<13.25	595	587	101	93	110	227776
$\chi^2 = 34.43$, 1df, p<0.001^a	13.25	38.49	499	605	83	75	90	236987
Townsend Score	−5.09	<0.33	1021	898	114	107	121	343607
$\chi^2 = 33.89$, 1df, p<0.001^a	0.33	13.66	776	899	86	80	93	352425

^{a} All χ^2 for trend
^{b} O – observed number of cases, E – expected number of cases, SIR – standardised incidence ratio, CI – confidence interval

3.2   Risk Factors

The univariate analysis results are given in Table 2 for the effect of mean nitrate level, Townsend score, ethnicity and person based population density on SIR. Mean nitrate levels showed a significant increase in SIR from 85 to 115 (χ^2 test for trend = 26.81, 1df, p <0.001), with a significant decrease in risk at the lowest level and significantly raised risk at the highest level. A significant decrease in SIR from 114 to 86 was observed with an increase in the proportion of non-whites in the population (χ^2 test for trend = 33.57, 1df, p<0.001). Areas of high person-based childhood population density were negatively associated with SIR which fell from 116 to 83 (χ^2 test for trend = 34.43, 1df, p<0.001) as were the most deprived WSZs (χ^2 test for trend = 33.89, 1df, p <0.001).

3.3   Poisson Regression Modelling

The significant factors from the univariate analysis, without nitrate, formed a base model and the effect of adding nitrate to the fit of the model was tested. There was an increase in relative IRR to 1.27 from a baseline level of 1 for mean nitrate levels in the highest third (>14.85 mg l^{-1} nitrate) and the likelihood-ratio test shows a highly significant effect of adding mean nitrate to the fitted model ($\chi^2 = 9.53$, 2df, p = 0.009) (Table 3). The goodness of fit chi-square for the 'with nitrate model' suggests the data is Poisson distributed. Using the overdispersion parameter α, derived from a negative binomial regression model of the data, confirmed there was no evidence of overdispersion ($\alpha = 0.005$, $\chi^2 = 0.395$, 1df, p = 0.53). An inspection of the residual deviance after model prediction displays no obvious spatial pattern. The

Table 3   *Comparison of incidence rate ratios (95% confidence intervals) for different mean nitrate levels in drinking water, Townsend Score, ethnicity, and person-based population density in a Poisson regression model.*

	Range		Model IRR (95% CI)	
			Without nitrate	*With nitrate*
Ethnicity (proportion	0.00	<0.02	1	1
non-white)	0.02	0.81	0.84 (0.76,0.94)	0.98 (0.85,1.13)
Population density	0.1	<8.27	1	1
(0–16 years per	8.27	<13.25	0.97 (0.86,1.09)	0.93 (0.82,1.05)
hectare)	13.25	38.49	0.88 (0.75,1.04)	0.85 (0.71,1.01)
Townsend score	−5.09	<0.33	1	1
	0.33	13.66	0.87 (0.77,0.99)	0.86 (0.75,0.98)
Mean nitrate level	1.48	<3.22	–	1
(mg l^{-1})	3.22	<14.85	–	1.11 (0.98,1.26)
	14.85	40.01	–	1.27 (1.09,1.48)
Model goodness of fit χ^2			175.02, 141df	165.49, 141df
Likelihood ratio test χ^2, = 9.53, 2df, p = 0.009				

standardised residuals were normally distributed and no single observation significantly altered the fit of the model.

A restricted analysis was performed using the 120 WSZs for which constant mean nitrate levels were recorded. The results were entirely consistent with the initial regression model, demonstrating that potential misclassification of nitrate exposure does not appear to affect the magnitude of the results.

4 Discussion

Our analysis of nitrate levels in Yorkshire drinking water and childhood diabetes shows a positive association between raised incidence of the disease and the nitrate content of domestic water.

Nitrate levels in drinking water vary according to the supply source which is in turn affected by climate and local geology. In addition, seasonal fluctuation in nitrate levels is observed. However, this variation affects the whole population of a WSZ equally although different WSZs may have different levels and frequency of fluctuation. Individuals ingest nitrate *via* drinking water and food and nitrates are hypothesised as being potentially toxic due to the formation of nitrosamines in the body. An estimated 5% of dietary nitrates are reduced to nitrite, the chemical precursor of nitrosamines (Choi, 1985). However, high intake of nitrate from drinking water may be more harmful than high intake from food as water does not contain nitrosation inhibitors such as Vitamin C (Virtanen *et al.*, 1994). A Spanish ecological study of nitrate and cancer mortality suggested high consumption of local citrus fruits was protective for cancer mortality, even in areas with high levels of nitrate in drinking water (Morales-Suarez-Varela *et al.*, 1995). No study has been conducted in the UK of variation in the consumption of bottled water at the level of water supply zones. However, a recent report has demonstrated that tap water consumption as a proportion of total liquid consumption has remained constant since 1978. For 0–5 year-olds, 67% of total liquid consumption is tap water (MEL Research 1996).

Nitrosamines such as STZ, or nitrosamine rich foods are recognised as specifically toxic to pancreatic b-cells in animal models (Dulin and Soret, 1978; LeDoux *et al.*, 1988), conferring a degree of biological plausibility to the hypothesised deleterious effects in humans. Studies investigating this issue in human populations are extremely limited. Two case control studies on children with IDDM in Scandinavia present conflicting results for intakes of nitrate and nitrite in food and drinking water (Helgason *et al.*, 1982; Dahlquist *et al.*, 1990). In Finland, where the levels of nitrate in domestic water are particularly low (<5 mg l^{-1}), there was no association between drinking water nitrate or nitrite levels and IDDM, although a relationship was found for dietary nitrite but not nitrate (Virtanen *et al.*, 1994). In Sweden, Dahlquist and colleagues in a case control study of dietary factors demonstrated a dose response for intake of foods with a high *nitrosamine* content in relation to increased risk of developing diabetes (Dahlquist *et al.*, 1990). Data from this study on intake of *nitrate* and *nitrite* was less clear cut and more difficult to interpret.

An ecological study from Colorado, US found a positive link between raised incidence of IDDM and levels of nitrate in drinking water (Kostraba *et al.*, 1992), lending support to the findings of the present study. The two studies are not directly comparable as despite similar methodology and equivalent numbers of cases, the areal units of analysis were of a different order of magnitude. In Colorado the public supply system was infrequently tested and the level of public water supply within counties varied between 16% and 100%. The 3.4 million population of Colorado was subdivided into 63 counties compared to the 148 WSZ in Yorkshire with a population of 2.8 million. The sampling rate in the counties of Colorado was annual whereas the frequency for the WSZ was minimally 12 samples per year. The Yorkshire study provides a more comprehensive base for investigation than any previous research.

The rates of IDDM in Yorkshire were 15% above that expected for the entire area in the WSZ where mean nitrate levels exceeded 14.85 mg l^{-1}. These levels were well within the acceptable maximum 50 mg l^{-1} recommended by the EC (EEC 1980), although it is also recommended that levels do not exceed 25 mg l^{-1}. The study results are based on an analysis of a large geographical area and a substantial number of cases but the heterogeneous distribution of the high level WSZ across the study area make individual identification of specific high risk areas inappropriate. The observations of our Yorkshire study find some support in the limited literature available but should be considered as hypothesis generating.

An ideal assessment of nitrate exposure would record measurements of an individual's intake. However, the underlying assumption that exposure of a population represents that of the individual is considered to apply in these circumstances for the following reasons. Firstly, domestic water is distributed to relatively small populations of approximately equal size (50,000 persons) and the chemical composition of the water supplied within each WSZ is homogeneous. Thus individual households within a WSZ will receive water of equivalent quality.

Secondly, the assumption that nitrate levels from measurements made between 1990 and 1995 represent likely exposure over the longer period of case collection (1978–1994) is based on the following evidence. Significant temporal trends in nitrate levels are absent in 81% of WSZs for the recent time period, 1990 to 1995. However, over the preceding two decades, the water supply companies report decreasing levels suggesting that recent measurements are conservative. For the remaining WSZs displaying either increasing (n = 12) or decreasing (n = 16) levels, misclassification of exposure does not influence our findings, as their removal from the Poisson regression model had no effect on the overall results. The climate and geology of the study area has not altered over the last 30 years and changing patterns of land use are more likely to have resulted in reductions of nitrate levels in recent years. Our measures of exposure are therefore considered to be conservative. An investigation of drinking water nitrate and cancer in urbanised areas of the UK, found minimal variation in nitrate levels over time (Beresford *et al.*, 1985). Close examination of the Yorkshire data, sample by sample for each WSZ, clearly showed the mean level of nitrate adequately expressed the exposure level for the area and was an appropriate parameter for analysis. A German study of primary brain tumours, comparing an area based survey of water nitrate levels and semi-quantitative measurement of nitrate in subjects' drinking water, showed correlations (Steindorf *et al.*, 1994) which add further support to the use of mean levels for an area reflecting the exposure to individuals.

The high quality of the register used, in conjunction with the systematically collected water nitrate data has presented a unique opportunity to investigate the relationship between the occurrence of IDDM in children and levels of nitrate across a large area comprising varied geology and land use. Children are located by their residence at the time of diagnosis and their previous length of stay in the area is unknown. However, it is established from a study of childhood leukaemia that migration of children in Yorkshire is generally restricted to small localities and few children will have left the area to be diagnosed elsewhere (Alexander *et al.*, 1993). The inclusion of cases who spent their lives outside the study area but moving in prior to diagnosis will counterbalance this effect. No large scale systematic migration pattern is apparent to suggest the majority of cases were not exposed to the measured nitrate levels.

The present study has compensated for known confounding factors as certain area based features of populations are known to be linked to IDDM incidence, however co-varying factors, such as underlying geology, for nitrate levels are impossible to assess in the context of the current study. The following 3 specific risk factors for IDDM were selected *a priori* to investigate as potential explanatory variables for variation in incidence – population density, ethnicity and deprivation. The proportion of non-white children was included as a variable as McKinney *et al.* (1996) have found reduced incidence of childhood onset IDDM in some local government districts in Yorkshire with high proportions of non-white children. Similarly, the effect of population density has been associated with incidence of IDDM (Patterson and Waugh, 1992). A higher risk of IDDM in rural compared to urban areas has been shown in Yorkshire (McKinney *et al.*, 1996) and elsewhere (Patterson *et al.*, 1996); however, the small geographical areas and similar population base comprising the WSZs prevented any meaningful analysis of the effect of urban/rural status. Low rates of IDDM are present in the Asian subcontinent and certain districts in the county of West Yorkshire have up to 28.2% of non-white children in the population. Recent geographical analyses have shown low rates in these areas and the significance of this has been confirmed by the clear association using the WSZ as the unit of analysis. Residence in affluent areas has been demonstrated as increasing risk at a small geographical scale and again this was shown to be the case in our study. Both ethnicity and deprivation are correlated in small census areas with population density and a person based measure of population density (Dorling and Atkins, 1995) was selected for analysis. All variables, significant at the univariate level were potentially explanatory in the Poisson regression analysis. However, the results of the multivariate approach clearly showed mean nitrate levels had an independent effect on the variation in incidence which was not explained by the associations with ethnicity, deprivation or population density.

5 Conclusion

The results of our ecological analysis makes a strong case for further investigation of the diabetogenic effect of nitrate in the diet and the contribution of nitrate levels in drinking water. The observations in Yorkshire are currently being tested in other areas of the UK where population-based registers of childhood diabetes are in existence. Questions that still need to be answered are the relative effect of consumption of nitrate in food and drinking water, age when exposure is most important, whether short-term exposure to high-levels are more important than mean exposure over time. More detailed analyses of nitrate in diet including drinking water are required to make more robust conclusions.

Acknowledgements

We are grateful to Yorkshire Water plc for providing digitised boundary data and nitrate sample data, to York Water Company for providing nitrate sample data and to Dr J Barrett for statistical advice. All paediatricians, clinicians and diabetes specialist nurses are thanked for assisting Ms C Stephenson, Ms H Lilley and Dr K Gurney with assiduous data collection. Funding: Children's Research Fund, Northern and Yorkshire NHS Research and Development, The General Infirmary at Leeds Special Trustees. Dr G Law was supported by a Leeds University Postgraduate Scholarship, Novo Nordisk.

References

Alexander FE, McKinney PA, Cartwright RA (1993) Migration patterns of children with leukaemia and non-Hodgkin's lymphoma in three areas of Northern England. Journal of Public Health Medicine 15(1): 9–15

Assan R, Larger E (1993) The Role of Toxins. In: Leslie RDG (ed) Causes of Diabetes: genetic and environmental factors. John Wiley, Chichester, pp 105–123

Bach J-F (1994) Insulin-dependent diabetes mellitus as an autoimmune disease. Endocrine Reviews 15: 516–542

Beresford SAA (1985) Is nitrate in the drinking water associated with the risk of cancer in the urban UK? International Journal of Epidemiology 14: 57–63

Breslow NE, Day NE (1987) Statistical methods in cancer research. Vol 2. The design and analysis of cohort studies. (IARC Scientific Publications No. 82), IARC, Lyon.

Chilvers C, Inskip H, Caygill C, Bartholemew B, Fraser P, Hill MA (1984) A survey of dietary nitrate in well-water users. International Journal of Epidemiology 13: 324–331

Choi BCK (1985) N-nitroso compounds and human cancer: a molecular epidemiolgic approach. American Journal of Epidemiology 121(5): 737–743

Dahlquist G, Blom LG, Persson LA, Sandström AIM, Wall SGI (1990) Dietary factors and the risk of developing insulin dependent diabetes in childhood. British Medical Journal 300: 1303–1306.

Dorling D, Atkins D (1995) Population density, change and concentration in Great-Britain 1971, 1981 and 1991. Studies on Medical and Population Subjects No. 58. HMSO, London.

Dulin WE, Soret MG (1978) Chemically and hormonally induced diabetes. In Volk BW, Welman KE (eds) The Diabetic Pancreas. Plenum Press, New York.

EEC (1980) Council directive on the quality of water for human consumption. Official Journal of the EEC 229: 11–29

Helgason T, Ewen SWB, Ross IS, Stowers JM (1982) Diabetes produced in mice by smoked/cured mutton. The Lancet 2: 1017–1021

HMSO (1991) Water Industry Act 1991. HMSO, London.

International Agency for Research on Cancer (1987). The relevance of N-nitroso compounds to human cancer: exposures and mechanisms. IARC Scientific Publications No. 84. IARC, Lyon.

Kostraba JN, Gay EC, Rewers M, Hamman RF(1992) Nitrate levels in community drinking waters and risk of IDDM. An ecological analysis. Diabetes Care 15: 1505–1508

LeDoux SP, Hall CR, Forbes PM, Patton NJ, Wilson GL (1988) Mechanisms of nicotinamide and thymidine protection from alloxan and streptozotocin toxicity. Diabetes 37: 1015–1019

McKinney PA, Law G, Bodansky HJ, Staines A, Williams DRR (1996) Geographical Mapping of Childhood Diabetes in the Northern County of Yorkshire. Diabetic Medicine 13(8): 734–740

M.E.L Research (1996) Tap water consumption in England and Wales: findings from the 1995 national survey. M.E.L Research Report 9448/01. Crown Copyright 1996, UK.

Morales-Suarez-Varela MM, Llopis-Gonzalez A, Terjerizo-Perez ML (1995) Impact of nitrates in drinking water on cancer mortality in Valencia, Spain. European Journal of Epidemiology 11: 15–21

Office of Population Censuses and Surveys OPCS (1996) Ethnicity in the 1991 Census. (Eds) Coleman D, Salt J. HMSO, London.

Parslow RC, McKinney PA, Law GR, Staines A, Williams R, Bodansky HJ. (1997) Incidence of childhood diabetes mellitus in Yorkshire, northern England, is associated with nitrate in drinking water: an ecological analysis. Diabetologia 40; 550–556.

Patterson CC, Waugh NR (1992) Urban/rural and deprivational differences in incidence and clustering of childhood diabetes in Scotland. International Journal of Epidemiology 21: 108–117

Patterson CC, Carson DJ, Hadden DR (1996) Epidemiology of childhood IDDM in Northern Ireland 1989–1994: Low incidence in areas with highest population density and most household crowding. Diabetologia 39: 1063–1069

Phillimore P, Beattie A, Townsend P (1994) Widening inequality of health in northern England, 1981–91. British Medical Journal 308: 1125–1128

Staines A, Bodansky HJ, Lilley HEB, Stephenson C, McNally RJQ, Cartwright RA (1993) The epidemiology of diabetes mellitus in the United Kingdom: the Yorkshire Regional Childhood Diabetes Register. Diabetologia 36: 1282–1287

StataCorp (1995) Stata Statistical Software: Release 4.0 College Station, TX: Stata Corporation.

Steindorf K, Schlehofer B, Becher H, Hornig G, Wahrendorf J (1994) Nitrate in drinking water. A case-control study on primary brain tumours with an embedded drinking water survey in Germany. International Journal of Epidemiology 23: 451–457

Verge CF, Howard NJ, Irwig L, Simpson JM, Mackerras D, Silink M (1994) Environmental Factors in Childhood IDDM: A population-based, case-control study. Diabetes Care 17: 1381–1389

Virtanen SM, Jaakkola L, Räsänen L *et al.* (1994) Nitrate and nitrite intake and the risk for Type 1 diabetes in Finnish children. Diabetic Medicine 11: 656–66.

Subject Index

Acetic acid 102, 103
Achlorohydria 254
Acid neutralising capacity 154
Acidification, 102, 157, 160, 282
Acidified nitrite 285, 295, 299, 311
Activation energy 124
Adirondacks 159
Adrenal morphometry 322
Adsorption equilibria 68
Aerosol 128, 129, 133, 136
Afforestation 170
Agricultural Development and Advisory
   Service (ADAS) 94, 95
Agro-industry 240
Agronomic efectiveness 11, 46, 49
Air-pollutant 123
Alanine 222
Albedo 136
Aldosteron 324, 325
Algae 157, 160, 165, 175
Algal bloom 156, 165
Algal scum 165
Algorithm 33
Alloxan 327
Aluminium 159, 160
Alveolar hyperinflation 318
Amide 176, 312
Amination 327
Amine 104, 271, 281, 319
Amino acid 74, 206, 210, 225
Amino nitrogen 206
Ammonia volatilisation 10, 11, 50
Ammonium 92, 102, 126, 129, 157,
   177
Ammonium sulfate 228
Amoxycillin 283, 310
Anaerobe 283
Anaerobic incubation 73

Analysis of Variance (ANOVA) 62, 100,
   263, 291
Angiotensin 324, 325
Animal manure 43, 47
Anthropogenic activity 110, 121
Antibacterial activity 300, 307, 331
Antibiotic therapy 267, 285, 286
Anticyclonic weather 130
Antifungal activity 312
Antimicrobial activity 260, 312, 313
Antioxidant 272
Apoplast 126
Apple Valley 140, 149, 150, 152
Aquatic
   biota 159, 160
   ecosystem 175
   environment 155
   flora 170
   habitat 68
   pollutant 175
Aquifer 150, 152, 191, 196, 199
Arable 135, 240
   crop 28, 47, 52
   rotation 63
   soil 65
Arginine 282, 285
Aromaticity 219
Ascorbic acid 255, 295, 300, 311
   oxidase 284
Asellus 178, 184
Aspartic acid 222
Assimilation 5
Atmosphere
   budget 133
   deposition 48, 166, 170
   global 121
   pollution 171
Auto immune process 327